METHODS IN MOLECULAR BIOLOGY™

Series Editor
John M. Walker
School of Life Sciences
University of Hertfordshire
Hatfield, Hertfordshire, AL10 9AB, UK

For further volumes:
http://www.springer.com/series/7651

Membrane Protein Structure and Dynamics

Methods and Protocols

Edited by

Nagarajan Vaidehi

Division of Immunology, Beckman Research Institute of the City of Hope, Duarte, CA, USA

Judith Klein-Seetharaman

Department of Structural Biology, University of Pittsburgh School of Medicine, Pittsburgh, PA, USA

 Humana Press

Editors
Nagarajan Vaidehi
Division of Immunology
Beckman Research Institute of the City of Hope
Duarte, CA, USA

Judith Klein-Seetharaman
Department of Structural Biology
University of Pittsburgh School of Medicine
Pittsburgh, PA, USA

ISSN 1064-3745   ISSN 1940-6029 (electronic)
ISBN 978-1-62703-022-9   ISBN 978-1-62703-023-6 (eBook)
DOI 10.1007/978-1-62703-023-6
Springer New York Heidelberg Dordrecht London

Library of Congress Control Number: 2012945407

Humana Press is a brand of Springer
Springer is part of Springer Science+Business Media (www.springer.com)

Preface

Membrane proteins play key roles in numerous cellular processes, in particular, mediating cell-to-cell communication and signaling events that lead to a multitude of biological effects. This functional diversity is achieved through a variety of different membrane proteins, encoded by some 30 % of genes in typical genome. Membrane proteins have also been implicated in many critical diseases such as atherosclerosis, hypertension, diabetes, and cancer. Therefore the three-dimensional structure and dynamics of membrane proteins and their relationship to the function of the proteins remain an important field of research in molecular biology, biochemistry, and biophysics. Researchers have worked relentlessly for many years to obtain sufficient quantities of purified proteins and structural information using several different biophysical techniques, such as X-ray crystallography, NMR, and other spectroscopic methods. In the last few years there has been a gold rush of structures published especially for the superfamily of membrane proteins known as the G-protein coupled receptors, providing the scientific community with insight into the inner workings of cell signaling through this largest class of membrane proteins. These structures also serve as the starting point for drug design targeting these receptors. However, individual experimental techniques often provide incomplete information on the structure and dynamics of membrane receptors, especially due to the difficulty in working with membrane proteins experimentally as a result of their hydrophobic nature. Thus, different biophysical and structural techniques are increasingly used together in exploiting their complementary information content. Increasingly sophisticated computational techniques are becoming available to bridge the gap in connecting the fragmented structural information obtained from different techniques, to generate and test biological hypotheses. In particular, mechanistic questions related to ligand binding and activation of membrane proteins tremendously benefit from computational analyses of membrane protein structures. The integrated cycle of experimental data generation, computational prediction, and experimental validation is now an established approach to rational drug design in both academia and pharmaceutical industries. It is the goal of this volume to highlight the numerous advances in both experimental and computational approaches to the study of structure, dynamics, and interactions of membrane proteins.

This volume is divided into two sections. Section A brings to light the details of the procedures used for measurements of structure and dynamics of membrane proteins using X-ray crystallography, nuclear magnetic resonance spectroscopy, mass spectrometry, ultraviolet/visible spectroscopy, infrared spectroscopy, and Fluorescence Resonance Energy Transfer methods. The elucidation of numerous examples of ligand protein dynamics as well as protein–protein interactions of the complexes that the receptors form upon activation using these approaches is highlighted to demonstrate the functional significance of these structural and spectroscopic studies. Section B of this volume contains a survey of the computational methods that have played a critical role in membrane protein structure prediction as well as in providing atomic level insight into the mechanism of the dynamics of membrane receptors. Thus, this section has been inspired by the usage of computational methods that tie several fragmented experimental data from structural and spectroscopic studies and provide an integrated visualization of the mechanism of activation of membrane receptors. It is intended to provide a resource on the current state-of-the-art techniques available in the computational area for studying membrane protein structure, dynamics, and drug design.

Numerous experts in this area have contributed chapters to this volume. We thank them all for taking their time writing the detailed procedures that would serve as a laboratory guide to all researchers in this field. Given the broad range of expertise covered by the authors and their authoritative status in each specific area, we are sure that this volume will be an invaluable resource to all readers interested in membrane protein structure and function.

We both thank our respective families for putting up with us working through times normally reserved for them, including the many lonely dinners they had without mom at the table. Editing this volume with chapters from varied fields has been a learning experience for both of us. We thank John Walker for giving us this opportunity and all at Humana Press, especially David Casey and Monica Beaumont, for their generous help offered to make this volume useful.

Duarte, CA, USA   *Nagarajan Vaidehi*
Pittsburgh, PA, USA   *Judith Klein-Seetharaman*

Contents

PART I   EXPERIMENTAL TECHNIQUES FOR MEMBRANE PROTEIN
STRUCTURE DETERMINATION

Contributors

RAVINDER ABROL • *Materials and Process Simulation Center, MC, California Institute of Technology, Pasadena, CA, USA*

SAYEH AGAH • *Department of Molecular Physiology and Biological Physics, University of Virginia, Charlottesville, VA, USA*

FARIBA M. ASSADI-PORTER • *Department of Biochemistry and NMRFAM, University of Wisconsin-Madison, Madison, WI, USA*

IVET BAHAR • *Department of Computational and Systems Biology, School of Medicine, University of Pittsburgh, Pittsburgh, PA, USA*

DOV BARAK • *Israel Institute for Biological Research, Ness Ziona, Israel*

MAIK BEHRENS • *Department of Molecular Genetics, German Institute of Human Nutrition Potsdam-Rehbruecke, Nuthetal, Germany*

OUTHIRIARADJOU BENARD • *Department of Neuroscience, Mount Sinai School of Medicine, New York, NY, USA*

SUPRIYO BHATTACHARYA • *Division of Immunology, Beckman Research Institute of the City of Hope, Duarte, CA, USA*

JENELLE K. BRAY • *Simbios NIH Center for Biomedical Computation, Stanford University, Stanford, CA, USA*

MICHAEL F. BROWN • *Department of Chemistry and Biochemistry, University of Arizona, Tucson, AZ, USA; Department of Physics, University of Arizona, Tucson, AZ, USA*

STEFANO COSTANZI • *NIDDK, National Institutes of Health, Berthesda, MD, USA*

XAVIER DEUPI • *Condensed Matter Theory Group and Laboratory of Biomolecular Research, Paul Scherrer Institut, Villigen, Switzerland*

SALEM FAHAM • *Department of Molecular Physiology and Biological Physics, Fontaine Research Park, University of Virginia, Charlottesville, VA, USA*

FRANCESCA FANELLI • *Department of Chemistry, Dulbecco Telethon Institute (DTI), University of Modena and Reggio Emilia, Modena, Italy*

ANTOINE GAUTIER • *Department of Biochemistry, University of Cambridge, Cambridge, UK*

CLEMENS GLAUBITZ • *Centre for Biomolecular Magnetic Resonance, Institute for Biophysical Chemistry, Goethe University Frankfurt, Frankfurt, Germany*

WILLIAM A. GODDARD III • *Materials and Process Simulation Center, California Institute of Technology, Pasadena, CA, USA*

ADAM R. GRIFFITH • *Materials and Process Simulation Center, California Institute of Technology, Pasadena, CA, USA*

JAMES GUMBART • *Biosciences Division, Argonne National Laboratory, Argonne, IL, USA*

BASAK ISIN • *Department of Pathology, University of Pittsburgh School of Medicine, Pittsburgh, PA, USA*

JUDITH KLEIN-SEETHARAMAN • *Department of Structural Biology, University of Pittsburgh School of Medicine, Pittsburgh, PA, USA*

J. KUBICEK • *QIAGEN GmbH, QIAGEN, Hilden, Germany*

J. LABAHN • *Institute for Structural Biology and Biophysics, Research Center Jülich, Jülich, Germany; Institute of Complex Systems, Jülich, Germany*

ANDREA LAKATOS • *Centre for Biomolecular Magnetic Resonance, Institute for Biophysical Chemistry, Goethe University Frankfurt, Frankfurt, Germany*

TZVETANA LAZAROVA • *Unitat de Biofísica, Departament de Bioquímica i de Biologia Molecular, Centre d'Estudis en Biofísica, Facultat de Medicina, Universitat Autònoma de Barcelona, Barcelona, Spain*

XAVIER LEÓN • *Centre de Biotecnologia Animal i Teràpia Gènica, Universitat Autònoma de Barcelona, Barcelona, Spain*

ANAT LEVIT • *Institute of Biochemistry, Food Science, and Nutrition, The Robert H. Smith Faculty of Agriculture, Food and Environment, The Hebrew University, Rehovot, Israel*

VÍCTOR A. LÓRENZ-FONFRÍA • *Unitat de Biofísica, Departament de Bioquímica i de Biologia Molecular, Facultat de Medicina, and Centre d'Estudis en Biofísica, Universitat Autònoma de Barcelona, Barcelona, Spain*

JOHN L. MARKLEY • *Department of Biochemistry, University of Wisconsin-Madison, Madison, WI, USA; National Magnetic Resonance Facility at Madison, University of Wisconsin-Madison, Madison, WI, USA*

MARIANNA MAX • *Department of Neuroscience, Mount Sinai School of Medicine, New York, NY, USA*

WOLFGANG MEYERHOF • *Department of Molecular Genetics, German Institute of Human Nutrition Potsdam-Rehbruecke, Nuthetal, Germany*

SARAH M. MOORE • *Department of Cell Biology, University of Pittsburgh School of Medicine, Pittsburgh, PA, USA*

KARSTEN MÖRS • *Centre for Biomolecular Magnetic Resonance, Institute for Biophysical Chemistry, Goethe University Frankfurt, Frankfurt, Germany*

DANIEL NIETLISPACH • *Department of Biochemistry, University of Cambridge, Cambridge, UK*

MASHA Y. NIV • *Institute of Biochemistry, Food Science, and Nutrition, The Robert H. Smith Faculty of Agriculture, Food and Environment, The Hebrew University, Rehovot, Israel*

ZOLTÁN N. OLTVAI • *Department of Pathology, University of Pittsburgh School of Medicine, Pittsburgh, PA, USA*

ESTEVE PADRÓS • *Unitat de Biofísica, Departament de Bioquímica i de Biologia Molecular, Facultat de Medicina, and Centre d'Estudis en Biofísica, Universitat Autònoma de Barcelona, Barcelona, Spain*

ABBY L. PARRILL • *Department of Chemistry, The University of Memphis, Memphis, TN, USA*

F. SCHÄFER • *QIAGEN GmbH, QIAGEN, Hilden, Germany*

CHRISTOFER S. TAUTERMANN • *Department of Lead Identification and Optimization Support, Boehringer Ingelheim Pharma GmbH & Co. KG, Ingelheim am Rhein, Germany*

DAMIEN THÉVENIN • *Department of Molecular Biophysics and Biochemistry, Yale University, New Haven, CT, USA*

KALYAN C. TIRUPULA • *Department of Structural Biology, University of Pittsburgh School of Medicine, Pittsburgh, PA, USA*

NAGARAJAN VAIDEHI • *Division of Immunology, Beckman Research Institute of the City of Hope, Duarte, CA, USA*

RANI PARVATHY VENKITAKRISHNAN • *Department of Biochemistry, University of Wisconsin-Madison, Madison, WI, USA*

SANTIAGO VILAR • *NIDDK, National Institutes of Health, Berthesda, MD, USA*

CHRISTINE C. WU • *Department Cell Biology, University of Pittsburgh School of Medicine, Pittsburgh, PA, USA*

Part I

Experimental Techniques for Membrane Protein Structure Determination

Chapter 1

Crystallization of Membrane Proteins in Bicelles

Sayeh Agah and Salem Faham

Abstract

The structural biology of membrane proteins remains a challenging field, partly due to the difficulty in obtaining high-quality crystals. We developed the bicelle method as a tool to aid with the production of membrane protein crystals. Bicelles are bilayer discs that are formed by a mixture of a detergent and a lipid. They combine the ease of use of detergents with the benefits of a lipidic medium. Bicelles maintain membrane proteins in a bilayer milieu, which is more similar to their native environment than detergent micelles. At the same time, bicelles are liquid at certain temperatures and they can be integrated into standard crystallization techniques without the need for specialized equipment.

   Key words: Phase transition, Crystal packing, Liquid crystals, Bilayer micelles, Crystallization technique, Lipidic cubic phase, Sponge phase

1. Introduction

1.1. Crystallization of Membrane Proteins

Despite many heroic efforts and some success stories, the number of membrane protein structures determined continues to lag far behind soluble proteins. One major hurdle is the difficulty in obtaining good-quality crystals. For many membrane proteins, detergents are the partner that they "cannot live with, or without." Detergents are necessary for the extraction and solubilization of membrane proteins, while at the same time they can interfere with crystal growth. By their nature, detergents cover much of the hydrophobic surface of a membrane protein leaving little surface area for crystal growth. In the absence of a sizable soluble domain, the protein contacts required for crystal growth are limited to mostly loop regions. This often leads to poor crystal quality since loops are typically the more flexible portion of the protein (1). This challenging task of growing well-ordered membrane protein crystals has prompted the development of a number of technical

Nagarajan Vaidehi and Judith Klein-Seetharaman (eds.), *Membrane Protein Structure and Dynamics: Methods and Protocols*, Methods in Molecular Biology, vol. 914, DOI 10.1007/978-1-62703-023-6_1, © Springer Science+Business Media, LLC 2012

advances. These developments include the antibody method (2), the protein fusion method (3, 4), and the lipidic cubic phase (LCP) method (5). These methods have proven useful, even critical in certain cases. Nonetheless, a universal solution appears out of reach as the unique features of each protein can considerably influence the crystallization process. The distinctive features of each protein may favor a certain method over others; thus the development and advancement of various approaches are beneficial. The success of the LCP method demonstrated that membrane proteins can be crystallized from lipid media (5), not just from detergent media. The lipid cubic phase forms a network of interconnected bilayers that is generated by the lipid monoolein (1-cis-9-octadecenoyl)-rac-glycerol. Due to inherent challenges of the LCP method, such as its high viscosity, we explored the utility of a different lipid phase, namely, bicelles, for the crystallization of membrane proteins, and developed the bicelle method described here.

1.2. Bicelles

Bicelles are bilayer micelles that are formed by specific mixtures of a lipid and a detergent (Fig. 1). Bicelles were originally used in solid-state NMR studies and continue to receive much attention in the NMR field, since they tend to partially align in a magnetic field (6). We were successful in crystallizing bacteriorhodopsin in bicelles (7), and determined its structure from two different bicelle-forming

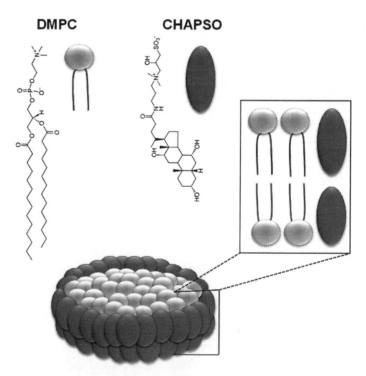

Fig. 1. Representation of the morphology of bicelles. Bicelles are bilayer discs. The bilayer is formed by the lipid (DMPC), with the detergent (CHAPSO) covering the edges of the hydrophobic bilayer.

compositions, demonstrating that bicelles can be a useful medium for the crystallization of membrane proteins (8). Since our initial report, additional membrane proteins have been crystallized from bicelles confirming that this method holds great promise. These are the structures of β2-adrenergic receptor (β2AR) (9), voltage-dependent anion channel (VDAC) (10), xanthorhodopsin (11), and rhomboid protease (12).

Bicelles are made from a mixture of 1,2-dimyristoyl-*sn*-glycero-3-phosphocholine (DMPC) and 1,2-dihexanoyl-*sn*-glycero-3-phosphocholine (DHPC) or a mixture of DMPC and 3-((3-cholamidopropyl)dimethylammonio)-2-hydroxy-1-propanesulfonate (CHAPSO) (13, 14). The long-chain lipid (DMPC) forms the bilayer portion of the bicelles, and the detergent molecules (DHPC, or CHAPSO) cover the edges (Fig. 1). Bicelles can form under a broad range of concentrations and lipid-to-detergent ratios. For DMPC/DHPC mixtures these ranges include (~3% to ~40%) in lipid concentration and (~1:2 to ~5:1) in lipid-to-detergent ratio (15, 16). Bicelles have been examined by NMR (17), electron microscopy (EM) (18), and small angle neutron scattering (SANS) (19). A significant aspect of the bicelle mixture is that its (temperature–concentration) phase diagram is complex involving a number of phase transitions (19). By plotting the concentration on the *x*-axis and temperature on the *y*-axis it is possible to represent where each of the phases occur and where the phase transitions happen. It is important to note that bicelles form in specific regions of the phase diagram. These additional phases include the perforated lamellar phase, and the multi-lamellar vesicles phase (Fig. 2). The fine details of the phase diagram are a matter of debate (20), and additional phases have been proposed such as the branched flat cylindrical micellar phase (16).

Phase transitions can occur as a result of variations in a number of parameters such as temperature, bicelle concentration, and bicelle composition (15). A significant aspect of the bicelle phase is that it is a liquid, thus easy to manipulate. The lamellar phase is a gel, and may be beneficial for crystal growth in some circumstances. The transition from a liquid to a gel occurs as temperature is raised. A possible explanation for this seemingly unusual behavior is that as the temperature is raised more of the detergent molecules are incorporated into the bilayer and fewer detergent molecules remain available to cover the edges; as a result the bilayer portion expands and at a certain point a phase transition occurs and the gel phase forms (20–22). The transition temperature is influenced by the bicelle composition. For example, DMPC:DHPC bicelles are a liquid at room temperature and form a gel over ~30°C. In contrast, bicelles prepared with 1,2-ditridecanoyl-*sn*-glycero-3-phosphocholine (DTPC) instead of DMPC have a lower transition temperature. DTPC:DHPC bicelles are liquid at ~12°C and form a gel at ~20°C (17). The morphology of bicelles can also vary. The diameter of

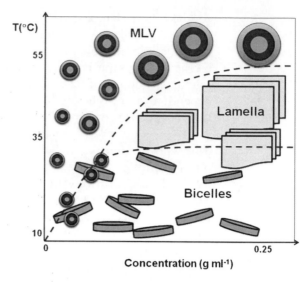

Fig. 2. The bicelle phase diagram. Temperature–concentration phase diagram for 3.2:1 DMPC:DHPC mixture, with total lipid concentration on the x-axis, and temperature on the y-axis. Although the actual phase diagram may be even more complex, this diagram provides a visual illustration on the effects of temperature and concentration on the bicelle phase. Temperature alone can influence the shift from the liquid bicelle phase to the gel lamellar phase. MLV represents multilamellar vesicles. This figure is adapted from (19).

bicelles has been reported to vary from 8 to 50 nm (23, 24). The size of the bicelles is influenced by the bicelle composition and the presence of divalent cations such as calcium or magnesium (25).

Membrane protein crystals grown in lipidic media including LCP and bicelles appear to have a similar type of packing that is distinct from the packing observed for crystals grown in detergents. Instead of having crystal contacts limited to loop regions, substantial crystal contacts are formed by the transmembrane hydrophobic region of the proteins. It was suggested early on that membrane proteins can produce two types of crystals based on their packing (26). In type I crystals two-dimensional layers stack on top of each other to produce a three-dimensional lattice. In each 2D layer the membrane proteins pack side by side positioned roughly as they would be in a bilayer, namely, with the planes that define the thickness of the transmembrane regions arranged in a parallel orientation (Fig. 3a). With this type of packing, substantial crystal contacts are formed by the transmembrane region of the protein. Crystals grown in lipid media mainly produce type I crystal packing (27, 28). In type II packing the majority of crystal contacts are formed by the water-exposed portion of the membrane protein, and it is mainly observed for membrane protein crystals grown from detergents (26) (Fig. 3b).

The phase diagrams of detergents, LCP, and bicelles can be influenced by the various precipitants commonly used in crystallization trials. At the same time each phase can significantly affect the

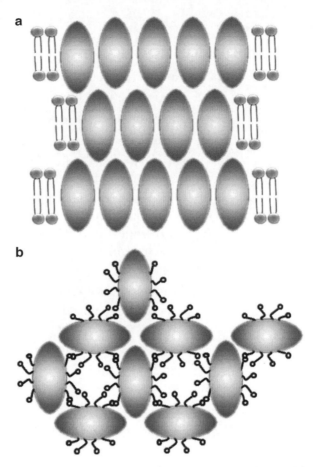

Fig. 3. Crystal packing for membrane proteins. (a) Type I crystal packing, protein molecules pack side by side forming a two-dimensional (2D) layer. The 2D layers align to form a three-dimensional (3D) crystal. Crystals grown from lipidic media mainly belong to type I. (b) Type II crystal packing. Most of the crystal contacts are formed by the soluble and loop regions. This is mainly observed for crystals grown in detergents.

crystallization process. It has been reported for detergents that some regions in the phase diagram are more favorable to crystal growth (29). For the LCP method, much attention has been paid to identifying compatible crystallization conditions that are able to maintain the LCP phase (30). However, it is recognized that crystal growth in LCP may require the formation of a lamellar phase (31). Careful analysis of the monoolein phase diagram has led to the lipidic sponge phase (LSP) method (32), which also relies on being in a specific region of the phase diagram. As for the bicelle method, results suggest that crystallization can occur in both the liquid phase, and in the gel phase. The process of crystal nucleation and growth may require unidentified local phase changes to accommodate the packing of protein molecules and to make room for growing protein crystals.

2. Materials

1. Lipids: DMPC and/or DTPC (Avanti Polar Lipids, Inc.).
2. Detergents: DHPC and/or CHAPSO (Sigma-Aldrich).
3. Sonicator (Heat systems ultrasonics w385).
4. Vortex (VWR).
5. Purified protein of interest at ~10 mg/ml. The choice of detergent is protein dependent (see Note 1).
6. A crystallization setup. Such 24 well trays (Hampton Research) for manual crystallization or a high throughput robot (mosquito from TTP labtech) for 96-well format. Both conventional hanging drops and sitting drops are suitable (see Note 2).
7. Lids or crystal clear tape for 96 well trays, and cover slides for 24 well trays (Hampton Research).
8. Standard commercially available crystallization screens (Hampton Research, or Qiagen).
9. A UV microscope that detects tryptophan fluorescence (Korima Inc., PRS1000) (33).
10. Cryoloops and tools for freezing crystals (Hampton Research).

3. Methods

Protein crystallization in bicelles can be described in four steps. First, the bicelle medium is prepared. Second, the protein is incorporated into the bicelles. Third, the crystallization trials are set up. Fourth, crystallization trials are monitored and crystals extracted.

3.1. Preparation of the Bicelle Mixture Stock Solution

Bicelles can be prepared from a number of ingredients and at different concentrations (~3–40%) (see Notes 3 and 4).

Here is an example for preparation of a 35% bicelle solution at 2.8:1 (DMPC:CHAPSO) (see Notes 5 and 6). The 2.8:1 ratio refers to the molar ratio; it is not a direct weight:weight ratio. The 35% is a weight/volume ratio. The weight referred to in the 35% case is the sum of the weights of (DMPC + CHAPSO). For this calculation we approximate the density of the DMPC/CHAPSO mixture to be close to ~1 mg/ml (34). Naturally, the density will change with temperature and as the composition transitions from a liquid phase to a gel phase. Therefore, to prepare 100 ml of a 35% bicelle solution the sum of the weights of DMPC and CHAPSO would be 35 g. The addition of 65 ml H_2O results in a final volume that is approximately the desired 100 ml. We find this approximation to be useful since the volumes we work with are much smaller than 100 ml.

The molecular weights used for this calculation are DMPC = 677.933 g/mol and CHAPSO = 630.877 g/mol. The molecular weight of a 2.8:1 bicelle is $1,898.2124 + 630.877 = 2,529.0894$. In order to prepare 0.5 ml (35%, 2.8:1) bicelles, 325 µl water is added to 175 mg bicelles. The amount of DMPC needed equals $(175 \times 1,898.2124)/2,529.0894 = 131.347$ mg, and the amount of CHAPSO needed is $(175 \times 630.877)/2,529.0894 = 43.653$ mg.

1. Add 325 µl water to 43.653 mg CHAPSO in a 1.5 ml Eppendorf tube.

2. Vortex until CHAPSO dissolves.

3. Weigh out 131.347 mg DMPC.

4. Add the dissolved CHAPSO to the DMPC.

5. Vortex. The final volume should be close to 0.5 ml.

6. Sonicate the mixture in short (1 s) bursts using a microtip sonicator.
   Sonication should be brief as to avoid froth formation. One to two bursts should be sufficient. Some recommend against sonication, in which case extended periods of vortexing will be required. Flash freezing with liquid nitrogen may help as well.

7. Go through cycles of cooling (4°C, or on ice), vortexing, and heating (~37°C) until the sample is homogeneous (for example, ~5 cycles and 1–2 min/cycle).
   An indication that the bicelle phase has formed is that the mixture will be a clear liquid when kept cool, and will form a clear gel when warmed up to 37°C. It is possible to transition between these two phases several times (see Note 7).

8. Keep bicelle solution cool (~4°C) for immediate use (see Note 8).

3.2. Preparation of the Protein/Bicelle Mixture

Addition of the protein to the bicelle mixture will bring along the detergent used in the purification. Detergents will likely influence the bicelle phase diagram. To minimize the effect of the detergent we try to maintain a low detergent concentration relative to the bicelle concentration. Proteins purified in DDM and LDAO have both been successfully crystallized in bicelles (see Notes 9 and 10). It is possible to purify membrane proteins from native membranes without the addition of detergents, and to crystallize successfully with the bicelle method as was the case with bacteriorhodopsin and xanthorhodopsin (7, 11).

1. Thaw the bicelle mixture if frozen, and keep cool (~4°C).

2. You may need to vortex the mixture to reform a homogeneous bicelle mixture, as precipitation might occur during freezing, thawing, or even as a result of prolonged periods on ice.

Bicelles prepared at lower concentrations (<35%), or at higher detergent-to-lipid ratio will not experience this precipitation as often, and may be easier and faster to prepare.

3. Add the protein of interest (~10 mg/ml) in a 4:1 protein:bicelle ratio while kept cold. If the initial bicelle concentration is 35%, then it will be diluted to 7%, which has been used successfully in crystallization. If the initial protein concentration is 10 mg/ml, it will only be diluted to 8 mg/ml (see Note 11).

4. Mix by pipetting up and down to homogeneity.

5. Incubate the protein/bicelle mixture on ice for 30 min with occasional mixing to allow for complete incorporation of the protein into the bicelles. NMR can be used to assess the incorporation of the membrane protein into bicelles (35, 36); however we do not experimentally validate the incorporation of the protein into the bicelles.

6. If protein activity assays are available, testing the activity of the protein sample after incorporation into bicelles can be very useful to confirm that the protein remains active and properly folded.

3.3. Crystallization of the Protein/Bicelle Mixture

One of the attractions and benefits of the bicelle method is that it is possible to carry out crystallization trials using standard tools and protocols. Nonetheless, it is sensible to check the liquidity of the protein/bicelle mixture by visual inspection during the pipetting procedure as bicelles can get more viscous at room temperature. In case gelling is observed, then extra caution should be taken by keeping the sample cool enough to be in the liquid phase. This should not be a problem in most cases; however if the bicelle composition is altered to one that forms a gel at low temperature (for example as a result of using DTPC instead of DMPC), then additional care may be required. Also setup of the crystallization trials in a warm room should be avoided.

1. Crystallization trials can be set up by standard methods such as pipetting the protein by hand or by a nanoliter dispensing robot (see Note 12).

2. Commercially available screens are convenient to use with bicelles. However, some of the conditions included in the usual screens may lead to phase separation. For example, we have observed phase separation in condition G6 of the classics screen from Qiagen (0.1 M NaCl, 0.1 M Bicine pH 9.0, 20% PEG550MME) (see Note 13). The appearance of nonprotein crystalline material is also a concern and care must be taken when analyzing the results of the crystallization trials. Given the limited number of structures solved from bicelles, along with the fact that bicelles can be prepared from different components, it is difficult to develop a complete crystallization

screen specific for the bicelle method. Nonetheless, lessons can be learned from the currently available information. Protein crystals have been formed under various conditions. Salt-based screens (phosphates, and sulfates) are the most successful so far, but crystals have been obtained in polyethylene glycols (PEG2K) (8), and organics (MPD) as well (10).

3. After the crystal trays have been prepared, the trays can be transferred to the temperature of interest (37°C, room temperature, or anywhere in between) (see Note 14). Protein crystallization in bicelles at 4°C has not yet been reported; it is plausible, although possibly tricky as it may lead to increased precipitation of the lipids (see Note 15).

The crystallization temperature can play an important role in the outcome, as temperature can strongly influence the phase behavior of the bicelle mixture. Small temperature changes of only 1–2° have been suspected to cause difficulty in reproducing crystallization results.

3.4. Visualization and Extraction of Crystals Grown in Bicelles

Identification of protein crystals in bicelles is not trivial as it may be difficult to distinguish between lipid features and protein crystals. To minimize confusion we recommend one of the two approaches:

(a) Set up double drops for each condition, one with bicelles alone, and one with protein and bicelles.

(b) Use a UV microscope (33). By relying on tryptophan fluorescence a UV microscope can help distinguish between protein crystals and nonprotein features (Fig. 4) (see Note 16).

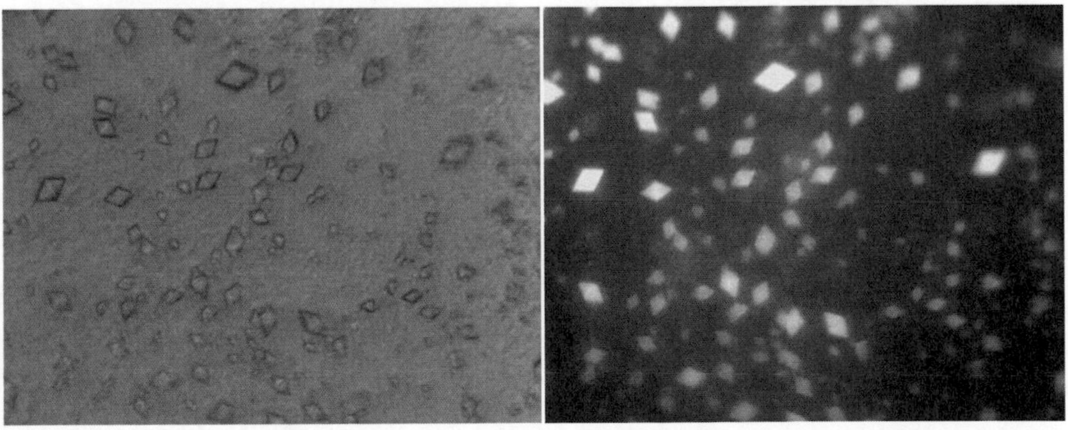

Visible microscope   UV microscope

Fig. 4. UV microscope. Protein crystals can be easier to identify with the assistance of a UV microscope. Tryptophan fluorescence can help distinguish protein from nonprotein features as long as other fluorescent compounds are not included in the crystallization experiment.

Table 1
List of cryo conditions used for freezing protein crystals grown with the bicelle method

Protein name	Cryo conditions
Bacteriorhodopsin	3.0 M phosphate, 10% bicelles
Bacteriorhodopsin	35% PEG 2K
β2-Adrenergic receptor	Reservoir solution, 20% glycerol
Xanthorhodopsin	Reservoir solution, 15% ethylene glycol
Voltage-dependent anion channel	Reservoir solution, 10% PEG400
Rhomboid protease	Reservoir solution, 25% glycerol

Crystal extraction from bicelle drops should be no different from standard methods. The gel phase that forms with the bicelle mixture is only mildly viscous and does not prevent handling the crystals or mounting them with crystal loops. Addition of a cryo solution can help loosen the crystals from the bicelle gel if necessary.

It is possible to screen for cryo conditions using standard approaches, without a need to remove the bicelles. The bicelle mixture may provide some additional cryo protection for the crystals. Cryo conditions that have been used with the bicelle method are listed in Table 1.

4. Notes

1. Detergents carried along with the protein can influence the bicelle behavior. If one desires to maintain a bicelle mixture as close to the original components as possible, one needs to be careful not to carry along too much detergent with the protein. Therefore, care must be taken when the protein sample is concentrated.

2. Standard crystallization setups used for typical protein crystallization experiments are suitable with the bicelle method. In contrast to the LCP method, no additional specialized tools are required.

3. Successful results have been reported from a number of bicelle-forming mixtures including DMPC:CHAPSO, DTPC:CHAPSO, and DMPC:nonyl-maltoside (NM). Protein crystals have also been observed in DMPC:DHPC bicelles. Additionally, bicelles can be prepared from other lipids or detergents. Including the

use of 1,2-dilauroyl-*sn*-glycero-3-phosphocholine (DLPC) or 1,2-dipalmitoyl-*sn*-glycero-3-phosphocholine (DPPC) instead of DMPC (37) or the use of 3-((3-cholamidopropyl) dimethylammonio)-1-propanesulfonate (CHAPS) instead of CHAPSO (38).

4. Lipid additives can be used in the process of screening for crystallization conditions. These lipids can be added at the first step, the bicelle preparation stage, or the second step, where the protein is incorporated into bicelles.

5. Protein stability analysis on opsin, the apo form of rhodopsin, in bicelles showed that the protein is more stable in DMPC:CHAPS bicelles than in DMPC:DHPC bicelles. This was demonstrated by the increased resistance of the protein to urea-induced denaturation (38). One possible explanation suggested is the increased rigidity of the DMPC:CHAPS bicelles.

6. Small amounts of hexadecyl(cetyl)trimethylammonium bromide (CTAB) or tetradecyltrimethylammonium bromide (TTAB) have been reported to stabilize bicelles (39, 40). It has been also reported that DMPC:DHPC bicelles can be stabilized by the addition of cholesterol (41) or cholesterol sulfate (42).

7. If the sample gets dispersed and stuck on the inner walls of the tube during the preparation process, centrifugation of the tube is recommended. This will maintain the proper concentration and ensure reproducibility.

8. For long-term storage bicelles can be kept at –20°C. When thawing the bicelle mixture, caution is required to ensure that the solution is homogeneous before use. Potentially, cycles of heating, cooling, and vortexing may be required.

9. The β2AR structure showed that bicelles can be used in conjunction with the bound Fab fragment method (9).

10. Both β-barrels and α-helical membrane proteins have been crystallized with the bicelle method (7, 9–12).

11. Preparing samples at a high initial bicelle concentration reduces the extent to which the protein is diluted. However, if one is able to concentrate the protein sample to high levels (~20 mg/ml), then it is not necessary to have the bicelles at a high initial concentration. Additionally, some proteins may crystallize at lower concentrations. Xanthorhodopsin crystallized in bicelles at ~4 mg/ml concentration.

12. It is possible to chill the platform of the nanoliter dispending robot (such as the mosquito) with ice prior to applying the protein sample. Only the plate position where the protein is loaded needs to be chilled. In case the bicelle composition

being used has a low gelling temperature (for example DTPC:CHAPSO), then chilling may help avoid sample gelling. The tray setup procedure is not long, and the cooler platform can help keep the sample in the liquid phase.

13. Many of the crystallization conditions may dramatically alter the bicelle phase diagram. We do not consider these conditions to be incompatible with the bicelle method. On the other hand, conditions that cause lipid precipitation are incompatible.

14. Phosphatidylcholine lipids that are used to form the bicelles may slowly hydrolyze at room temperature. This process apparently did not interfere with the growth of bacteriorhodopsin crystals even at low pH and over extended periods of time. However, it should be noted that phosphatidylcholine lipids with an ether linkage instead of an ester linkage are able to form bicelles and are more resistant to hydrolysis (43, 44).

15. Setting up crystallization trials at 4°C can lead to precipitation of the lipids in some of the typical crystallization conditions that have high precipitant concentrations.

16. The observation of fluorescent crystals in the UV microscope does not guarantee that they are protein crystals. The protein may precipitate on a salt crystal and cover it completely.

Acknowledgments

The authors would like to thank Professors Jochen Zimmer and Michael Wiener for critical review of the manuscript.

References

1. Carpenter EP, Beis K, Cameron AD, Iwata S (2008) Overcoming the challenges of membrane protein crystallography. Curr Opin Struct Biol 18:581–586

2. Ostermeier C, Iwata S, Ludwig B, Michel H (1995) Fv fragment-mediated crystallization of the membrane protein bacterial cytochrome c oxidase. Nat Struct Biol 2:842–846

3. Cherezov V, Rosenbaum DM, Hanson MA, Rasmussen SG, Thian FS, Kobilka TS, Choi HJ, Kuhn P, Weis WI, Kobilka BK, Stevens RC (2007) High-resolution crystal structure of an engineered human beta2-adrenergic G protein-coupled receptor. Science 318:1258–1265

4. Prive GG, Verner GE, Weitzman C, Zen KH, Eisenberg D, Kaback HR (1994) Fusion proteins as tools for crystallization: the lactose permease from *Escherichia coli*. Acta Crystallogr D: Biol Crystallogr 50:375–379

5. Landau EM, Rosenbusch JP (1996) Lipidic cubic phases: a novel concept for the crystallization of membrane proteins. Proc Natl Acad Sci U S A 93:14532–14535

6. Sanders CR, Prosser RS (1998) Bicelles: a model membrane system for all seasons? Structure 6:1227–1234

7. Faham S, Bowie JU (2002) Bicelle crystallization: a new method for crystallizing membrane proteins yields a monomeric bacteriorhodopsin structure. J Mol Biol 316:1–6

8. Faham S, Boulting GL, Massey EA, Yohannan S, Yang D, Bowie JU (2005) Crystallization of bacteriorhodopsin from bicelle formulations at room temperature. Protein Sci 14:836–840

9. Rasmussen SG, Choi HJ, Rosenbaum DM, Kobilka TS, Thian FS, Edwards PC, Burghammer M, Ratnala VR, Sanishvili R, Fischetti RF, Schertler GF, Weis WI, Kobilka BK (2007) Crystal structure of the human beta2 adrenergic G-protein-coupled receptor. Nature 450:383–387

10. Ujwal R, Cascio D, Colletier JP, Faham S, Zhang J, Toro L, Ping P, Abramson J (2008) The crystal structure of mouse VDAC1 at 2.3 A resolution reveals mechanistic insights into metabolite gating. Proc Natl Acad Sci U S A 105(46):17742–17747

11. Luecke H, Schobert B, Stagno J, Imasheva ES, Wang JM, Balashov SP, Lanyi JK (2008) Crystallographic structure of xanthorhodopsin, the light-driven proton pump with a dual chromophore. Proc Natl Acad Sci U S A 105:16561–16565

12. Vinothkumar KR (2011) Structure of rhomboid protease in a lipid environment. J Mol Biol 407:232–247

13. Sanders CR 2nd, Prestegard JH (1990) Magnetically orientable phospholipid bilayers containing small amounts of a bile salt analogue, CHAPSO. Biophys J 58:447–460

14. Sanders CR 2nd, Schwonek JP (1992) Characterization of magnetically orientable bilayers in mixtures of dihexanoylphosphatidylcholine and dimyristoylphosphatidylcholine by solid-state NMR. Biochemistry 31:8898–8905

15. Harroun TA, Koslowsky M, Nieh MP, de Lannoy CF, Raghunathan VA, Katsaras J (2005) Comprehensive examination of mesophases formed by DMPC and DHPC mixtures. Langmuir 21:5356–5361

16. van Dam L, Karlsson G, Edwards K (2004) Direct observation and characterization of DMPC/DHPC aggregates under conditions relevant for biological solution NMR. Biochim Biophys Acta 1664:241–256

17. Ottiger M, Bax A (1998) Characterization of magnetically oriented phospholipid micelles for measurement of dipolar couplings in macromolecules. J Biomol NMR 12:361–372

18. van Dam L, Karlsson G, Edwards K (2006) Morphology of magnetically aligning DMPC/DHPC aggregates-perforated sheets, not disks. Langmuir 22:3280–3285

19. Nieh MP, Glinka CJ, Krueger S, Prosser RS, Katsaras J (2002) SANS study on the effect of lanthanide ions and charged lipids on the morphology of phospholipid mixtures. Small-angle neutron scattering. Biophys J 82:2487–2498

20. Triba MN, Warschawski DE, Devaux PF (2005) Reinvestigation by phosphorus NMR of lipid distribution in bicelles. Biophys J 88:1887–1901

21. Jiang Y, Wang H, Kindt JT (2010) Atomistic simulations of bicelle mixtures. Biophys J 98:2895–2903

22. Nieh MP, Raghunathan VA, Pabst G, Harroun T, Nagashima K, Morales H, Katsaras J, Macdonald P (2011) Temperature driven annealing of perforations in bicellar model membranes. Langmuir 27:4838–4847

23. Vold RR, Prosser RS, Deese AJ (1997) Isotropic solutions of phospholipid bicelles: a new membrane mimetic for high-resolution NMR studies of polypeptides. J Biomol NMR 9:329–335

24. Barbosa-Barros L, De la Maza A, Walther P, Estelrich J, Lopez O (2008) Morphological effects of ceramide on DMPC/DHPC bicelles. J Microsc 230:16–26

25. Arnold A, Labrot T, Oda R, Dufourc EJ (2002) Cation modulation of bicelle size and magnetic alignment as revealed by solid-state NMR and electron microscopy. Biophys J 83:2667–2680

26. Ostermeier C, Michel H (1997) Crystallization of membrane proteins. Curr Opin Struct Biol 7:697–701

27. Caffrey M (2009) Crystallizing membrane proteins for structure determination: use of lipidic mesophases. Annu Rev Biophys 38:29–51

28. Faham S, Ujwal R, Abramson J, Bowie JU (2009) Chapter 5 Practical aspects of membrane proteins crystallization in bicelles. In: Larry D (ed) Current topics in membranes, vol 63. Academic, San Diego, CA, p 109

29. Koszelak-Rosenblum M, Krol A, Mozumdar N, Wunsch K, Ferin A, Cook E, Veatch CK, Nagel R, Luft JR, Detitta GT, Malkowski MG (2009) Determination and application of empirically derived detergent phase boundaries to effectively crystallize membrane proteins. Protein Sci 18:1828–1839

30. Cherezov V, Fersi H, Caffrey M (2001) Crystallization screens: compatibility with the lipidic cubic phase for in meso crystallization of membrane proteins. Biophys J 81:225–242

31. Nollert P, Qiu H, Caffrey M, Rosenbusch JP, Landau EM (2001) Molecular mechanism for the crystallization of bacteriorhodopsin in lipidic cubic phases. FEBS Lett 504:179–186

32. Wadsten P, Wohri AB, Snijder A, Katona G, Gardiner AT, Cogdell RJ, Neutze R, Engstrom S (2006) Lipidic sponge phase crystallization of membrane proteins. J Mol Biol 364:44–53

33. Judge RA, Swift K, Gonzalez C (2005) An ultraviolet fluorescence-based method for identifying and distinguishing protein crystals. Acta Crystallogr D: Biol Crystallogr 61:60–66

34. Hianik T, Haburcak M, Lohner K, Prenner E, Paltauf F, Hermetter A (1998) Compressibility and density of lipid bilayers composed of

polyunsaturated phospholipids and cholesterol. Colloids Surf 139:189–197

35. De Angelis AA, Opella SJ (2007) Bicelle samples for solid-state NMR of membrane proteins. Nat Protoc 2:2332–2338

36. Matsumori N, Murata M (2010) 3D structures of membrane-associated small molecules as determined in isotropic bicelles. Nat Prod Rep 27:1480–1492

37. Lind J, Nordin J, Maler L (2008) Lipid dynamics in fast-tumbling bicelles with varying bilayer thickness: effect of model transmembrane peptides. Biochim Biophys Acta 1778:2526–2534

38. McKibbin C, Farmer NA, Edwards PC, Villa C, Booth PJ (2009) Urea unfolding of opsin in phospholipid bicelles. Photochem Photobiol 85:494–500

39. Losonczi JA, Prestegard JH (1998) Improved dilute bicelle solutions for high-resolution NMR of biological macromolecules. J Biomol NMR 12:447–451

40. Fleming K, Matthews S (2004) Media for studies of partially aligned states. Methods Mol Biol 278:79–88

41. Ghimire H, Inbaraj JJ, Lorigan GA (2009) A comparative study of the effect of cholesterol on bicelle model membranes using X-band and Q-band EPR spectroscopy. Chem Phys Lipids 160:98–104

42. Shapiro RA, Brindley AJ, Martin RW (2010) Thermal stabilization of DMPC/DHPC bicelles by addition of cholesterol sulfate. J Am Chem Soc 132:11406–11407

43. Ottiger M, Bax A (1999) Bicelle-based liquid crystals for NMR-measurement of dipolar couplings at acidic and basic pH values. J Biomol NMR 13:187–191

44. Aussenac F, Lavigne B, Dufourc EJ (2005) Toward bicelle stability with ether-linked phospholipids: temperature, composition, and hydration diagrams by 2H and 31P solid-state NMR. Langmuir 21:7129–7135

Chapter 2

Vapor Diffusion-Controlled *Meso* Crystallization of Membrane Proteins

J. Labahn, J. Kubicek, and F. Schäfer

Abstract

The presented method to crystallize membrane proteins combines the advantages of the meso-phase crystallization method and the classical vapor diffusion crystallization. It allows fast screening of crystallization conditions employing automated liquid handlers suited for the 96-well crystallization format.

Key words: Membrane protein crystallization, In meso crystallization, Lipidic cubic phase, Meso phase, Vapor diffusion, Monoolein

1. Introduction

The fact that there is only a small number of membrane protein structures known can be directly traced back to the common problems in obtaining membrane protein crystals for structural investigations. Methods developed for soluble protein crystallization are inefficient for membrane proteins. Landau and Rosenbusch (1) used lipidic meso-phases to accommodate the specific needs of membrane proteins in a way compatible with crystallization: the lipidic component monoolein is an amphiphile and therefore self-organizes in water into complex structures, the meso-phases (2) (Fig. 1). The cubic phase Pn3m consists of a bi-continuous bilayer, that separates two channel systems of aqueous phase (depicted as circles in picture of Pn3m phase in Fig. 1). The bilayer locally appears two dimensional similar to a cell membrane and therefore allows the incorporation of membrane proteins. However, it extends continuously through space and therefore supports diffusion of the protein in three dimensions and crystallization upon dehydration.

Dehydration of the meso-phase can be achieved by lowering the level of humidity. In vapor diffusion experiments this is

Nagarajan Vaidehi and Judith Klein-Seetharaman (eds.), *Membrane Protein Structure and Dynamics: Methods and Protocols*,
Methods in Molecular Biology, vol. 914, DOI 10.1007/978-1-62703-023-6_2, © Springer Science+Business Media, LLC 2012

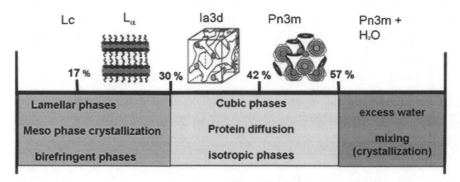

Fig. 1. The monoolein–water isotherm at 22°C (2). With increasing water content the layered birefringent lamellar phases (Lc, Lα), the optically isotropic cubic phases (Ia3d, Pn3m), and the swollen cubic phase in the presence of excess water (sponge) phase are formed reversibly. The added amount of water determines the maximal hydration level. Upon dehydration by vapor diffusion the maximally hydrated system reverts back to the final meso phase which is determined by the dilution of the added screening solution (Fig. 2).

experimentally realized by enclosing the wetted monoolein together with a reservoir solution that takes up water from the gas phase that separates the two condensed phases (3).

In contrast to (1) in meso crystallization in batch (1), no weighing of mg amounts of monoolein or dehydrating salt is required. Furthermore, in contrast to (2) sponge phase crystallization (4), we are not limited to a small number of screening solutions. Also, in contrast to (3) the active mixing approach (5), we find the protein–monoolein ratio in the crystallization experiment not to be limited by the solubility of the protein in detergent solution when targeting a certain meso-phase because excess water can be removed via vapor diffusion. Finally, the vapor diffusion-controlled *in meso* crystallization experiment can be set up with the same equipment used for the crystallization of soluble proteins.

Similar to the crystallization of soluble proteins the likelihood of crystallization will depend on the presence of a crystallizable species, which may be controlled by the addition of specific lipids to the protein solution or other stabilizing compounds to the diluted screening solution (6).

In the typical setup, 132 µg solid monoolein is hydrated with 900 nl aqueous solution in total: (1) the membrane protein in 450 nl protein solution will dissolve in the meso-phase matrix that acts as a solvent. This matrix will also separate excess detergent from the protein. (2) The content of undiluted screening solution in the added 450 nl of diluted screening solution will determine the final dehydration level that can be reached upon equilibration against the undiluted reservoir and thereby the final meso-phase, as well as the protein concentration in the aqueous phase above the monoolein (Fig. 2).

Crystals may be obtained with excess water conditions, in the cubic phase or in the presence of a lamellar phase (Fig. 3).

Effects of using diluted reserviors

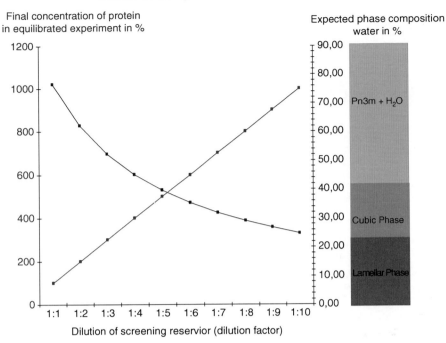

Final concentration of protein
in equilibrated experiment in %

Expected phase composition
water in %

Pn3m + H₂O

Cubic Phase

Lamellar Phase

Dilution of screening reservior (dilution factor)

Fig. 2. Targeting final hydration levels and meso-phases using dilutions. *Red curve*: Expected final protein concentration (stock solution 100%) depending on dilution of screening solution for a vapor diffusion experiment without transfer of protein or water from aqueous solution into monoolein, if a mixture of equal volumes of protein solution and (diluted) screening solution is equilibrated against undiluted screening solution. *Black curve*: Expected final hydration level of meso-phase depending on dilution of screening solution, if the final water volume of the ideal vapor diffusion experiment (*red curve*) is available for meso-phase formation. *Right*: Targeted type of meso-phase as a function of expected hydration level (*black curve*) based on the phase isotherm of the monoolein–water system at 22°C.

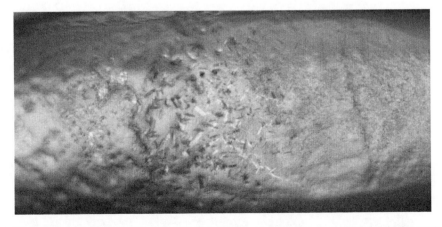

Fig. 3. Crystallization of a naturally colored rhodopsin. The largest yellow protein crystals appear in the *center* where the protein droplet was set. At the right side of the experiment birefringent speckles of lamellar phase can be detected in the overall isotropic cubic phase.

2. Materials

1. Monoolein pre-coated crystallization plates (see Note 3b)

   Either 132 μg monoolein prefilled evolution microplate (see Note 14) or 29 μg prefilled crystallization plate in MRC or Intelli format (Qiagen GmbH, Germany).

2. Screening solutions (see Note 2b).

   Cubic Phase I and Cubic Phase II (Qiagen GmbH, Germany) or other commercially available screening solutions in Deep Well Blocks.

3. Plate sealing.

   AMP Liseal, Transparent Microplate Sealer (Greiner bio-one, Germany).

4. Anti-evaporation layer.

   Perfluoropolyether (PFO-X175/08, Hampton Research, USA).

5. Other

   Deep Well Blocks.

6. Instruments and equipment

   (a) Polarization microscope.

   (b) Incubator (22–24°C).

   (c) Fluorescence microscope (optional).

   (d) Automatic liquid handler (optional).

3. Methods

(a) Sample preparation

1. Prepare a membrane protein solution with a concentration of 2–35 mg/ml of known detergent content (see Note 1).

2. Prepare the 96 screening solutions that are to be used for the crystallization plate in a 96-well Deep Well Block or use a block of premade solutions. Prepare the required (50–100 μl each) dilution of the screening solutions in a second block (see Note 2).

(b) Setup of the crystallization experiment

1. The ready-to-use microplate, pre-coated with 132 μg (29 μg) monoolein, is unfrozen for 10 min at 22°C (see Note 3).

2. To the experimental wells with monoolein 450 nl (100 nl) of protein solution is added (see Note 4).

    3. The plate is sealed and incubated for 3 h at 22°C (see Note 5).

    4. Fill the reservoir wells with 75 μl of undiluted screening solution (see Note 6).

    5. Add 450 nl (100 nl) of diluted screening solution to the experimental well (see Note 7).

    6. Reseal the plate and incubate at 22°C (see Note 8).

(c) Monitoring progress

    1. Monitor the formation of cubic meso-phase initially every day until the meso-phase becomes optically isotropic when viewed with a polarization microscope (see Note 9).

    2. Monitor for dehydration as indicated by loss of optical isotropy when viewed with a polarization microscope and for protein crystal formation twice a month for 3–4 months (see Note 10).

(d) Harvesting crystals

    1. Add 0.5–2 μl of screening solution or a water-immiscible liquid (e.g., perfluoropolyether) to the experiment to minimize evaporation (see Note 11).

    2. Remove monoolein from the crystal with a mounting loop without actually touching the crystal (see Note 12).

    3. Transfer the crystal with a mounting loop into suitable cryoprotectant and freeze the crystal without monoolein (see Note 13).

4. Notes

1. (a) Though protein purity appears to be a less important factor in the case of meso-phase crystallization when compared to crystallization of soluble proteins (7), detergent concentration is an additional search parameter: if the protein is concentrated by ultrafiltration it should *not be assumed* that detergent micelles pass through the membrane into the filtrate. If the detergent content of the protein sample after concentration is high, the compatibility of a protein-free detergent solution with meso-phase formation should be tested to avoid waste of protein. Generally the concentration of detergent should be controlled, e.g., by thin-layer chromatography to insure the reproducibility of crystallization experiments (8).

    (b) The presence of glycerol will reduce dehydration and lengthen the equilibration time.

(c) Organism-specific lipids may be added after concentrating the protein.

2. (a) For initial experiments dilutions of 1:1 (undiluted), 1:4, and 1:7 are reasonable starting points to target different meso-phases.

(b) Commercially available screening solutions for membrane proteins are a good starting point.

(c) Often a screening for an additive may be required to succeed in crystallization. Therefore a salt and a PEG screen are recommended to us as dehydrating screens and mix these (90%, v/v) with other screens (10%, v/v) (e.g., anion screen, cation screen, etc.) when searching for additives. The addition of a strongly binding ligand that stabilizes a certain protein conformation is recommended in any case.

3. (a) Pre-coated plates should be stored at –20°C to avoid oxidation of monoolein.

(b) Only the evolution plate prefilled with 132 µg monoolein (see Subheading 2) is suitable for manual dispensing of solutions (see Note 14).

4. (a) If the protein concentration is very low (<2 mg/ml) or a high ratio of protein/monoolein is targeted, the dispensed volume of protein solution can be increased.

(b) If the protein is dispensed by an automated liquid handler the actual release of the protein droplets should be monitored. Problems with droplet release can be solved by (1) changed ejection speed and (2) change of detergent or salt content of protein solution.

5. This incubation step can be omitted for experiments with 29 µg monoolein (see Note 14).

6. If it is desired to slow the kinetics of dehydration, some µl of a low density, low vapor pressure, with water-immiscible liquid can be added on top of the screening solution in the reservoir well (9).

7. If a different final hydration level is required, another dilution of the screening solution can be used.

8. (a) If evaporation of reservoir solution or salt crystal formation is observed the plate was not properly sealed.

(b) Incubation temperature should be between 22 and 24°C. During incubation any variation of temperature should be avoided.

9. (a) The time for the complete transformation of solid monoolein into cubic phase varies between 20 min and 3 days, depending on the composition of the aqueous phase.

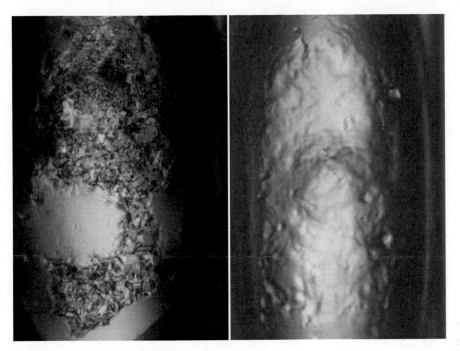

Fig. 4. Effect of screening solution on meso-phase formation upon hydration of solid monoolein. *Left:* Arrest of phase transformation at strongly polarizing lamellar phase due to incompatible screening solution. *Right:* Complete transformation into isotropic cubic phase with cubic phase salt screen I.

    (b) If no formation of optically isotropic cubic phase is observed (Fig. 4), the protein or the screening solution contains components or concentration of components incompatible with meso-phase crystallization: check compatibility of detergent concentration and/or concentration of screening solution without protein to establish a proper concentration range for your new screen or detergent.

10. (a) If an increased volume of protein solution has been used (see Note 4a) crystallization of optically isotropic crystals of monoolein may be observed.

    (b) Formation of lamellar phase from cubic phase upon dehydration occurs typically in form of birefringent speckles or whiskers (Fig. 3).

    (c) Discrimination between very small birefringent protein crystals and local transformation of isotropic cubic phase into birefringent lamellar phase may be difficult using a polarization microscope. Monoolein phases and protein phases can be distinguished by tryptophan-fluorescence (10).

11. If complete removal of monoolein from the protein crystal is desired, better results will be obtained with the anti-evaporation layer, e.g., of PFO (11).

12. (a) Meso-phase turns opaque on freezing. Typically protein crystals frozen with large excess of monoolein become undetectable.

(b) In the case crystals stick to the loop when removing the meso-phase, it will be necessary to have a second loop at hand.

13. Meso-phases frozen at 100 K show pulver diffraction rings upon data collection which may be treated like ice rings.

14. The addition of 450 nl protein solution to 132 μg monoolein and subsequent incubation generate a protein gradient in the meso-phase. Under this condition all protein concentrations down to almost zero are screened in the same experiment at the same time (Fig. 3).

References

1. Landau EM, Rosenbusch JP (1996) Lipidic cubic phases: a novel concept for the crystallization of membrane proteins. Proc Natl Acad Sci U S A 93:14532–14535
2. Qiu H, Caffrey M (2000) The phase diagram of the monoolein/water system: metastability and equilibrium aspects. Biomaterials 21:223–234
3. Benvenuti M, Mangani S (2007) Crystallization of soluble proteins in vapour diffusion for X-ray crystallography. Nat Protoc 2:1633–1651
4. Wöhri AB et al (2008) A lipidic-sponge phase screen for membrane protein crystallization. Structure 16:1003–1009
5. Caffrey M, Cherezov V (2009) Crystallizing membrane proteins using lipidic mesophases. Nat Protoc 4:706–731
6. McPherson A, Cudney B (2006) Searching for silver bullets: an alternative strategy for crystallizing macromolecules. J Struct Biol 156:387–406
7. Kors CA et al (2009) Effects of impurities on membrane-protein crystallization in different systems. Acta Crystallogr D 65:1062–1073
8. Misquitta Y, Caffrey M (2003) Detergents destabilize the cubic phase of monoolein: implications for membrane protein crystallization. Biophys J 85:3084–3096
9. Chayen NE (1997) A novel technique to control the rate of vapour diffusion, giving larger protein crystals. J Appl Crystallogr 30:198–202
10. Li F, Robinson H, Yeung ES (2005) Automated high-throughput nanoliter scale protein crystallization screening. Anal Bioanal Chem 383:1034–1041
11. Brumshtein B et al (2008) Control of the rate of evaporation in protein crystallization by the microbatch under oil method. J Appl Crystallogr 41:969–971

Chapter 3

Solution NMR Studies of Integral Polytopic α-Helical Membrane Proteins: The Structure Determination of the Seven-Helix Transmembrane Receptor Sensory Rhodopsin II, pSRII

Antoine Gautier and Daniel Nietlispach

Abstract

About 30% of the proteins encoded in the genome are expressed as membrane proteins but these represent <1% of all the structures solved today. In view of the physiological and pharmaceutical significance of membrane proteins it is clear that a better and more comprehensive understanding of their three-dimensional (3D) structures at atomic resolution is required. α-Helical integral membrane proteins are generally more difficult to work with than β-barrel-type proteins and this has particularly been true for the polytopic members such as the large family of seven-helical proteins. In this chapter we describe the practical aspects of the solution-state NMR spectroscopy structure determination of the seven-helical transmembrane (7-TM) protein receptor sensory rhodopsin pSRII from the haloalkaliphilic archaeon *Natronomonas pharaonis* reconstituted in detergent micelles. This is the first time that a three-dimensional structure of a 7-TM protein has been determined by NMR.

Key words: Integral membrane proteins, α-Helical, NMR spectroscopy, 7-TM receptor, Sensory rhodopsin, Protein expression, Isotope labeling, Deuteration, Selective methyl protonation, Structure determination, Detergent solubilization

1. Introduction

Structure determination of integral membrane proteins (IMPs) is challenging for a range of reasons that cause difficulties for all available structure determination methods (1, 2). Consequently, despite their large abundance IMPs are structurally underrepresented and account for <0.5% of the deposited structures in the protein data bank (PDB) (3). Solution-NMR spectroscopy is rapidly gaining importance in the study of structure and dynamics of integral membrane proteins but the majority of these investigations have

Nagarajan Vaidehi and Judith Klein-Seetharaman (eds.), *Membrane Protein Structure and Dynamics: Methods and Protocols*, Methods in Molecular Biology, vol. 914, DOI 10.1007/978-1-62703-023-6_3, © Springer Science+Business Media, LLC 2012

concentrated on the more amenable β-barrel-type proteins (1, 4–7). The larger group of α-helical IMPs is more challenging to work with and so far, representative of the larger polytopic proteins, only a few monomeric structures with up to four helices had been solved by solution NMR. In addition several oligomeric structures have been determined (8–12). Structures of seven-helical transmembrane (7-TM) proteins, which include the G protein-coupled receptors (GPCRs) as their largest family, have been particularly inaccessible and the development of structure determination methods in this area can be considered as especially valuable. The aim of the work presented here was to establish methodology for solution-state NMR spectroscopy three-dimensional structure determination of 7-TM proteins and to validate the suitability of the developed approach using pSRII as a model 7-TM protein (13, 14). To some extent pSRII can be considered as a structural analogue of smaller GPCRs with whom it shares some of the common features of the unique heptahelical architecture (15–17). Working with pSRII combined several advantages as the protein could be expressed in *E. coli* and the wild-type form was sufficiently thermally stable. This allowed the focus of our efforts to be directed towards the development of the NMR methodological aspects without being impaired by low expression levels, short protein lifetimes, or similar problems that would have been encountered with GPCRs.

Optimization of expression and purification protocols yielded very pure and homogeneous protein in sufficient milligram quantities, allowing the preparation of several isotopically labeled protein samples for NMR studies. The successful solubilization of pSRII in detergent micelles resulted in an NMR accessible size of the protein–detergent complex of ~70 kDa and enabled the recording of high-quality NMR spectra using high levels of deuteration in combination with TROSY techniques (18). The sufficient stability of pSRII in diheptanoylphosphatidylcholine (c7-DHPC) at elevated temperatures (50°C) for many days allowed the extensive recording of 3D and 4D experiments which generated sufficient structural restraints to result in a high-quality three-dimensional structure with well-defined backbone and side chain conformations for protein and the retinal chromophore (14).

2. Materials

2.1. Host Strain, Expression Vector, and Antibiotics

1. Chloramphenicol resistant *E. coli* Tuner (DE3)pLacI cells (Novagen).
2. pSRII cDNA.
3. Kanamycin resistant pET-28b(+) plasmid (Novagen).

4. Chloramphenicol (Melford Laboratories, UK): Used as a stock solution of 34 mg/mL in absolute ethanol. The solution was filter-sterilized and stored at –20°C.

5. Kanamycin (Melford Laboratories, UK): Used as a stock solution of 50 mg/mL in water. The solution was filter-sterilized, aliquoted, and stored at –20°C.

2.2. Cell Cultures and Isotope Labeling

1. M9 minimal medium: Sterile water, 1× M9 salts: 0.002 mM $FeCl_3$, 1× M9 salt mix, 5 mM $MgSO_4$, 0.1 mM $CaCl_2$, 0.5% D-glucose, 0.1% NH_4Cl. 10× M9 salts: 0.477 M Na_2HPO_4, 0.22 M KH_2PO_4, 0.086 M NaCl. 1,000× M9 salt mix: 4 mM $ZnSO_4$, 1 mM $MnCl_2$, 0.7 mM H_3BO_3, 0.7 mM $CuSO_4$. All solutions were prepared using distilled water sterilized by filtration through a 0.22 μm pore size syringe filter (Sartorius).

2. Stable isotopes and chemicals for NMR sample preparation: ^{15}N-ammonium chloride (98% ^{15}N, Aldrich), $^{13}C_6$-D-glucose (99% ^{13}C, Aldrich), $^2H_7,^{13}C_6$-D-glucose (97% ^2H, 99% ^{13}C, Aldrich), ^2H_2O (99.8% ^2H, Cambridge Isotope Laboratories, MA), 41.7 mg/mL 2-keto-3-methyl-d$_3$-butyric acid-1,2,3,4-$^{13}C_4$,3-d$_1$ sodium salt (Isotec, USA; Cat. No. 637858) in ^2H_2O, 50 mg/mL 2-ketobutyric acid-$^{13}C_4$,3,3-d$_2$ hydrate sodium salt (Cambridge Isotope Laboratories, USA; Cat. No. CDLM-4611) in ^2H_2O.

3. Stock solutions of 1 M isopropyl β-D-1-thiogalactopyranoside (IPTG) (Melford Laboratories, UK) in sterile water and 10 mM all-*trans* retinal (Sigma) in absolute ethanol.

4. Expressions were performed in 2 L baffled flasks containing 500 mL M9 minimum medium. Starter cultures were grown in 250 mL conical flasks.

2.3. Protein Purification, NMR Sample Preparation, and Detergent Screening

1. Buffer A: 50 mM Tris–HCl (pH 8.0), 5 mM $MgCl_2$; Buffer S: 300 mM NaCl, 50 mM MES–NaOH (pH 6.5), 5 mM imidazole; Buffer W1: 300 mM NaCl, 50 mM MES–NaOH (pH 6.5), 25 mM imidazole; Buffer W2: 300 mM NaCl, 50 mM MES–NaOH (pH 6.5), 50 mM imidazole; Buffer E: 300 mM NaCl, 50 mM Tris–HCl (pH 7.0), 150 mM imidazole; Buffer N: 50 mM NaCl, 28.85 mM Na_2HPO_4, 21.15 mM NaH_2PO_4 (pH 7.0), 0.5% NaN_3; Buffer P: 50 mM NaCl, 6 mM Na_2HPO_4, 44 mM NaH_2PO_4 (pH 6.0), 0.5% NaN_3. All buffers were prepared using distilled water and sterilized by filtration.

2. One bottle of protease inhibitor cocktail (PIC) (Sigma, Cat. No. P2714: AEBSF, 2 mM; Aprotinin, 0.3 μM; Bestatin, 130 μM; EDTA, 1 mM; E-64, 14 μM; Leupeptin, 1 μM) in 100 mL sterile water. The solution was filter-sterilized, aliquoted, and stored at –20°C.

3. Emulsiflex-C5 High Pressure Homogeniser (Avestin, Canada).

4. Dounce homogeniser (Jencons-PLS).

5. Detergents: 10% (w/v) n-dodecyl-β-D-maltoside (DDM) (Melford Laboratories, UK) in sterile water, 10% (w/v) 1-myristoyl-2-hydroxy-*sn*-glycero-3-phosphocholine (LMPC) (Avanti Polar Lipids, USA) in sterile water, 10% (w/v) 1-myristoyl-2-hydroxy-*sn*-glycero-3-phosphoglycerol(LMPG) (Avanti Polar Lipids, USA) in sterile water, 10% (w/v) 1-palmitoyl-2-hydroxy-*sn*-glycero-3-phosphoglycerol (LPPG) (Avanti Polar Lipids, USA) in sterile water, 10% (w/v) 1-oleoyl-2-hydroxy-*sn*-glycero-3-phosphoglycerol (LOPG) (Avanti Polar Lipids, USA) in sterile water, 10% (w/v) 1,2-dihexanoyl-*sn*-glycero-3-phosphocholine (c6-DHPC) (Avanti Polar Lipids, USA) in sterile water, 20% (w/v) 1,2-diheptanoyl-*sn*-glycero-3-phosphocholine (c7-DHPC) (Avanti Polar Lipids, USA) in sterile water.

6. Ni-NTA resin (Novagen).

7. Empty chromatography columns (Bio-Rad).

8. Vivaspin 20 mL centrifugal concentrators, 10 kDa MWCO (Sartorius-Stedim).

2.4. NMR Spectroscopy, Data Analysis, and Structure Calculations

1. NMR experiments were recorded at 50°C on Bruker DRX600 and DRX800 spectrometers equipped with 5 mm triple-resonance HCN TXI/z probes (Department of Biochemistry, University of Cambridge, UK).

2. All data were processed using the *Azara* suite of programs. (W. Boucher, unpublished).

3. All backbone and side chain NMR assignments, intensity measurements, and data fitting were done using the CCPNMR *Analysis* software suite (19).

4. Structures were calculated using CNS 1.1 (20) and ARIA 1.2 (21).

5. The structures obtained from the calculations were viewed in MOLMOL (22) to obtain approximate RMSD values.

6. Structures were validated using ARIA 1.2 diagnostics in addition to PROCHECK (23) and PROCHECK-NMR (24).

3. Methods

In the description of the methods below emphasis is placed on the following aspects: (1) the expression, purification, detergent screening, and NMR sample preparation of pSRII; (2) the NMR experiments required for backbone and side chain assignments and to obtain structural restraints; (3) the structure calculations and structure validation of pSRII.

3.1. Expression and Purification of Isotopically Labeled pSRII for NMR Analysis

3.1.1. Molecular Biology

The plasmid pET28b(+) (Novagen) containing the *psopII* gene (TrEMBL accession number Q3IMZ8) was used to transform *E. coli* Tuner (DE3)pLacI cells (Novagen) by the calcium chloride technique (25). The gene was inserted between the NcoI and XhoI restriction sites. The clone was obtained from Prof. J. Navarro (University of Texas Medical Branch, USA). The final construct encoded amino acid residues 1–241 of *Natronomonas pharaonis* plus a carboxy-terminal His$_6$ purification tag.

3.1.2. Uniform ^{15}N- and ^{13}C,^{15}N-Labeling

Three different variants of minimal medium for the preparation of labeled samples were evaluated with respect to cell growth and protein expression. These were M9 medium (26), MOPS medium (27), and high cell density medium (HCDM). Growth and expression were found to be similar in all three media, and due to its lower cost and relative ease of preparation, M9 medium was used for the final expressions of isotopically labeled pSRII. Expression levels were similar in M9-H$_2$O medium to those in rich medium, and over 30 mg of labeled pSRII was obtained from 6 L of culture.

1. Transformed cells were grown at 37°C in M9 minimal medium (see Note 1) supplemented with 50 μg/mL kanamycin and 34 μg/mL chloramphenicol. The 50 mL cultures were grown until they reached stationary growth.

2. 25 mL of this starter culture was used to inoculate 500 mL of M9 minimal medium culture containing ^{15}NH$_4$Cl and unlabeled or ^{13}C-labeled D-glucose. The cultures were grown at 37°C in a shaking incubator at 200 rpm until an absorbance at 600 nm of ~1.0 was achieved.

3. The culture was then induced with 1 mM IPTG, after which 10 μM all-*trans* retinal was added and incubated at 25°C in an orbital shaker for 10 h. Cultures were supplemented every 2 h with 10 μM all-*trans* retinal.

4. Cells were harvested by centrifugation for 30 min at 6,400 × g and 4°C and then frozen for storage at −20°C.

3.1.3. Uniform ^{2}H,^{13}C, ^{15}N-Labeling

1. Transformed cells were adapted for growth in perdeuterated medium by subculturing the cultures in M9 minimal liquid media with progressively higher D$_2$O content (50, 70, 90 (twice), and 100% (three times)). Each step in the adaptation process lasted for 12 h. M9 was supplemented with 34 μg/mL chloramphenicol and 50 μg/mL kanamycin.

2. The final 25 mL cultures were grown with shaking to an A_{600} of approximately 3.0 at 37°C and 1:10 dilution was used to inoculate 500 mL of the M9 medium containing ^{15}NH$_4$Cl, sterile filtered D$_2$O, and ^{2}H$_7$,^{13}C$_6$-D-glucose as the sole carbon

source (see Note 2). This culture was grown under the same conditions to an A_{600} of approximately 1.0 (ca. 12 h).

3. Induction conditions and protein expression were the same as already described previously.

3.1.4. Uniform (^2H, ^{13}C, ^{15}N), ($^{13}CH_3$)-Ile($\delta 1$)($^{13}CH_3$/$^{12}CD_3$) Leu, Val Labeling

The uniformly deuterated ^{13}C,^{15}N-labeled ILV methyl protonated pSRII sample was produced following exactly the same protocol as for the expression of deuterated ^{13}C,^{15}N-labeled pSRII except that 83.3 mg/L of 2-keto-3-methyl-d$_3$-butyric acid-1,2,3,4-$^{13}C_4$,3-d$_1$ and 50 mg/L 2-ketobutyric acid-$^{13}C_4$,3,3-d$_2$ hydrate were added 1 h before induction (see Note 3).

3.1.5. Purification of pSRII

1. Thawed cells were resuspended in 1/25th of the original culture volume of lysis Buffer A with added PIC. The lysate was passed twice through an Emulsiflex-C5 High Pressure Homogeniser.

2. Crude membrane fragments were sedimented by centrifugation 100,000×g, 90 min, and 4°C and resuspended in 1/40th of the original culture volume of Buffer S with added PIC (see Note 4).

3. The detergent DDM was added to a final concentration of 1.5% to the resuspension and the mixture was incubated overnight in the dark with gentle agitation at 4°C.

4. After solubilization, the lysate was centrifuged for 1 h at 4°C and 100,000×g. The supernatant was added to Ni-NTA agarose beads.

5. The resin was then packed into a gravity-flow column and washed with 25 column volumes of Buffer S containing 0.06% DDM. Impurities were then eluted with 25 column volumes of Buffer W1 and 7 column volumes of Buffer W2. These two washing Buffers contained 0.06% DDM.

6. pSRII was eluted from the beads in Buffer E containing 0.1% DDM. The quantity and purity of pSRII were determined by UV-vis spectroscopy (see Note 5).

3.1.6. Detergent Screening

To allow structural studies of membrane proteins using solution-NMR spectroscopy, the protein needs to be reconstituted using a membrane mimetic. Different media are currently available of which at the moment detergent micelles still seem to produce the most consistent results. When evaluating suitable conditions a compromise has to be found where the overall size of the protein–detergent complex is kept small enough to provide NMR spectra of sufficient quality, where sufficient protein can be solubilized and the protein is stable over several days to allow extensive use of the sample for NMR data recording, and where the protein is able to maintain its functional integrity. In general the suitability of a particular detergent is difficult to predict requiring screening of a wide range of detergents and conditions. In the case of pSRII a first round of detergent screening was done using the characteristic

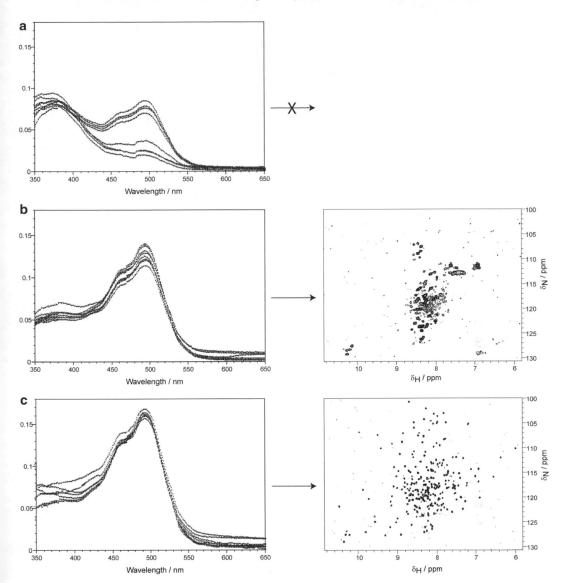

Fig. 1. Detergent screening for NMR studies of pSRII. UV/vis spectra (*left*) with the retinal absorbance at 498 nm and (^{15}N,^{1}H)-TROSY NMR spectra (*right*) are shown for three different types of detergents: (**a**) DPC, (**b**) LMPG, and (**c**) c7-DHPC. To monitor the protein stability at 25°C, UV/vis spectra were recorded every 6 h. DPC was not able to maintain the structural integrity of pSRII and was therefore not further assessed by NMR. LMPG kept the protein stable but the NMR spectra recorded at pH 6 and 40°C were only of modest quality showing insufficient number of peaks. c7-DHPC kept pSRII stable over several days and provided the necessary high-quality NMR spectra at temperatures between 30 and 50°C.

UV/vis absorbance of the protein and retinal moiety to assess the thermal stability of the detergent-solubilized protein at 25°C. Subsequently, the suitability of potential detergent candidates for NMR was then further evaluated by 2D (^{15}N,^{1}H)-TROSY spectroscopy and by pulsed field gradient translational diffusion measurements (Fig. 1). While NMR spectra were recorded at a range of temperatures (see below), for a more consistent comparison of the UV/vis spectra in different detergents all measurements were done at the lowest temperature of 25°C.

The pSRII protein was initially solubilized in DDM. The substitution of DDM for a new detergent took place when pSRII was bound to nickel beads. Several detergents such as c7-DHPC, c6-DHPC, LMPC, LMPG, LPPG, and LOPG showed promising thermal stability in our UV/vis absorbance monitoring assay at 25°C and were then tested for their suitability to provide NMR spectra of sufficient quality (Fig. 1). To this purpose ^{15}N-labeled pSRII was solubilized using varying protein detergent ratios and $(^{15}N,^{1}H)$-TROSY spectra were recorded to assess the overall spectral appearance and linewidths (see Note 6). Temperatures were varied between 25 and 50°C, depending on the ability of the detergent to successfully stabilize the protein.

1. The pSRII bound nickel agarose beads were thoroughly washed with 50 column volumes of Buffer N and no detergent (see Note 7).

2. The tested detergent was added to a concentration equal to ten times its CMC. The column containing pSRII bound to the beads and the new detergent was sealed and incubated for 16 h in the dark at 4°C.

3. Remaining DDM was washed out with 50 column volumes of Buffer N and the new detergent added to a concentration close to 4× CMC.

4. pSRII was eluted with Buffer P containing 300 mM imidazole and the new detergent at a concentration of 4× CMC.

5. Imidazole was discarded by alternating concentration and dilution in Buffer P with a detergent concentration equal to the CMC. The diluent Buffer P was prepared with a detergent concentration equal to the CMC by pre-filtration using the same concentrator (see Note 8).

6. Probing the thermal stability, UV/vis spectra were recorded in 6-h intervals and the intensities of the protein and retinal signals were monitored at 280 and 498 nm.

7. 1D ^{1}H NMR measurements were carried out to monitor the complete removal of DDM as assessed by the absence of the anomeric proton signal.

8. 2D $(^{15}N,^{1}H)$-TROSY spectra were recorded at different temperatures between 25 and 50°C to check whether the new detergent would yield a high-quality NMR spectrum (0.4 mM pSRII, experiment time 2 h). After completion of each NMR experiment the sample integrity was validated and quantified through UV/vis spectra. Examples of suitable and inadequate detergent choices are shown in Fig. 1.

9. 1D ^{1}H NMR translational diffusion measurements were then used to assess the size of the protein–detergent complex.

10. The procedure was then repeated for several detergent concentrations.

3.1.7. NMR Sample
Preparation

1. Solutions of purified pSRII in detergent micelles were concentrated to 450 µL and D_2O was added to a final concentration of 10% (v/v). Final sample conditions were 0.5 mM pSRII in Buffer P and 60 mM DHPC.

2. The sample was transferred to a 5 mm NMR tube (Wilmad), which was then sealed with parafilm.

3.2. NMR Experiments
for Backbone and Side
Chain Assignments

The backbone and side chain assignments and structural restraints were obtained from 11 protein samples which were labeled accordingly either with ^{15}N (three samples); $^{15}N,^2H$ (two samples); $^{15}N,^{13}C,^2H$ (two samples); and $^{15}N,^{13}C,^2H$ and selective methyl protonation of isoleucine (Ile), leucine (Leu), and valine (Val) residues (two samples) or $^{15}N,^{13}C$ (two samples) (Table 1). A total of 21 different NMR experiments were recorded. All spectra were recorded at 50°C at 600 or 800 MHz. Specific spectral parameters of the most important experiments are given in Table 1. In the following the general peak assignment strategies are outlined.

3.2.1. Sequential
Backbone Assignment

1. Sequential assignments were derived from six out-and-back amide detected experiments through pairwise matching of the frequencies for ^{13}Ca nuclei in HNCA/HN(CO)CA, ^{13}Cb nuclei in HN(CA)CB/HN(COCA)CB, and $^{13}C'$ nuclei in HN(CA)CO/HNCO (see Note 9). All experiments were recorded on a perdeuterated sample using ($^{15}N,^1H$)-TROSY implementations of the triple-resonance experiments to benefit from the reduced transverse relaxation rates for ^1HN and ^{15}N spin states. Most experiments were repeated several times and after confirming the sample integrity added together to improve the signal-to-noise ratio. Nonuniform sampling was used for all indirectly detected dimensions (see Note 10) to increase the resolution per unit time (28). Maximum entropy processing as implemented in *Azara* was employed to convert the time domain signals in the indirect dimensions into frequency domain data (Azara is freely available and can be downloaded from http://www.ccpn.ac.uk/azara). Backbone assignment ambiguities were reduced through sequential matching using all three carbon nuclei.

2. Assignments were confirmed or where incomplete extended through the use of sequential HN-HN NOE information obtained from 3D and 4D ^{15}N NOESY experiments (see Note 11). Sequential NOEs ($i/i\pm 1, 2$) were in most cases consistently observed within the individual seven α-helices and were particularly prominent when using a perdeuterated sample. The same experiments were also recorded on a ^{15}N-labeled sample and residue characteristic NOE pattern between side chain and amide protons were used to confirm the assignments made. A total of 460 inter-amide NOEs were observed of which 359 were $i/i\pm 1$. The quality of the data was dramatically improved at the higher field of 800 MHz.

Table 1

Heteronuclear NMR experiments used for assignment and structure determination of pSRII solubilized in c7-DHPC detergent micelles: isotope-labeled samples and acquisition parameters

Experiment	Sample	^1H (MHz)	Expected time	Scans	Recovery delay (s)	Water flip-back	Dim1/ pnts	sw (Hz)/ t_1^{max} (ms)	Dim2/pnts (proc)	sw (Hz)/ t_1^{max} (ms)	Dim3/pnts (proc)	sw (Hz) / t_1^{max} (ms)	Dim4/pnts (proc)	sw (Hz)/ t_1^{max} (ms)
3D HNCA (NUS)	C, D	600	3 days 17 h	184	1.1	g	H/512	10,000/51	N/16 (M)	1,898/21.1	C/16 (M)	4,386/8.2		
3D HN(CO)CA (NUS)	C	600	5 days 12 h	272	1.1	z	H/512	10,000/51	N/16 (M)	1,898/21.1	C/16 (M)	4,386/8.2		
3D HN(CA)CB (NUS)	C, D	600	5 days 4 h	256	1.0	g	H/512	10,000/51	N/16 (M)	1,898/21.1	C/16 (M)	10,000/4.6		
3D HN(COCA)CB (NUS)	C	600	6 days 12 h	320	1.0	z	H/512	10,000/51	N/16 (M)	1,898/21.1	C/16 (M)	10,000/4.6		
3D HN(CA)CO (NUS)	C	600	2 days 15 h	128	1.0	z	H/512	10,000/51	N/16 (M)	1,898/21.1	C/16 (M)	2,000/18.0		
3D HNCO (NUS)	C	600	1 days 1 h	48	1.0	z	H/512	10,000/51	N/16 (M)	1,898/21.1	C/16 (M)	2,000/18.0		
2D ^{13}C-CT-HMQC	D	800	1 day	288	1.1	g	H/512	10,000/51	C/114 (L)	4,025/28.3				
2D ^{13}C-CT-HSQC	E	800	1 day	288	1.1	g	H/512	10,000/51	C/114 (L)	6,579/17.3				
3D NOESY ^{15}N-HSQC	A, B, D	800	9 days 12 h	28	1.35	z	H/512	10,000/51	H/110	8,403/13.1	N/40	2,530/15.8		
3D Val-(HM) CM(CBCA)NH	D	600	8 days 8 h	192	1.2	g	H/512	10,000/51	C/19 (L)	1,358/14.0	N/32 (L)	1,898/16.9		
3D Leu-(HM) CM(CGCBCA)NH	D	600	8 days 8 h	192	1.2	g	H/512	10,000/51	C/19 (L)	1,358/14.0	N/32 (L)	1,898/16.9		
3D Ile-(HM) CM(CGCBCA)NH	D	600	7 days	160	1.2	g	H/512	10,000/51	C/19 (L)	1,358/14.0	N/32 (L)	1,898/16.9		

3D Leu-HM(CMCGCBCA)NH	D	600	7 days 13 h	224	1.2	g	H/512	10,000/51	H/16	1,000/16.0	N/32 (L)	1,898/16.9		
3D Ala-HMCM(CG)CBCA	E	600	5 days 10 h	120	1.2	g	H/512	10,000/51	C/20	1,509/13.3	C/36 (L)	3,019/11.9		
3D HMCM(CG)CBCA	D	600	6 days 15 h	64	1.1	g	H/512	10,000/51	C/48	9,058/5.3	C/36 (L)	3,019/11.9		
3D (H)CCH COSY	E	600	7 days 16 h	32	1.2	g	H/512	10,000/51	C/60 (L)	12,255/4.9	C/60 (L)	12,255/4.9		
3D H(C)CH COSY	E	600	7 days 18 h	40	1.2	g	H/512	10,000/51	C/60 (L)	12,255/4.9	H/48	3,731/12.9		
3D NOESY 13C-HMQC	D	800	7 days 17 h	24	1.1	g	H/512	10,000/51	H/80	8,000/10.0	C/24	4,025/6.0		
3D NOESY 13C-HSQC	E	800	7 days 17 h	24	1.1	z,g	H/512	10,000/51	H/80	8,000/10.0	C/40	6,576/6.1		
4D 13C,13C-HMQC-NOESY-HMQC	D	600	9 days 10 h	12	1.0	g	H/512	10,000/51	H/16	1,200/13.3	C/20 (F,M)	3,012/6.6	C/20 (F,M)	3,012/6.6
3D 15N,15N-HMQC-NOESY-HMQC	B	600	6 days 12 h	64	1.3	z	H/512	10,000/51	N/48	1,898/25.2	N/32	1,898/16.9		

NMR samples are isotope labeled as follows: (A) ^{15}N; (B) $^2H,^{15}N$; (C) $^2H,^{13}C,^{15}N$; (D) u-($^2H,^{13}C,^{15}N$) Ile d1-($^{13}CH_3$) Leu,Val-($^{13}CH_3$, $^{12}CD_3$); (E) $^{13}C,^{15}N$. Sample concentrations varied between 0.4 and 0.8 mM in pSRII. All data were recorded on Bruker spectrometers equipped with conventional (600 MHz) or cryoprobe (800 MHz) HCN z-gradient triple-resonance probes. NMR raw data were processed using the software package Azara: (L) constant-time data dimensions were extended using mirror-image linear prediction; (M) maximum entropy processing was used for non-uniformly sampled (NUS) spectra. Typically 256 complex NUS data coordinates were recorded. An exponentially weighted sampling scheme was applied for decaying dimensions while a random sampling scheme was used for constant-time periods. NOE mixing times in ^{15}N NOESY were 120 and 250 ms and in ^{13}C NOESY were 80 and 100 ms. Experiments were recorded using either (z) water-flipback or (g) gradient PFG-coherence order selection implementations. All samples were recorded at 50°C in the presence of non-deuterated c7-DHPC

3. The combination of triple-resonance experiments and NOE information allowed assignment of 98% of all the amide-containing residues. Reducing the pH to 5.4 for one of the samples revealed the remaining four missing residues: two serine (Ser)–glycine (Gly) repeats located in inter-helical loop regions, which at pH 6 were in intermediate exchange. With this the sequential backbone assignments were completed to 100%.

3.2.2. Side Chain Assignment and NOE Distance Restraints

1. Highly deuterated while selectively methyl protonated samples were used to assign the ^{1}H and ^{13}C shifts of the methyl groups of residues Ile(δ^1), Leu(δ^1,δ^2), and Val(γ^1,γ^2) based on the sensitive 3D HMCM(CG)CBCA experiment. In order to improve the experimental sensitivity Leu and Val side chains contained linearized ^{13}C spin systems (29). The individual side chains were linked to the sequential main chain assignments via ^{13}Ca and ^{13}Cb resonance matching in the backbone out-and-back experiments (Fig. 2). The number of proton chemical shift assignments were increased using a combination of ^{13}C-CT-HMQC and the less sensitive 3D HM(CMCGCBCA)NH experiment. Ambiguity and overlap in the corresponding ^{13}C version of the latter experiment were reduced by recording three 3D variants, which were individually optimized for Ile and Leu as (HM)CM(CGCBCA)NH using band-selective ^{13}C pulses and as (HM)CM(CBCA)NH for Val residues. A total of 137 methyl groups out of 141 for Ile, Leu, and Val were assigned.

2. Additional methyl groups for alanine (Ala) and threonine (Thr) residues were assigned from 3D HCCH COSY experiments or for Ala using an Ala-selective HMCMCA experiment. In total 40 out of 47 methyl groups were assigned. Further assigned were nine Ile(γ^2) groups (Fig. 3) and the seven methionine (Met) methyl positions. In all the side chain experiments pulsed field gradient-based coherence order selection was necessary to reduce the intense detergent signals and to reduce artifact levels (see Note 12).

3. The identified methyl groups served as starting points to extend assignments onto the neighboring side chain positions within the same methyl containing residues through a combination of 3D (H)CCH COSY and H(C)CH COSY, 3D ^{13}C NOESY-HMQC and 3D ^{15}N NOESY experiments recorded on ^{13}C,^{15}N- and ^{15}N-labeled samples, respectively, and a 4D ^{13}C HMQC-NOESY-HMQC recorded on the selectively methyl protonated sample (see Note 13) (Fig. 4). The large amount of signals in the side chain experiments recorded on the protonated sample together with the increased linewidth made assignments a demanding task. Some assignments were obtained iteratively in conjunction with the structure calculations in progress. To reduce ambiguities as much as possible the experiments were recorded at 800 MHz.

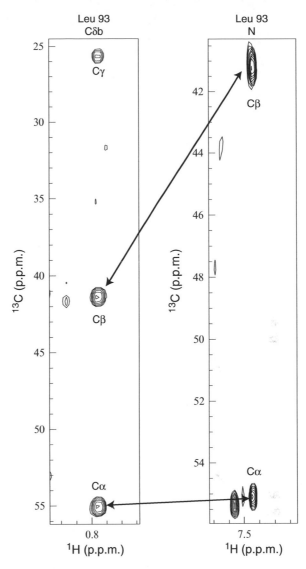

Fig. 2. Sequential methyl resonance assignment in selectively methyl protonated Ile, Leu, and Val residues. ω_1(^{1}H), ω_3(^{13}C) cross sections are shown from 3D HMCM(CG)CBCA (*left*) and 3D HNCA and 3D HN(CA)CB (*right*), respectively (the two ^{1}HN backbone strips on the *right* are superimposed). The correlation of the Leu93 ^{13}Cδb methyl group with the amide position is obtained via matching ^{13}Cα and ^{13}Cβ chemical shifts.

4. Once assignments had been propagated within the methyl-containing amino acids these side chain positions were then used as new starting points to identify adjacent residues. A combination of 3D NOESY spectra and HCCH COSY experiments allowed to progress with the proton and carbon assignments of those residues. Figure 5 shows the extent of the proton side chain assignment for the residues with methyl groups and those residues that are in the vicinity of Ile, Leu, Val, Ala, Thr, or Met residues.

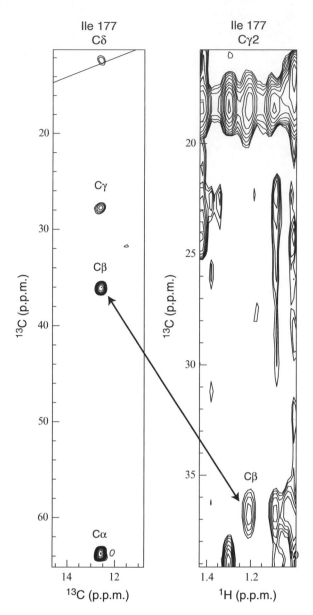

Fig. 3. Side chain assignment of isoleucine residues. On the *left* is shown the $\omega_2(^{13}C)$, $\omega_3(^{13}C)$ section extracted at the position of Ile177Hδ1 from a 3D HMCM(CG)CBCA recorded on a U-(^{2}H,^{13}C,^{15}N), (^{13}CH$_3$)-Ile(δ1)(^{13}CH$_3$/^{12}CD$_3$)-Leu,Val-labeled sample. The complementary $\omega_1(^1H)$, $\omega_2(^{13}C)$ strip at the position of Ile177Cγ2 from a 3D (H)CCH COSY recorded on a protonated ^{13}C,^{15}N-labeled pSRII is shown on the *right*. Matching ^{13}Cβ resonances in both experiments allowed the assignment of 14 out of 15 isoleucine Cγ2, Hγ2 positions.

5. Methyl-HN and methyl–methyl NOEs were obtained from 3D ^{15}N and ^{13}C separated NOESY spectra. A 4D HCCH NOESY experiment helped to increase the number of inter-methyl NOEs (Fig. 4). In the 3D spectra this region was often strongly overlapped. Further, the 4D helped in situations where due to the limited dispersion in the methyl region many

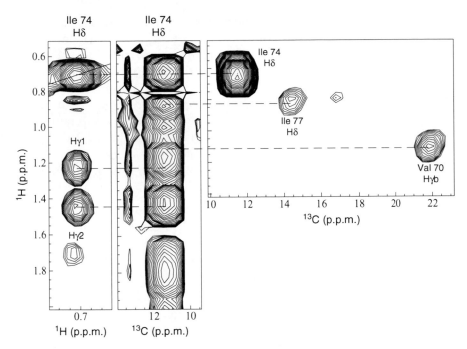

Fig. 4. Resonance assignment in ¹³C NOESY spectra. 3D ¹³C-separated NOESY (*left*) and HCCH COSY (*middle*) experiments were recorded on a protonated ¹³C,¹⁵N-labeled pSRII sample in order to extend assignments beyond the methyl groups. The large number of cross peaks in those spectra in the methyl region between 0 and 1.5 ppm makes this a very demanding task that is exacerbated by the presence of NOEs between pSRII and the detergent. A 4D ¹³C HCCH NOESY (*right*) was used in addition to reduce signal overlap near the diagonal. The ω_1(¹H), ω_3(¹³C) cross section from the 4D NOESY extracted at the ¹H,¹³C methyl position of Ile74δ1 shows further assignments that are not accessible in the 3D NOESY spectrum due to the intense overlap.

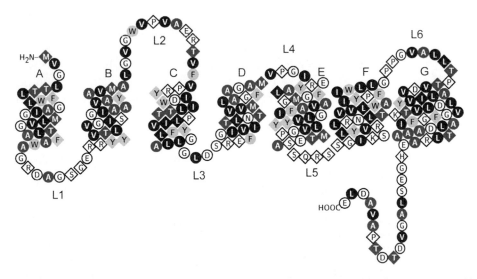

Fig. 5. Extent of side chain assignments. Simplified topology diagram of pSRII showing the primary sequence with helices A–G and the loop regions L1–L6. *Circles* and *diamonds* represent residues with full proton assignments and partial proton side chain assignments, respectively. Isoleucine, leucine, and valine residues are in *black*, whereas methionine, alanine, and threonine residues are in *dark gray*. Residues with aromatic side chains are in *light grey*. All the remaining residue types are in *white*.

NOEs were found in close proximity to or buried by the diagonal. The successful assignment of many Ala and Thr residues increased the number of medium- and long-range NOEs and their complementary positioning to Ile, Leu, and Val residues in the tertiary structure provided a more uniform distribution of distance restraints over the protein. Assignments of these additional methyl groups resulted in a substantially increased inter-helical total of 121 $HN-CH_3$ and 89 CH_3-CH_3 NOEs using protonated and selectively methyl protonated samples, while the latter sample on its own only gave 45 and 36 NOEs, respectively.

6. The aromatic side chains were assigned based on 3D ^{13}C and ^{15}N NOESY spectra, ^{13}C HSQC, and a set of 3D HCCH COSY experiments optimized for the aromatic side chains. A total of 198 NOEs were obtained between the aromatic side chains and amide or methyl groups.

7. The all-*trans* retinal moiety was assigned from 2D DQF COSY and 2D NOESY experiments and through the comparison of 3D ^{13}C NOESY spectra where the latter experiment was recorded twice, with and without ^{13}C decoupling in the indirect ^1H dimension. The proton chemical shifts of retinal are indicative of an all-*trans* configuration. With the exception of the three ring CH_2 positions and the two geminal methyl groups all of the retinal protons could be assigned.

3.3. Structure Calculations and Structure Validation

To determine the structure of pSRII by NMR, the NOE peaks that were observed and were partially or fully assigned had to be converted into distance restraints. These conformational restraints, along with other information about the primary sequence, dihedral angles, and hydrogen bonds provided the main input for the structure calculations. The crystallography and NMR system (CNS) was used to carry out the molecular dynamics, energy minimization, and simulated annealing steps. The program ARIA with its iterative structure calculation protocol was employed to reduce the amount of ambiguous NOE distance restraints.

1. ARIA was supplied with ambiguous and unambiguous NOE-based restraints, along with peak intensities converted by the program into ambiguous distance restraints. Maximum peak heights of cross peaks in NOESY spectra were measured first and then converted into peak intensities using a macro from the CCPNMR *Analysis* program. In our study ARIA was provided with three combined NOE tables representative of 4D and two 3D NOESY spectra. Information on rotational correlation time, spectrometer frequencies, and mixing times was used for the relaxation matrix-based spin diffusion correction during the calibration of the NOEs in ARIA.

2. Backbone chemical shifts were used to predict the backbone torsion angles ϕ and ψ using the program TALOS+ (30, 31). A total of 190 predictions that were qualified as good were included as dihedral angle restraints in the structure calculations. They proved to be crucial for the prediction of the secondary structure.

3. Based on the results of solvent exchange experiments such as CLEANEX (32) in combination with the temperature dependence of amide proton chemical shift changes a total of 132 amide hydrogen bonds were included in the structure calculations. Hydrogen bond restraints were used provided the residues involved were in α-helical or the β-sheet region.

4. A three-dimensional model of all-*trans* retinal was built using the program PRODRG (33). The retinal model was then covalently attached to the Nε of lysine (Lys) 205. During the calculations the position of the retinal was adjusted through experimental NOE restraints that were found between the retinal moiety and pSRII.

5. The finalized ARIA simulated annealing protocol used for the structure calculation of pSRII consisted of 180 ps of torsion angle dynamics (TAD) (20,000 steps at 10,000 K) followed by two cooling stages: 72 ps of slow cooling from 10,000 to 1,000 K (24,000 steps) using TAD and then 72 ps of slow cooling from 2,000 to 50 K (24,000 steps) using Cartesian dynamics (CD) (see Note 14).

6. In the last set of ARIA structure calculations, a total of eight iterations were performed. For the first seven iterations 20 structures were calculated whereas 100 were calculated for the last iteration. From the final ensemble the 30 lowest energy structures were selected for further analysis. This family and the structure closest to the mean were analyzed with respect to their coordinate precision, energy, and deviations from the experimental restraints and idealized geometry. The ensemble had an RMSD of 0.48 Å over the backbone atoms and 0.86 Å over the heavy atoms of the ordered residues (1–221) with no dihedral or distance restraint violations in any of the 30 structures. The structure closest to the mean and a superposition of the 30 lowest energy structures are shown in Fig. 6.

7. Using PROCHECK a Ramachandran analysis of the ensemble and the structure closest to the mean was performed. 93.9% of the residues in the ensemble were in the most favorable regions whereas 5.3, 0.2, and 0.6% of the residues were in additionally allowed, generously allowed, and disallowed regions, respectively.

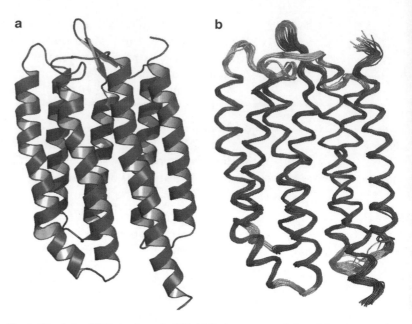

Fig. 6. 3D solution-NMR structure of pSRII. (**a**) Cartoon representation of the structure closest to the mean viewed across the membrane plane with the extracellular side of the protein at the top. (**b**) Overlay of the complete backbone traces for the 30 lowest energy structures shown in the same orientation as for (**a**). Loop regions are indicated in *light gray.*

4. Notes

1. M9 minimal medium in D_2O: The 2 L baffled flasks were first autoclaved empty and then dried in an oven until the condensation droplets on the inside surface of the flasks disappeared. M9 medium containing D_2O instead of H_2O was prepared using the same protocol except that D-glucose, NH_4Cl, and the M9 salts were prepared in D_2O while $FeCl_3$, the M9 salt mix, $MgSO_4$, and $CaCl_2$ were prepared in H_2O.

2. Starter cultures for the expression of [15]N- and ^{13}C,[15]N-labeled pSRII were grown in M9 minimal medium culture containing ^{15}N-ammonium chloride. Starter cultures for the expression of uniformly deuterated samples were grown in solutions that contained progressively higher levels of D_2O with ^{15}N-ammonium chloride.

3. ILV residue labeling was chosen due to the high abundance of these hydrophobic residues in pSRII (32%). The incorporation of the methyl-containing residues Ala (12%) and Thr (7%) could have extended the NOE network between the helices. However, to improve performance of the ILV-targeted experiments, Ala and Thr protonation was excluded. In return, methyl group assignments for Ala and Thr were derived from a protonated ^{13}C labeled sample.

4. In order to avoid excessive frothing during the resuspension, it is best to resuspend the membranes in Buffer S, followed by the addition of DDM. For this purpose, the Buffer S needs to be made up in 17/20th of the desired final volume—the remaining 3/20th are made up when a 10% DDM solution is added to a final concentration of 1.5%.

5. In this work the UV/vis absorbance of protein and retinal was used as the main criterion to assess quantity and purity of the prepared pSRII samples embedded in micelles. The amount of protein was calculated according to the following equation:

$$\text{protein quantity} = \frac{A_{498}MV}{\varepsilon_{498}},$$

where A_{498} is the absorbance at 498 nm, M is the molecular weight of pSRII, V is the volume, and ε_{498} is the extinction coefficient at 498 nm taken as 48,000 M^{-1} cm^{-1}.

The purity of the pSRII preparations was estimated as following:

$$\text{Purity} = \frac{\varepsilon_{280} / \varepsilon_{498}}{A_{280} / A_{498}}.$$

Based on the primary sequence, the extinction coefficient of pSRII at 280 nm (ε_{280}) was calculated to be 49,000 M^{-1} cm^{-1}, leading to a ratio $\varepsilon_{280}/\varepsilon_{498}$ of 1.02.

6. Frequently, an initial improvement in NMR spectral quality can be observed upon addition of more detergent as this results in increase in the number of micelles that are singly occupied with protein. Eventually, at higher concentrations the viscosity of the solution becomes too big so that the overall spectral quality deteriorates again.

7. The protein remains coated with a minimal amount of detergent, which delays protein denaturation when washing in the absence of additional detergent.

8. In order to obtain a detergent solution with a concentration similar to the CMC, an ~0.06% DDM solution was filtered using centrifugal concentrators with an MWCO of 10 kDa. While micelles could not pass through the membrane detergent monomers were found in the flow through at a concentration close to the CMC. The filtered detergent solution was then used to dilute the protein sample during the different rounds of dilution/concentration without increasing or decreasing the overall detergent concentration in the sample.

9. While the 3D HNCA and HN(CA)CB experiments benefit from the TROSY effect for ^1HN and ^{15}N at 800 MHz the experiments that involve transfer via the carbonyl nucleus

perform less well due to the rapid $^{13}C'$ transverse relaxation at higher fields. For consistency all experiments initially used for the backbone assignment were recorded at 600 MHz.

10. A typical non-uniformly sampled (NUS) experiment sampled 256 complex data points in the two indirect dimensions, which corresponds to ca. 17% of a fully sampled time domain matrix. Accordingly, to improve the overall sensitivity more transients were recorded.

11. 3D ^{15}N separated NOESY experiments were recorded as HSQC implementations, giving better sensitivity but also as TROSY-type experiments to benefit from the reduced ^{1}HN linewidth in the acquisition dimension.

12. In contrary to c6-DHPC at the time of our study c7-DHPC was not commercially available in deuterated form. The intense detergent signals interfered with the assignment procedure and were reduced using pulsed field gradient coherence order selection.

13. For the methyl groups the ^{13}C HMQC versions of the experiments were more sensitive than the ^{13}C HSQC implementations. All experiments were used as gradient selected experiments.

14. TAD uses fewer degrees of freedom only allowing rotation about bonds, whereas cartesian dynamics allows variation of the bond lengths and angles.

References

1. Kim HJ, Howell SC, Van Horn WD, Jeon YH, Sanders CR (2009) Recent advances in the application of solution NMR spectroscopy to multi-span integral membrane proteins. Prog Nucl Magn Reson Spectrosc 55:335–360

2. Vinothkumar KR, Henderson R (2010) Structures of membrane proteins. Quart Rev Biophys 43:65–158

3. Berman HM, Westbrook J, Feng Z, Gilliland G, Bhat TN, Weissig H, Shindyalov IN, Bourne PE (2000) The protein data bank. Nucleic Acids Res 28:235–242

4. Arora A, Abildgaard F, Bushweller JH, Tamm LK (2001) Structure of outer membrane protein A transmembrane domain by NMR spectroscopy. Nat Struct Biol 8:334–338

5. Fernandez C, Adeishvili K, Wuthrich K (2001) Transverse relaxation-optimized NMR spectroscopy with the outer membrane protein OmpX in dihexanoyl phosphatidylcholine micelles. Proc Natl Acad Sci U S A 98: 2358–2363

6. Hwang PM, Choy WY, Lo EI, Chen L, Forman-Kay JD, Raetz CRH, Prive GG, Bishop RE, Kay LE (2002) Solution structure and dynamics of the outer membrane enzyme PagP by NMR. Proc Natl Acad Sci U S A 99: 13560–13565

7. Hiller S, Garces RG, Malia TJ, Orekhov VY, Colombini M, Wagner G (2008) Solution structure of the integral human membrane protein VDAC-1 in detergent micelles. Science 321:1206–1210

8. Oxenoid K, Chou JJ (2005) The structure of phospholamban pentamer reveals a channel-like architecture in membranes. Proc Natl Acad Sci U S A 102:10870–10875

9. Schnell JR, Chou JJ (2008) Structure and mechanism of the M2 proton channel of influenza A virus. Nature 451:U512–U591

10. Zhou YP, Cierpicki T, Jimenez RHF, Lukasik SM, Ellena JF, Cafiso DS, Kadokura H, Beckwith J, Bushweller JH (2008) NMR solution structure of the integral membrane enzyme DsbB: functional insights into DsbB-catalyzed disulfide bond formation. Mol Cell 31:896–908

11. Van Horn WD, Kim HJ, Ellis CD, Hadziselimovic A, Sulistijo ES, Karra MD, Tian CL, Sonnichsen FD, Sanders CR (2009) Solution nuclear magnetic resonance structure of membrane-integral diacylglycerol kinase. Science 324:1726–1729

12. http://www.drorlist.com/nmr/MPNMR. html

13. Gautier A, Kirkpatrick JP, Nietlispach D (2008) Solution-state NMR spectroscopy of a seven-helix transmembrane protein receptor: backbone assignment, secondary structure, and dynamics. Angew Chem Int Ed 47: 7297–7300

14. Gautier A, Mott HR, Bostock MJ, Kirkpatrick JP, Nietlispach D (2010) Structure determination of the seven-helix transmembrane receptor sensory rhodopsin II by solution NMR spectroscopy. Nat Struct Mol Biol 17:768–774

15. Royant A, Nollert P, Edman K, Neutze R, Landau EM, Pebay-Peyroula E, Navarro J (2001) X-ray structure of sensory rhodopsin II at 2.1-angstrom resolution. Proc Natl Acad Sci U S A 98:10131–10136

16. Luecke H, Schobert B, Lanyi JK, Spudich EN, Spudich JL (2001) Crystal structure of sensory rhodopsin II at 2.4 angstroms: insights into color tuning and transducer interaction. Science 293:1499–1503

17. Gordeliy VI, Labahn J, Moukhametzianov R, Efremov R, Granzin J, Schlesinger R, Buldt G, Savopol T, Scheidig AJ, Klare JP, Engelhard M (2002) Molecular basis of transmembrane signalling by sensory rhodopsin II-transducer complex. Nature 419:484–487

18. Pervushin K, Riek R, Wider G, Wuthrich K (1997) Attenuated T-2 relaxation by mutual cancellation of dipole-dipole coupling and chemical shift anisotropy indicates an avenue to NMR structures of very large biological macromolecules in solution. Proc Natl Acad Sci U S A 94:12366–12371

19. Vranken WF, Boucher W, Stevens TJ, Fogh RH, Pajon A, Llinas P, Ulrich EL, Markley JL, Ionides J, Laue ED (2005) The CCPN data model for NMR spectroscopy: development of a software pipeline. Proteins: Struct Funct Bioinform 59:687–696

20. Brunger AT, Adams PD, Clore GM, DeLano WL, Gros P, Grosse-Kunstleve RW, Jiang JS, Kuszewski J, Nilges M, Pannu NS, Read RJ, Rice LM, Simonson T, Warren GL (1998) Crystallography & NMR system: a new software suite for macromolecular structure determination. Acta Crystallogr Section D: Biol Crystallogr 54:905–921

21. Linge JP, O'Donoghue SI, Nilges M (2001) Automated assignment of ambiguous nuclear overhauser effects with ARIA. Nucl Magn Reson Biol Macromol Part B 339:71–90

22. Koradi R, Billeter M, Wuthrich K (1996) MOLMOL: a program for display and analysis of macromolecular structures. J Mol Graphics 14:51–55

23. Laskowski RA, Macarthur MW, Moss DS, Thornton JM (1993) PROCHECK—a program to check the stereochemical quality of protein structures. J Appl Crystallogr 26:283–291

24. Laskowski RA, Rullmann JAC, MacArthur MW, Kaptein R, Thornton JM (1996) AQUA and PROCHECK-NMR: programs for checking the quality of protein structures solved by NMR. J Biomol NMR 8:477–486

25. Dagert M, Ehrlich SD (1979) Prolonged incubation in calcium chloride improves the competence of *Escherichia coli* cells. Gene 6:23–28

26. Sambrook J, Russell D (2001) Molecular cloning: a laboratory manual. Cold Spring Harbor Laboratory Press, Cold Spring Harbor, NY

27. Neidhard FC, Bloch PL, Smith DF (1974) Culture medium for enterobacteria. J Bacteriol 119:736–747

28. Rovnyak D, Frueh DP, Sastry M, Sun ZYJ, Stern AS, Hoch JC, Wagner G (2004) Accelerated acquisition of high resolution triple-resonance spectra using non-uniform sampling and maximum entropy reconstruction. J Magn Reson 170:15–21

29. Tugarinov V, Kay LE (2003) Ile, Leu, and Val methyl assignments of the 723-residue malate synthase G using a new labeling strategy and novel NMR methods. J Am Chem Soc 125: 13868–13878

30. Cornilescu G, Delaglio F, Bax A (1999) Protein backbone angle restraints from searching a database for chemical shift and sequence homology. J Biomol NMR 13:289–302

31. Shen Y, Delaglio F, Cornilescu G, Bax A (2009) TALOS plus: a hybrid method for predicting protein backbone torsion angles from NMR chemical shifts. J Biomol NMR 44:213–223

32. Hwang TL, Mori S, Shaka AJ, vanZijl PCM (1997) Application of phase-modulated CLEAN chemical EXchange spectroscopy (CLEANEX-PM) to detect water-protein proton exchange and intermolecular NOEs. J Am Chem Soc 119:6203–6204

33. Schuttelkopf AW, van Aalten DMF (2004) PRODRG: a tool for high-throughput crystallography of protein-ligand complexes. Acta Crystallogr Section D: Biol Crystallogr 60: 1355–1363

34. Isaacson RL, Simpson PJ, Liu M, Cota E, Zhang X, Freemont P, Matthews S (2007) A new labeling method for methyl transverse relaxation-optimized spectroscopy NMR spectra of alanine residues. J Am Chem Soc 129: 15428–15429

Chapter 4

Use of NMR Saturation Transfer Difference Spectroscopy to Study Ligand Binding to Membrane Proteins

Rani Parvathy Venkitakrishnan, Outhiriaradjou Benard, Marianna Max, John L. Markley, and Fariba M. Assadi-Porter

Abstract

Detection of weak ligand binding to membrane-spanning proteins, such as receptor proteins at low physiological concentrations, poses serious experimental challenges. Saturation transfer difference nuclear magnetic resonance (STD-NMR) spectroscopy offers an excellent way to surmount these problems. As the name suggests, magnetization transferred from the receptor to its bound ligand is measured by directly observing NMR signals from the ligand itself. Low-power irradiation is applied to a ^{1}H NMR spectral region containing protein signals but no ligand signals. This irradiation spreads quickly throughout the membrane protein by the process of spin diffusion and saturates all protein ^{1}H NMR signals. ^{1}H NMR signals from a ligand bound transiently to the membrane protein become saturated and, upon dissociation, serve to decrease the intensity of the ^{1}H NMR signals measured from the pool of free ligand. The experiment is repeated with the irradiation pulse placed outside the spectral region of protein and ligand, a condition that does not lead to saturation transfer to the ligand. The two resulting spectra are subtracted to yield the difference spectrum. As an illustration of the methodology, we review here STD-NMR experiments designed to investigate binding of ligands to the human sweet taste receptor, a member of the large family of G-protein-coupled receptors. Sweetener molecules bind to the sweet receptor with low affinity but high specificity and lead to a variety of physiological responses.

Key words: Saturation transfer difference (STD), Saturation transfer double difference (STDD), Membrane-bound receptors, G-protein-coupled receptor (GPCR), Sweet taste receptor, T1R2, T1R3, Neotame, Dextrose

1. Introduction

Nuclear magnetic resonance (NMR) is a powerful tool for directly monitoring ligand binding to protein receptors. Saturation transfer difference (STD) NMR can be used to monitor weak ligand binding ($K_d \sim$ mM–µM) via non-scalar magnetization transfer from a large protein (receptor, >20 kDa, example used here ~190 kDa for the

Nagarajan Vaidehi and Judith Klein-Seetharaman (eds.), *Membrane Protein Structure and Dynamics: Methods and Protocols*, Methods in Molecular Biology, vol. 914, DOI 10.1007/978-1-62703-023-6_4, © Springer Science+Business Media, LLC 2012

heterodimeric sweet receptor) to smaller ligands (1–3). STD-NMR does not require expensive stable isotope or radioisotope labeling, and the experiment requires only a low receptor concentration (nM–pM) in the presence of 20–1,000 times excess ligand in a sample of ~200 μl. Hence it offers an economical method to assay the function of proteins in an isolated form or when expressed on a cell surface (2, 4). Membrane proteins studied by this approach can be present in a cell surface membrane or in isolated membranes or liposomes. Transferred Nuclear Overhauser Enhancement (TrNOE) (5, 6) and saturation transfer methods (3), which both utilize transfer of magnetization between receptor and bound ligand, have been applied successfully in vivo (7, 8) to monitor oligomerization (9) and to measure dissociation constants for complexes (10). Because signals from small ligands are monitored, the molecular weight of the protein is not an issue with STD. In fact, spin diffusion is more efficient for a large or membrane-bound protein than for a small protein in solution. These experiments are particularly useful for monitoring weak (mM scale) binding. Further, when a protein (or receptor) has more than one binding site, competition STD experiments can be used to distinguish between competitive and noncompetitive ligand binding (11). The binding of multiple ligands can be analyzed simultaneously as long as signals from the different ligands do not overlap.

An alternative approach to investigating ligand binding by NMR spectroscopy is to observe signals from the protein itself as ligands are added. Binding is followed by monitoring changes in the chemical shifts of signals from the protein. This approach requires a high concentration of protein (>100 μM) labeled with stable isotopes. In addition, membrane proteins studied by this approach in solution must be investigated in detergent micelles or small bicelles so that they tumble rapidly enough to yield resolvable NMR signals.

Membrane proteins are notoriously difficult to produce and isolate in functional form. Over-expression in bacterial systems often requires refolding of the membrane proteins, and a functional assay is needed to determine whether the protein has refolded properly. The STD experiment is sensitive enough to detect ligand binding at protein concentration levels frequently found in cells (pM–nM). This relieves the need for high protein yields from expression systems making the use of bacterial expression systems unnecessary. Furthermore, samples of membrane-bound proteins from cells can be obtained under nondestructive conditions without unfolding and refolding the protein.

The initial step in the STD experiment is the saturation of proton spins on the receptor, which is achieved by applying a selective radiofrequency (RF) signal to a spectral region that contains signals from the receptor but not the ligand(s). Magnetization transferred from the receptor to the bound ligand serves to saturate its NMR signals. Upon dissociation, the saturated ligand contributes a decrease in the intensity of the NMR signal from the pool of free ligand until the ligand loses the saturation condition by longitudinal (R_1) relaxation. Because the R_1 relaxation rate is much slower than the rate of dissociation of ligands from the saturated protein, one protein has the capacity to saturate many ligand molecules. The k_{on}, k_{off}, and R_1 rates in conjunction with the power and duration of the selective RF determine the observed strengths of the ligand signals. Several experimental parameters can be varied to achieve optimal signal-to-noise ratios (S/N). If k_{off} is slow (tight binding, nM), the net saturation transferred to the pool of free ligand may be low. If k_{off} is very fast (extremely weak binding, i.e., >mM), the residence time of the ligand in the complex may be insufficient to achieve full saturation. Excess ligand is required to ensure that the receptor molecules are saturated sufficiently with ligand, but very high ligand concentrations dilute out the effects of saturation. Temperature and pH can be varied to modify k_{on} and k_{off}. As a rule of thumb, when the ligand:receptor ratio is increased from 20:1 to 1,000:1, the magnitude of observed saturation follows a sigmoidal curve. This information can be used to choose the optimal conditions for the STD experiment. To test for competitive binding to a single site on the receptor, the concentration of one ligand is changed at a time. In order to correct for effects of nonspecific binding, a control solution lacking the receptor or containing a receptor with blocked binding site can be used in a parallel STD experiment. Subtraction of the control STD from the experimental STD yields a double difference STD (STDD) spectrum (Figs. 5 and 6).

We have successfully used STD and STDD methods to monitor ligand binding to the sweet receptor (2, 12). By optimizing concentrations of receptor-to-ligand ratios and the temperature, we were also able to distinguish between ligands that bind to mutant receptors but fail to activate and those that do not bind at all (see Note 1) (Figs. 5 and 6).

The sweet receptor is a member of the family of Class C G-protein-coupled receptors. The heterodimeric sweet receptor consists of the two subunits: T1R2 and T1R3. Each T1R subunit contains a large extracellular domain called the Venus FlyTrap Module (VFTM), linked via a Cysteine-Rich Domain (CDR) to a seven helix Trans-Membrane Domain (TMD). A schematic of the sweet receptor is shown in Fig. 1. A large number of small molecule ligands are known to interact with the sweet receptor, and these interactions can be investigated in cells transfected with sweet receptors by means of a calcium flux assay. This has enabled comparison of the properties of sweet receptors from different species, such as mouse and human. Results have shown that dipeptide sweeteners, aspartame and its analog neotame, while active with human receptor, do not activate the mouse receptor (12). On the other

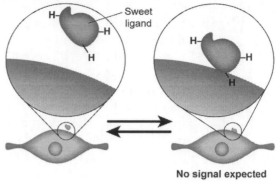

1) Parental HEK cells (negative control)

Sweet ligand

No signal expected
or small at noise level
(non-specific interaction)

2) T1R2/T1R3 HEK cells with transfected receptors

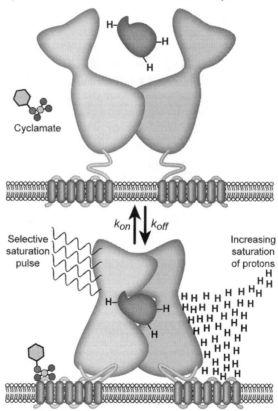

Cyclamate

k_{on} k_{off}

Selective
saturation
pulse

Increasing
saturation
of protons

Signal expected

Fig. 1. Schematic illustrating the use of STD NMR for monitoring binding interactions of non-expressing cells and cell-expressed receptors (2): (1) negative control: due to possible nonspecific binding of ligand to parental membrane background parental cells that are not transfected and do not contain receptor of interest, e.g., parental HEK cells, are used for preparing membranes from these cells as negative control membranes; (2) ligand binding to the sweet receptor is expected for T1R2/T1R3 transfected HEK cells where the receptors are expressed and displayed leading to STD signals. STD signals depend on both equilibrium constants (k_{on}/k_{off} rates), which describe kinetic interactions between ligand and receptor. A selective saturation NMR pulse is applied to the receptor to transfer magnetization from receptor through spin diffusion to the nearby (bound) ligand (*bottom*). Effects are detected as *STD signals* on the pool of free ligands by rapid exchange of the saturation transfer only if binding occurs.

hand, both human and mouse receptors are activated by natural sugars, such as dextrose and some small artificial sweeteners such as sucralose or sweet amino acids such as D-tryptophan. We prepared membranes from transfected cells expressing the sweet receptor and from the parental cell line for negative control (Fig. 1). The control parental cell line was used to detect and correct for nonspecific binding in the STDD spectrum.

We describe here our protocol for STD-NMR studies of neotame and dextrose binding to human sweet receptor subunits, which includes experimental procedures used in acquiring and analyzing STD and STDD data.

2. Materials

Most of the chemicals were purchased from Sigma Aldrich (http://www.sigmaaldrich.com/sigma-aldrich/home.html). Deuterated water was obtained from Cambridge Isotopes Limited (http://www.isotope.com/cil/index.cfm?123&CFID=25580810&CFTOKEN=89808295). Shigemi tubes were obtained from Shigemi Inc. (http://www.shigeminmr.com/main.html). Neotame was obtained from NutraSweet Company (http://www.nutrasweet.com/).

3. Methods

In the following four sections, we describe (Subheading 3.1) our method for preparing membrane samples containing proteins of interest, (Subheading 3.2) sample preparation for NMR, (Subheading 3.3) NMR data collection methods, and (Subheading 3.4) data analysis.

3.1. Preparation of Membranes for STD-NMR

This method uses transfected human embryonic kidney 293 E cells, grown in 175 cm^2 flasks. Transfect 293 E cells grown to 60–70% confluence in 175 cm^2 flasks ($\sim 1.2 \times 10^7$ cells/flask) with Lipofectamine-2000, per Invitrogen protocol, http://tools.invitrogen.com/content/sfs/manuals/lipofectamine2000_man.pdf; 48 h after transfection, harvest the cells using trypsin-free cell dissociation buffer (0.5 mM EDTA in phosphate-buffered saline—PBS). Suspend the cells in PBS after dissociating them from the flask and spin at $800 \times g$ for 10 min to pellet the cells as described below.

1. Day 1: 24 h prior to transfection, seed 6.2×10^6 cells in OPTI-MEM containing 10% fetal bovine serum with no antibiotics in a 175 cm^2 flask.

2. Day 2: Make sure that cells are 60–70% confluent before transfection. Dilute 75 µg DNA (37.5 µg DNA each of T1R2 and
T1R3 plasmid) in 2.0 ml OPTI-MEM (serum free) and mix
gently. Make sure that Lipofectamine 2000 is completely suspended by gently shaking it before use and then dilute 150 µl
of it into 2.0 ml serum-free OPTI-MEM. Mix gently and let sit
for 5 min at room temperature. Combine the diluted
Lipofectamine 2000 with the diluted DNA and incubate for
30 min at room temperature. Add 4 ml DNA-Lipofectamine
2000 mixture to each flask. Distribute the DNA–lipofectamine
solution over the cells by gently rocking the plate back and
forth. Incubate the cells at 37°C in a CO_2 incubator.

3. Day 3: Replace growth medium with regular OPTI-MEM containing antibiotics and serum. This is an optimized protocol
and works well with 293 cells. Usually, transfection makes cells
come off the plate more easily, resulting in loss. Gentle aspira-
tion and addition of new media are required for good yields.

4. Day 4: Remove media completely from the flasks and add 5 ml
trypsin-free cell dissociation rinse buffer (0.5 mM EDTA in
PBS). Rock the rinse buffer over the cells thoroughly but only
once, and then gently aspirate the medium. Leave at room
temperature for 5 min to allow the cells to detach. Gentle tap-
ping on the side of flask will help to dislodge them. Once
detached, add 10 ml PBS lacking Ca^{2+}/Mg^{2+} and harvest the
cells using a broken, fire-polished pipette. Avoid harsh handling of the cells, because if they lyse at this point, nuclear
DNA can be released, spoiling the membrane preparation. To
pellet the cells, spin at $800 \times g$ for 10 min. Do not spin any
faster, to avoid breaking the cells (see Note 1).

3.2. Membrane Preparation from Cell Pellet

Add 4 ml homogenization buffer, 20 mM Tris–HCl pH 7.4, 10%
glycerol, and complete protease inhibitor (Roche, Indianapolis,
IN) to the cell pellet and homogenize with a Polytron® homogenizer. This buffer composition and protease inhibitor cocktail work
very well. Allow only eight strokes of the Polytron at half-maximal
speed while the cells are kept on ice (see Note 2).

1. Centrifuge homogenate at $1,500 \times g$ for 15 min at 4°C to
remove unbroken cells, cell debris, and nuclei. The resulting
supernatant contains cytosol and total cellular membranes.
Transfer the supernatant to a new centrifuge tube.

2. Ultracentrifuge the supernatant at $100,000 \times g$ for 1 h at 4°C.
The resulting pellet is referred as the membrane pellet. Remove
the supernatant without disturbing the membrane pellet.

3. Add 8 ml homogenization buffer (20 mM Tris–HCl pH 7.4,
10% glycerol lacking protease inhibitor) to further wash, and
gently resuspend the membrane pellet using a 1 ml pipette tip.

Ultracentrifuge at $100,000 \times g$ for 30 min at 4°C, and carefully remove the supernatant without disturbing the resulting membrane pellet.

4. To the pellet add 200 µl homogenization buffer lacking protease inhibitor and resuspend by 20 passages through a 25-gauge needle. This renders an even suspension of membranes in the buffer (see Note 3). Protease inhibitor-free homogenization buffer is preferred to avoid interference from inhibitors in future experiments.

5. Membranes can be stored in this buffer at –80°C. Total membrane protein concentration is checked by the Lowry protein assay (13). Total membrane protein (including the receptor) is found to be 10–13 µg/µl when following this protocol.

6. For STD-NMR studies, prepare 50–75 µg total protein/160 µl NMR PBS consisting of 137 mM NaCl, 10 mM Na_2HPO_4, 2.7 mM KCl, and 1 mM KH_2PO_4 at pH 7.2–7.6 in D_2O (see Note 4).

3.3. Preparation of the Receptor–Ligand Complex (the Same Procedure is Used to Prepare Control Membrane Without Receptor)

The total amount of membrane protein to be used in an NMR sample is 50–75 µg. We found that membrane protein preparations were stable in PBS buffer. Buffers for small molecule ligands were prepared in 99.98% D_2O; buffers for peptide or small protein ligands were prepared in H_2O (see Note 5).

1. To 50 µl of membrane (with its incorporated receptor), 150 µl of ice-cold PBS buffer is added, and the membrane is resuspended by gentle pipetting (using 250 µl tip size) up and down 2–3 times. The sample is spun in a tabletop centrifuge at $800 \times g$ for 3 min.

2. The supernatant is removed; to the washed pellet containing the receptor, 160 µl of ligand solution (at the appropriate concentration: e.g., 5 mM for neotame, obtained from http://www.nutrasweet.com/) in deuterated PBS buffer is added, and the pellet is resuspended by gentle vortexing at $5,000 \times g$ for 5 min (see Note 6).

3. A well-dispersed sample has an opaque color (Fig. 2). If membranes are not prepared correctly, they can include genomic DNA, which will prevent proper dispersal. The sample turns to a solid clump without any dispersion (see Note 7). In such a situation, the sample should be discarded and replaced with freshly made membrane.

4. The ligand concentrations varied from 1 mM for the sweet proteins (brazzein and other protein ligands) to 5–10 mM for small molecule ligands (see Note 8).

5. The membrane sample (150 µl with opaque color) is placed into a 3 mm outer diameter (o.d.) Shigemi tube (Fig. 2), and

Fig. 2. (*Left*) A 1.7 mm o.d. capillary tube (30 ml) used in a 1.7 mm cryogenic probe when membrane or protein samples are severely limited. (*Right*) Membrane sample (150 ml) placed into a 3 mm o.d. Shigemi tube with opaque color showing the appearance of a properly dispersed membrane sample prepared for NMR. We have used 3 and 5 mm Shigemi tubes in 5 mm cryogenic probes.

NMR data are recorded immediately. We also use NMR tubes of other sizes (with different volume requirements): 5 mm o.d. regular or Shigemi tube (500 μl), or 1.7 mm o.d. NMR tube (30 μl) (see Note 9).

3.4. STD NMR Spectroscopy

3.4.1. NMR Data Collection

NMR data are collected at 25°C on a Varian (Agilent Technologies) Avance 600 NMR spectrometer equipped with a cryogenic probe (see Note 10). The pulse programs are written in house and can be obtained from nmrfam.wisc.edu.

1. Selective saturation of the receptor is achieved by a train of Gaussian-shaped pulses of about 30 ms each, saturating a bandwidth of about 20 Hz, at –2 ppm (where the receptor has signals, but the ligands do not) for a saturation time of approximately 3 s. This ensures full saturation of the receptor (see Note 11). The on-resonance irradiation of the protein can be

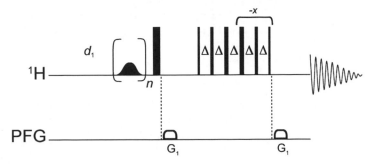

Fig. 3. Pulse sequence used in the STD-NMR experiment. The shaped pulse in the beginning is for saturation; the 90° ¹H pulse is followed by gradients and 3-9-19 pulses for water suppression by Watergate.

varied depending on the type of ligand present (normally around −1.0 ppm to 1.0 ppm).

2. Watergate 3-9-19 pulse sequence with gradients is employed prior to acquisition of the FID signal for solvent water suppression (see Note 12).

3. Data are collected with 12 k complex points in the direct dimension for 1.5 s (see Note 13). Small molecules relax slowly and are in a dynamic equilibrium in solution between the free and in complex states. A delay of 3 s is applied between each FID to ensure complete relaxation of the ligand.

4. 256–1,024 scans are accumulated for each of the STD experiments depending on signal intensity. The pulse sequence used to record the experiments is shown in Fig. 3. STD pulse sequence is available in Varian or Bruker pulse libraries which require their own specific pulse sequences set up for this section please consult your NMR pulse library and its setup. For an experiment of 256 scans, the experiment time is about 160 min.

5. Off-resonance irradiation is set at ~50 ppm, a region where no receptor or ligand signals are present. The spectra for both on-resonance and off-resonance saturation are collected interleaved (see Note 14). The decrease in signal intensity, resulting from the transfer of saturation from the protein to the ligand, is evaluated by subtracting the on-resonance spectrum from the off-resonance spectrum. This subtraction yields a positive signal from a bound ligand.

6. An in-house script is used to separately process and add on- and off-resonance FIDs in Varian (Agilent Technologies) VNMRJ software (http://www.varianinc.com/cgi-bin/nav? products/nmr/apps/corner&cid=LPHKNLKJFL), and the peak intensities are manually inspected. Usually the difference between on and off reference spectra is between 10 and 200

times less than the ^{1}H signal intensity. If the STD signal is very weak or negligible it can be at noise level or 1,000 times less than the 1D intensity. If any FID has values differing significantly from the average, the FID is excluded from further analysis (see Note 15).

7. To remove any possible nonspecific binding contribution of the ligand to the parental membrane itself, the experiment is repeated with a membrane preparation lacking the receptor of interest; this is referred to as a negative control. The STD signal from the negative control is subtracted from that of the receptor-containing preparation to yield the STDD spectrum (see Notes 16 and 17).

8. With a ^{15}N-labeled peptide or ^{15}N–^{2}H-labeled protein samples, either 1D or 2D STD edited ^{1}H–^{15}N HSQC NMR spectra are collected, with 80–100 increments in the indirect dimension and with 2 k sampling points in the direct dimension. The in-house written pulse program is available by request at nmrfam. wisc.edu. The resulting data are processed and analyzed by NMRPipe   (http://spin.niddk.nih.gov/NMRPipe/)   and SPARKY (http://www.cgl.ucsf.edu/home/sparky/) software.

3.4.2. NMR Data Analysis and Interpretation

The subtraction of signals arising from nonspecific interaction with the parental membrane from total specific and nonspecific binding signals to membrane expressing the receptor protein can be achieved in the time domain or the frequency domain (see Note 18). VNMR provides simple and elegant steps that enable time domain subtraction. All spectra are processed and analyzed using the VNMRJ (version 2.1B) software in the time domain mode (see Note 19). The steps involved in data processing using VNMRJ are the following:

1. Process the interleaved 1D data for receptor and parental membrane samples to obtain the resulting STD signals. By default, all the operations are done on arbitrary experiment number 5 (see Note 20). Save the 1D STD spectrum, which is the *third increment* with the appropriate file name.

2. In two new experiments other than number 5, load the two saved spectra.

3. Go to the experiment containing receptor STD data. Process data with the "wft" command. Apply shifted sine multiplication and line broadening parameters as per requirement. Routinely sbs = sb and lb = 5 are used for processing. Reference the peaks of interest (see Note 20).

4. Command "clradd add" deletes and adds the experiment on experiment number 5.

a	b
``#!/bin/csh
var2pipe -in ./fid -noaswap \
 -xN   24038 \
 -xT   12019 \
 -xMODE   Complex \
 -xSW   8012.821 \
 -xOBS   599.708 \
 -xCAR   4.773 \
 -xLAB   H1 \
 -ndim   1 \
 -out ./test.fid -verb -ov``` | ```#!/bin/csh

#
Basic 2D Phase-Sensitive Processing:
Cosine-Bells are used in both dimensions.
Use of "ZF -auto" doubles size, then rounds to power
of 2.
Use of "FT -auto" chooses correct Transform mode.
Imaginaries are deleted with "-di" in each dimension.
Phase corrections should be inserted by hand.

nmrPipe -in test.fid \
| nmrPipe -fn SP -off 0.5 -end 1.00 -pow 1 -c 1.0   \
| nmrPipe -fn ZF -auto   \
| nmrPipe -fn FT -auto   \
| nmrPipe -fn PS -p0 74.00 -p1 0.00 -di -verb   \
| nmrPipe -fn EXT -x1 0.4ppm -xn 1.2ppm \
  -ov -out test.ft2``` |

Fig. 4. (**a**) Example script for Varian (Agilent) to NMRPipe format conversion; (**b**) example script for processing the converted NMRPipe data.

5. Go to the experiment containing the parental (negative control) STD data. Process as in step 3. Use command "add(−1.0)" to subtract the data on experiment number 5.

6. If normalization is required for the parental membrane, use the coefficient c in *add(c)* instead of −1.

7. Go to experiment number 5 and process with command "wft."

Data processing and subtractions in the frequency domain can be achieved by a combination of NMRPipe software for data processing and in-house-developed Newton software for subtractions. Newton is an application for performing spectral deconvolution. The program contains libraries for performing generalized addition and subtraction of NMR matrices in NMRPipe and SPARKY data formats and is available at nmrfam.wisc.edu. The frequency domain data are matched by chemical shifts in the two spectra, by grid interpolation before subtraction. First, the data from the spectrometer are converted to NMRPipe format. The data are further Fourier transformed with required processing parameters and correct phases (see details in VNMRJ manual; http://www.varianinc.com/). Samples of NMRPipe conversion scripts are given in Fig. 4. To add spectra, the command "newton add-matrix −m1 file1.ft2 −m2 file2.ft2 −c2 coefficient 2" is used, where coefficient 2 is the scaling factor for the second matrix.

Examples of STD spectra produced by time domain analysis are shown in Fig. 5. The STD results show binding of dextrose to human (hT1R2) and mouse (mT1R2) sweet receptor subunits.

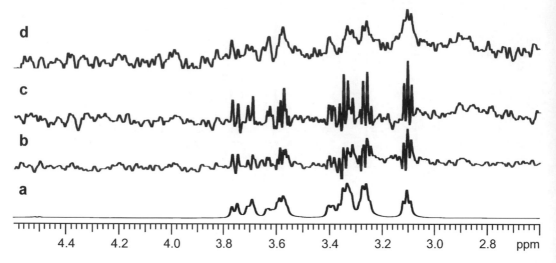

Fig. 5. (**a**) ^{1}H 1D spectrum of dextrose; (**b**) dextrose binding to parental membrane; (**c**) dextrose binding to mT1R2; (**d**) dextrose binding to hT1R2. Data sets were processed by VNMRJ software in the time domain.

A control 1D ^{1}H spectrum of dextrose is shown in *panel a*. Dextrose binds weakly to parental membrane alone, as seen in *panel b*. This signal is subtracted to obtain the STDD signal for receptor binding. Dextrose binding to mT1R2 is shown in *panel c*, and dextrose binding to hT1R2 is shown in *panel d*. Notice the significant line broadening of the ligand peaks upon binding to the receptor.

Previous mutational and chimeric studies have reported several binding sites on the heterodimeric sweet receptor (T1R2 + T1R3) (14). Jiang et al. have demonstrated that the TMD of hT1R3 is the binding site for lactisole (a sweet receptor antagonist) (15), whereas the extracellular domain of hT1R2 is the binding site for neotame (a dipeptide sweet receptor agonist) (Max and Maillet, submitted). However, binding evaluation of the T1R2 (and T1R3) subunit has remained elusive because homodimeric form or interspecies form of the receptor mT1R2 + hT1R3 cannot be activated by the sweet ligand (16). Previous studies indicated that neotame does not activate to mT1R2 + mT1R3 (12); however, it is unknown whether this reflects lack of binding or binding that does not lead to activity. STD experiments may prove to be useful to test the binding properties of sweet ligands such as dipeptide sweeteners (for example: neotame) to domains for which species differences in activity cannot be used.

Examples of STD spectra produced by frequency domain processing are illustrated in Fig. 6, which shows results for neotame binding to the purified VFTM (the extracellular domain) of hT1R2 as is consistent with indirect biological chimera studies (12). This example shows that STD NMR also can serve as a powerful method to detect binding to isolated domains of receptor proteins which cannot be determined from activity assays.

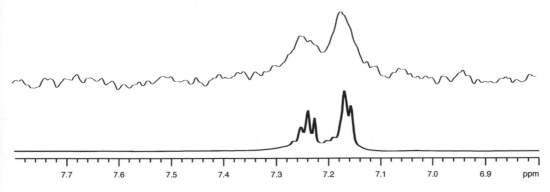

Fig. 6. (a) ^{1}H 1D spectrum of neotame; (b) STD signal resulting from binding of neotame to the VFTM of T1R2. Only resonances at lower frequencies from the water signal are shown; these signals arise from binding of neotame to the purified amino terminal domain of the hT1R2 subunit. Data were processed by NMRPipe software, and subtraction was performed by Newton (in-house) software in the frequency domain.

4. Notes

1. $800 \times g$ for 10 min is just sufficient to spin the cells down and will not lyse cells. Centrifugal speeds or longer times may result in cell lysis, so take care in maintaining the recommended limits.

2. To protect the receptor-expressing cells from trypsin digestion during cell harvest, it is important to harvest the cells in PBS containing 0.5 mM EDTA. While homogenizing the cells, great care has to be taken to restrict homogenization to only eight strokes. Exceeding this number may result in DNA breaks resulting in contamination of the preparation. DNA contamination leads to string-like clumps in the supernatant and ruins the membrane preparation.

3. At the stage of final suspension, the membrane has to be thoroughly mixed by pulling up and down with a syringe. Incomplete suspension may result in clumping of suspension. Aliquot membrane preparations in smaller volumes in Eppendorf tubes and keep frozen at −80°C for long-term storage.

4. To make PBS buffer for 1-l volume, weigh the following: 8 g NaCl (137 mM), 1.15 g Na_2HPO_4 (10 mM), 0.2 g KCl (2.7 mM), and 0.2 g KH_2PO_4 (1 mM). Dissolve in 1-l double-distilled water; the pH should be around 7.4.

5. Small molecule ligands are normally detected by conventional one-dimensional (1D) ^{1}H STD-NMR experiments and do not require labeling. With ligands that are peptides or small proteins, uniform ^{15}N-labeling (or uniform ^{2}H, ^{15}N-labeling with back exchange in H_2O to protonate NH-groups) can be used along with ^{1}H–^{15}N edited 2D-STD detection to determine what parts of the ligand bind to the receptor (12).

6. When the receptor is membrane-spanning, it is crucial for the success of the experiment to obtain stable membrane preparations at optimal pH and ionic strength. In small molecule ligand-binding experiments, the signal-to-noise ratio of the detected signal can be improved by exchanging the solvent from H_2O to D_2O to improve the suppression of the solvent signal.

7. It is easier and cleaner to transfer samples to and from NMR tubes with a pipette tip that reaches the bottom of the tube. Commercially such tips are not available for purchase. For transfers to and from 3-mm NMR tubes, we added a 1 mm polyethylene capillary tube (Intramedic, http://vwrlabshop.com/intramedic-polyethylene-tubing-clay-adams/p/0013744/) running along the length of the NMR tube.

8. However, only 7–10% D_2O solvent (as required for the NMR frequency lock) should be used for ^{15}N-labeled peptide or protein ligands to minimize the loss of backbone amide signals. Further, if a dual sample spinner of 2.7 mm is available, samples can be prepared in 100% H_2O to increase S/N. In this case, the first tube contains the sample and the second tube is filled with deuterated buffer to provide the frequency lock for acquisition.

9. The bottom of the Shigemi tube is constructed of glass whose magnetic susceptibility matches that of an aqueous solution. The sample is placed in the tube, and a plunger made of susceptibility-matched glass is lowered to the top of the solution. Care must be taken to avoid air bubbles to ensure good shimming quality. The solid base and the plunger together limit the sample volume to the most sensitive region of the NMR receiver coil. We used 3 mm Shigemi tubes to reduce the sample volume to 150 μl. If sample volume is not limiting, a 5 mm Shigemi tube be used, which requires 250–300 μl sample volume. Although the 3 mm Shigemi tube reduces the sample volume, the S/N is lower than that from a solution of the same concentration in a 5 mm Shigemi tube.

10. Cryogenic probes increase S/N by about two when compared to room-temperature probes. If the spectrometer is not equipped with a cryogenic probe, data collection times are nearly four times longer to obtain similar S/N.

11. It should be noted that if the receptor or the ligand is completely deuterated, there is no saturation transfer, because there are no protons present to saturate. If the ligand is a larger protein, it may not be possible to selectively saturate the receptor, and artifacts will result from joint saturation of receptor and ligand.

12. Without water suppression, the spectrum may exhibit a large baseline roll. This can give rise to anomalies in subtraction resulting in peaks in the vicinity of water. For small ligands, the

use of >98% deuterated PBS buffer removes most of the water signal. However, the residual (1–2%) water peak is still larger than the ligand signals. Thus, water suppression ensures cleaner subtraction of the on- and off-resonance spectra.

13. The long acquisition time ensures detection of magnetization that is transferred upon binding to small molecules. If the ligand is a protein with larger molecular weight, shorter acquisition time may be employed.

14. Owing to long data collection times and low S/N, any instability during a single FID is sufficient to cause discrepancies in the subtracted data. Hence, on-resonance and off-resonance data are collected in an interleaved fashion.

15. A ^{1}H 1D spectrum is to be collected to decide the required offset for on-resonance irradiation. The irradiation is performed at least 1–2 ppm away from the nearest ligand resonances. If the on-resonance irradiation is too close to ligand resonances, ligand signals will be excited and will be seen in the spectrum even if there is no binding. Several NMR standardization experiments are required to ensure that the STD NMR experiment is set up properly on a given spectrometer. Examples of such experiments consist of ligand only in the sample buffer to ensure no excitation of the ligand by itself and no STD signal. Further, a sample of protein–ligand complex known to exhibit an STD can be used as a positive standard control. We recommend using 1:20 aldehyde dehydrogenase:NADH as positive STD standard control. This sample can be used to optimize the S/N ratio by varying ligand concentration. Some of the ligands under investigation exhibited STD signals in the presence of parental membrane (negative control) in which no receptor was expressed. In this case the STD spectrum of the parental membrane preparation is subtracted from that for membrane containing the expressed receptor to obtain the double difference saturation transfer (STDD spectrum containing specific signals arising from receptor–ligand binding).

16. It is important to have the receptor sample and the negative control membrane under similar concentrations and buffer conditions. All experimental conditions, including protein and ligand concentrations, sample volume, temperature, and number of scans, should be kept the same for both samples.

17. In time-domain subtraction of data we assume that the two data sets have the same offset, sweep width, and number of points. If there is a shift in the frequencies between the sample and control spectrum, it can be adjusted in the frequency domain before subtraction.

18. With data collected on Bruker spectrometers running TopSpin™ or XWIN-NMR, built-in processing scripts can be

used for data processing and analysis. For instructions on using the above programs, consult the manuals that come with the specific version of the program and the spectrometer. Currently, the pointers to manuals can be reached at http://www.bruker-biospin.com/software_nmr.html.

19. If the VnmrJ version used has a default experiment number other than 5, then the processed data usually go to that experiment number. The user should verify this experiment number in advance. In our example, we set the default experiment to number 5.

20. In VnmrJ, *wft* refers to Fourier transformation of the raw NMR data (to transform time domain data to frequency domain). A window function multiplication is normally used after transformation. Routinely, a sine-bell window function is used and is denoted as *sb*. Other window functions applied prior to Fourier transformation are *sbs* (shifted sine bell shifted) and *lb* (line broadening). The command "clradd" clears any previously added data from experiment 5, and "add" adds the present data to newly created default experiment 5. Because the control experiment is to be subtracted, a coefficient of –1.0 is added in parentheses and the command "add (–1.0)" is used to subtract the control signals from the sample. For referencing, 0.5–1 mM DSS may be used as internal standard.

Acknowledgement

This research was supported by NIH grants R21 DC008805 to MM and FAP, R01 DC009018 and Wisconsin Institute of Discovery Grant (WID-135A039) to FAP, R01 DC006696, R01 DC008301 to MM, and P41 RR02301 which funds the National Magnetic Resonance Facility at Madison. We thank Drs. Marco Tonelli and Roger Chylla for their helpful discussions and assistance with the Newton program developed by Dr. Chylla. We thank the NutraSweet Company for providing the sample of neotame.

References

1. Wang YS et al (2004) Competition STD NMR for the detection of high-affinity ligands and NMR-based screening. Magn Reson Chem 42:485–489

2. Assadi-Porter FM et al (2008) Direct NMR detection of the binding of functional ligands to the human sweet receptor, a heterodimeric family 3 GPCR. J Am Chem Soc 130: 7212–7213

3. Mayer MMayer B (1999) Characterization of ligand binding by saturation transfer difference NMR spectroscopy. Angew Chem Int Ed 38:1784–1788

4. Assadi-Porter FM et al (2010) Interactions between the human sweet-sensing T1R2-T1R3 receptor and sweeteners detected by saturation transfer difference NMR spectroscopy. Biochim Biophys Acta 1798:82–86

5. Balaram P et al (1972) Negative nuclear Overhauser effects as probes of macromolecular structure. J Am Chem Soc 94:4015–4017

6. Megy S et al (2005) STD and TRNOESY NMR studies on the conformation of the oncogenic protein beta-catenin containing the phosphorylated motif DpSGXXpS bound to the beta-TrCP protein. J Biol Chem 280:29107–29116

7. Claasen B et al (2005) Direct observation of ligand binding to membrane proteins in living cells by a saturation transfer double difference (STDD) NMR spectroscopy method shows a significantly higher affinity of integrin alpha(IIb)beta3 in native platelets than in liposomes. J Am Chem Soc 127:916–919

8. Mari S et al (2004) 1D saturation transfer difference NMR experiments on living cells: the DC-SIGN/oligomannose interaction. Angew Chem Int Ed Engl 44:296–298

9. Di Micco S et al (2005) Differential-frequency saturation transfer difference NMR spectroscopy allows the detection of different ligand-DNA binding modes. Angew Chem Int Ed Engl 45:224–228

10. Angulo J et al (2010) Ligand-receptor binding affinities from saturation transfer difference (STD) NMR spectroscopy: the binding isotherm of STD initial growth rates. Chemistry 16:7803–7812

11. Feher K et al (2008) Competition saturation transfer difference experiments improved with isotope editing and filtering schemes in NMR-based screening. J Am Chem Soc 130:17148–17153

12. Assadi-Porter FM et al (2010) Key amino acid residues involved in multi-point binding interactions between brazzein, a sweet protein, and the T1R2-T1R3 human sweet receptor. J Mol Biol 398:584–599

13. Lowry OH et al (1951) Protein measurement with the Folin phenol reagent. J Biol Chem 193:265–275

14. Cui M et al (2006) The heterodimeric sweet taste receptor has multiple potential ligand binding sites. Curr Pharm Des 12:4591–4600

15. Jiang P et al (2005) Lactisole interacts with the transmembrane domains of human T1R3 to inhibit sweet taste. J Biol Chem 280:15238–15246

16. Jiang P et al (2004) The cysteine-rich region of T1R3 determines responses to intensely sweet proteins. J Biol Chem 279:45068–45075

Chapter 5

How to Investigate Interactions Between Membrane Proteins and Ligands by Solid-State NMR

Andrea Lakatos, Karsten Mörs, and Clemens Glaubitz

Abstract

Solid-state NMR is an established method for biophysical studies of membrane proteins within the lipid bilayers and an emerging technique for structural biology in general. In particular magic angle sample spinning has been found to be very useful for the investigation of large membrane proteins and their interaction with small molecules within the lipid bilayer. Using a number of examples, we illustrate and discuss in this chapter, which information can be gained and which experimental parameters need to be considered when planning such experiments. We focus especially on the interaction of diffusive ligands with membrane proteins.

Key words: Solid-state NMR, Membrane proteins, Protein–ligand interactions

1. Introduction

Membrane proteins are responsible for mediating cross talk between extra- and intracellular events and they enable transport across the membrane. This makes membrane proteins major targets for pharmaceuticals. Unfortunately, this interest is not matched by the availability of protein structures to enable rational drug design. With more than 66,993 total protein structures in the pdb archive (www.pdb.org) (1), as of April 2011, membrane proteins with or without ligand are represented by less than 1%. While solid-state NMR (ssNMR) might be difficult and too expensive to fully integrate it into high-throughput approaches that are used to find new drugs it can certainly help to optimize possible candidates, for example by characterizing structure, dynamics, protein- and lipid interactions (2). The method is especially strong in cases of larger or hydrophobic ligands, which are difficult to approach by other

---

Andrea Lakatos and Karsten Mörs contributed equally to this chapter.

Nagarajan Vaidehi and Judith Klein-Seetharaman (eds.), *Membrane Protein Structure and Dynamics: Methods and Protocols*, Methods in Molecular Biology, vol. 914, DOI 10.1007/978-1-62703-023-6_5, © Springer Science+Business Media, LLC 2012

techniques. In the absence of high-resolution crystallographic data, ssNMR is able to provide unique site-specific protein information about the binding event. But even when crystallographic data are available, ssNMR offers highly complementary data on structure, dynamics, and specific interactions within a more native like lipid environment. Changes in structure and dynamics within the protein and within the ligand upon complex formation can be followed.

This chapter addresses the needs of the membrane protein biochemist who is considering ssNMR studies for investigating a protein–ligand complex and would like to have an overview about opportunities and challenges before approaching an NMR colleague for setting up a collaboration. We describe different experimental conditions and scenarios on how to investigate the binding of small molecules to large membrane proteins. A number of excellent reviews explaining in detail technical, theoretical and application aspects of such studies have been published during the last years and we point the reader's attention to those publications (e.g., (3, 4)). Therefore, we restrict ourselves to the discussion of ssNMR studies of ligands noncovalently bound to membrane proteins. We pay special attention to experimental parameters, which need to be considered when planning such studies. Using a number of examples, we show under which conditions (1) ligand binding parameters, (2) ligand structures in the protein binding pocket, (3) the structural response of the protein upon ligand binding, and (4) molecular dynamics of bound ligands can be obtained.

2. Which Sample Preparation Criteria Have to be Fulfilled for ssNMR Studies on Membrane Protein–Ligand Complexes?

2.1. Membrane Proteins

Expression and isotope labeling: One of the key criteria and often limiting factor is the amount of sample needed for ssNMR experiments which is usually in the range of nanomole to micromole, which means milligrams of protein within a sample volume of tens of microliters. Therefore, suitable prokaryotic expression systems like *Escherichia coli* or *Lactococcus lactis* or eukaryotic systems such as yeast (*Saccharomyces. cerevisiae* and *Pichia pastoris*), insect and mammalian cells are chosen. Due to the wide variety and easy available cloning and expression tools as well as scalability, *E. coli* is probably the most used expression system. Also in terms of ^{13}C or ^{15}N isotope labeling, the *E. coli* system provides a variety of options for uniform or selective labeling of membrane proteins. For uniformly labeled samples, growth media are usually supplemented by ^{13}C-glucose or ^{13}C-glycerol as well as ^{15}N ammonium chloride. Since most membrane proteins consist of α-helical bundles, it might be necessary to decrease the number of resonances to reduce spectral overlap. Here, the use of (1,3-^{13}C)-glycerol and

(2-^{13}C)-glycerol (5, 6) or 10% glucose (7) has been shown to work well. To reduce spectral crowding further or when functional aspects or specific interactions are of interest, extensive labeling is not always preferable. Residue specific labeling can be achieved for many amino acid types (8, 9) by adding the labeled amino acids to the otherwise unlabeled medium (6) or vice versa (10). Problems due to metabolic scrambling of some amino acids can be overcome by the use of auxotrophic strains (11) or optimized protocols (12). Therefore, almost any amino acid can be labeled. So-called unique pair labeling (13, 14) takes advantage of selective labeling strategies. Here, pairs of amino acids that only occur once in the primary structure are selected: as a result, residue i is ^{13}C labeled and residue $i+1$ is ^{15}N enriched. By transferring magnetization between ^{15}N and ^{13}C only those pairs will be observed and can therefore be unambiguously assigned. A number of computational tools have been developed to optimize the selection of unique pairs in a given protein (14, 15). These approaches allow for minimizing the amount of samples needed for an unambiguous assignment of the whole protein or selected regions of interest such as ligand binding sites.

A major obstacle in ssNMR experiments is the considerable background from naturally abundant isotopes. This usually prevents the observation of substrate resonances that overlap with resonances from lipids or proteins. To reduce background signals it is possible to grow cells in minimal medium with ^{13}C-depleted D-glucose (16). This allows ^{13}C detection of low molecular weight substrates bound to large membrane proteins.

Novel approaches to label only parts of a protein are protein *trans*-splicing (17) or chemical ligation (18). These approaches enable the labeling of protein segments by separate expression and consecutive joining of the expressed fragments. Although not yet shown for membrane proteins these techniques should enable the possibility to study larger systems while reducing the spectral overlap.

In cases in which *in vivo* expression of a desired protein does not provide the yield necessary for NMR or the protein is toxic for the cells, cell-free protein synthesis is a valuable tool to overcome these hurdles (19). For cell-free expression a variety of sources can be used, e.g., rabbit reticulocytes, wheat germ (20), *Leishmania tarentolae* (19), or *E. coli* cells (13, 21–23). The cell free system allows for complete control of the expression system which makes it possible to adjust the conditions according to the protein of choice. This includes the addition of detergents (24) or lipids (micelles or even liposomes) to provide a lipophilic environment for the membrane protein (25). Cell-free expression offers large flexibility with respect to isotope labeling.

Purification and reconstitution: Some membrane proteins can be expressed in high amounts so that they constitute up to 70% of the total membrane proteins in the inner bacterial membrane (26). In those cases the purification of the inner membrane can be sufficient for functional studies, which represents the most native state for ssNMR. An experimental limitation is given by the background signals from other proteins. Therefore, a high ligand binding specificity has to be assured. For structure determination, or in cases where the expression is less efficient, the protein has to be solubilized by detergent. A detergent should efficiently solubilize the protein while maintaining its structure and function. Although significant progress has been made in finding new detergents (27–30), defining a solubilization protocol for a particular membrane protein remains a matter of trial and error (31).

ssNMR allows to study membrane proteins in a lipid environment that resembles their native environment. Reconstituting solubilized membrane proteins into lipid bilayers is critical. The detergent has to be removed completely in order to allow for liposome formation but slow enough preventing protein aggregation. Typical methods to achieve this are extensive dialysis or treatment with nonpolar polystyrene beads (32, 33), as well as rapid dilution or on-column buffer exchange (34). The amount of protein in the NMR sample has to be on the order of nanomole to micromole. To be able to get this high amount of protein into a standard MAS-NMR rotor with <70 μL volume, a low lipid to protein ratio is required. While biochemical assays are often done with a lipid to protein ratio of, for example, 8,000 (mol/mol) (35) Hellmich et al. demonstrated that the right choice of lipid allows to use a ratio of 150:1 (mol/mol), while the protein stays fully active even under MAS conditions (36). Even lower lipid-protein ratios have been reported. Shi et al. were able to reconstitute green proteorhodopsin at a ratio as low as 20:1 (mol/mol) (37). To reduce the amount of lipids even further, two-dimensional crystallization might be a valuable approach (38–40). Two-dimensional crystals could in principle also help to improve the spectral quality if sample homogeneity is improved by crystallization. The success of the reconstitution protocol has to be monitored with respect to protein activity in the lipid bilayer as well as with respect to spectral quality.

Special consideration has to be given to the titration of ligands to reconstituted membrane protein samples. Usually, MAS-NMR studies are done on dispersions of multilamellar vesicles. This could cause problems with binding site accessibility. Therefore, depending on the specific protein–ligand complex, it needs to be decided at which preparation stage the ligand is added. An option to achieve full binding site accessibility while maintaining a lipid bilayer environment is given by the use of nanodiscs (41, 42).

2.2. Ligands

The key criterion for detecting a ligand bound to a membrane protein by ssNMR is the existence of an NMR active nucleus. Protons are not very suitable for direct detection in ssNMR due to their strong homonuclear dipolar coupling network which causes severe line broadening. Ideal nuclei are ^{13}C, ^{15}N, ^{2}H, ^{19}F, and ^{31}P.

The first three are the most commonly used isotopes especially for peptide ligands, but in some cases metal binding (e.g., ^{113}Cd) has been observed directly as well (43).

Isotope labeled nonpeptidic organic ligands can be obtained via chemical synthesis. The cost of this method depends on the efficiency of the reactions used and the type of labeling. A relatively inexpensive way to introduce NMR-active sites in an organic molecule is to prepare its fluorinated derivatives. ^{19}F can easily be introduced in an organic compound via chemical synthesis by replacing hydrogen. However, due to its high electronegativity, it can slightly modify the properties of the original organic compound and hence its affinity to the protein (44). Many small organic molecules are commercially available in uniformly or partially isotope labeled form. These molecules are either potential substrates for membrane proteins (i.e., sugars, amino acids) or can serve as building blocks or starting compounds in the synthesis of such substrates.

Short peptidic ligands such as hormones are usually prepared from isotope labeled protected amino acids by standard solid phase peptide synthesis using the Fmoc or Boc procedure (45, 46). The method is fully flexible for isotope labeling but relatively expensive. An alternative way for producing isotope labeled peptides is overexpression in *E. coli* (*see* for example (47)) or cell free synthesis (48). The peptide is fused to a carrier fusion protein which increases its stability and solubility. The expressed fusion protein is purified with affinity chromatography. The peptide is cleaved from the fusion protein by chemical or enzymatic cleavage and purified by HPLC.

The exact isotope labeling pattern strongly depends on the desired information and the NMR technique used. For K_d determinations one NMR active nucleus present at any site of the ligand is sufficient to conduct the measurements. For distance determinations usually two isolated spins have to be introduced carefully choosing the site to be labeled. Extensive isotopically labeled ligands are usually required for detailed conformational studies using chemical shift mapping.

An important issue that has to be considered is the hydrophobicity of the ligand. Water soluble substrates present a favorable case, since partitioning of the ligand into the membrane can be neglected. In case of hydrophobic ligands, additional control experiments are essential, which allow discriminating NMR signals between the membrane-bound and the protein-bound forms.

3. ssNMR Tools Needed for Membrane Protein–Ligand Interaction Studies

ssNMR is a well-established method for the study of membrane proteins within the lipid bilayer and for general membrane biophysics. There are three different basic variants of ssNMR within this field: (1) Magic angle sample spinning (MAS) is used for obtaining highly resolved spectra, (2) utilizing the two-dimensional character of lipid bilayers, highly ordered samples can be prepared from which orientational information is obtained without sample spinning, and (3) wide-line spectra of nonordered samples have been used especially for extracting dynamic information. The most universal and most widely used approach for larger membrane proteins is MAS-NMR, and we focus here only on examples which are based on this concept. A number of excellent reviews have been published introducing basic concepts of ssNMR (3) as well as discussing membrane protein specific methods in detail (2, 4, 49). We would like to refer the interested reader to those publications. Here, we only introduce the terminology needed to understand the examples discussed further below.

Magic angle sample spinning (MAS)-NMR: This concept is based on fast sample rotation at rates between approximately 5–25 kHz around an axis which is inclined to the magnetic field by the magic angle ($\theta = 54.7°$). This trick emulates fast isotropic molecular tumbling as it would occur in solution. In principle, all anisotropic interactions (dipole couplings, chemical shift anisotropy, quadrupole couplings) are affected and can be partially or completely eliminated, as all of them show a ($3\cos^2\theta - 1$) angle dependency. Usually, the detection of nuclei such as ^{13}C, ^{15}N, ^{19}F by MAS-NMR is combined with high power proton decoupling since MAS alone is not sufficient to remove the strong homo- and heteronuclear proton couplings. For these experiments, samples are placed in rotors which have usually diameters between 7 and 2.5 mm. Which rotor to choose depends on the available sample amount, the required volume per sample and requirements imposed by the NMR experiments. The most commonly used rotors for membrane proteins have diameters of 3.2 or 4 mm with active sample volumes of 20–70 µL. This offers a good compromise between the amount of sample which can be used and sample spinning speeds which can be achieved (10–25 kHz).

Cross-polarization (CP): This is a method routinely used for signal enhancement in ssNMR. Protons are the most abundant nuclei with the largest magnetic moment that causes high detection sensitivity which is compromised by the strong dipolar proton–proton coupling network in solids resulting in severe line broadening. Therefore, other nuclei such as ^{13}C or ^{15}N are detected, which are less sensitive. CP enables transfer of magnetization from protons to those nuclei hence enhancing their signal intensity. The transfer efficiency depends on dipolar couplings. Fast motions (microseconds or faster) scale down dipolar couplings resulting in reduced magnetization transfer. Therefore, CP can also be used as a filter technique to suppress signals of mobile compounds in the sample.

Recoupling: Under MAS, high resolution is achieved at the cost of averaging all anisotropic interactions. This means that, for example, valuable dipole couplings from which internuclear distances could be obtained are lost. The process of selectively reintroducing such couplings while maintaining MAS resolution by particular pulse sequences has been termed "recoupling" in correspondence to the opposite process of "decoupling".

Rotational resonance is a homonuclear dipolar recoupling experiment which is usually applied to isolated ^{13}C spin pairs. It is intriguingly simple to perform, as homonuclear couplings are reintroduced by adjusting the sample spin rate to the chemical shift difference of both spins (50). Precise distances can be obtained in a range of up to 0.6 nm. Data analysis requires numerical simulations. An extension to multispin systems has been demonstrated (51).

Proton-driven spin diffusion (PDSD) is often used to establish through-space ^{13}C–^{13}C correlation in multidimensional experiments performed on extensively or uniformly ^{13}C labeled samples (52). This is the simplest variant of a whole range of ^{13}C–^{13}C correlation experiments. Many correlations can be obtained and used for chemical shift mapping experiments. As in the NOESY experiments, only qualitative conclusions can be drawn from the cross peak size, but the occurrence of a cross peak means that both correlated nuclei are found within a range of approximately 0.6 nm.

Rotational echo double resonance (REDOR) is the best established most widely used heteronuclear dipolar recoupling experiment (53). Precise distances between pairs of, for example, ^{13}C and ^{15}N (up to 0.5 nm) or ^{13}C and ^{19}F (up to 1.0 nm) can be measured (54). The averaging effect of MAS is selectively disturbed by the application of a series of rotor synchronized 180° pulses. Transfer echo double resonance (TEDOR) is a variant of REDOR mainly used to record multiple ^{13}C–^{15}N correlations in multidimensional experiments (55, 56).

Double-quantum filtering methods are applied for suppressing the ^{13}C natural abundance background. This is of great importance especially in cases, when small ^{13}C labeled ligands are studied bound to large membrane proteins. The 1.1% natural abundance of ^{13}C is already sufficient to completely obscure the ligand resonances. In these experiments, double quantum transitions, which can only occur for coupled spins, are excited. The most widely used version of this experiment is the C7 pulse sequence (57, 58).

4. Observing Ligands Interacting with Membrane Proteins

In the following section, we discuss experiments which rely on the detection of bound ligands and do not require isotope labeled membrane protein. Using different examples, we show that binding parameters as well as structure, dynamics, and protonation state of ligands can be determined.

4.1. Determination of Ligand Binding Parameters

ssNMR experiments detecting soluble ligands bound to membrane proteins can be challenging due to overlapping signals from bound and nonbound ligand populations, especially in cases of low affinity. An elegant way of filtering signals arising from bound ligands is given by the use of CP (59). The magnetization transfer characteristics of CP depend on the molecular dynamics. Therefore, a ligand immobilized in the protein binding pocket will give rise to a CP signal while it is not visible in solution. Quantitative binding data can be extracted from this basic experimental concept through titration studies (60). This has been beautifully demonstrated on the *E. coli* GalP, which is a sugar–proton symporter of roughly 50 kDa related to Glut1 in higher organisms. The protein can be overexpressed to 50–60% of total inner membrane protein which enabled the authors to use the inner membrane for their experiments without further purification (59). ^{13}C-CP MAS NMR experiments were carried out on nonfrozen samples using ^{13}C-labeled glucose. For the bound sugar species two signals with different intensities from both anomeric forms of glucose could be detected. The larger peak stems from the β-anomer indicating its preferential binding to GalP. The authors have estimated from relaxation measurements and cross-polarization properties that the lifetime for the substrate-transporter complex must be in the millisecond range. Similar studies have also been carried out by Spooner et al. (61) on the 48 kDa L-fucose-H$^+$ symport protein, FucP, from *E. coli*. Overexpressed FucP accounts for 40% of the inner membrane protein, this also allowed using the purified inner *E. coli* membrane for their studies.

The approach described above can also be applied to ions bound to membrane proteins, if this particular ion is an NMR active isotope. A very interesting case has been demonstrated by Rahman et al. (43) who investigated the zinc transporter ZitB from *E. coli*. The authors were using ^{113}Cd CP MAS NMR to directly study metal-ion binding to ZitB overexpressed to 15% of total protein in native membranes. ^{113}Cd is often used as NMR-active nucleus for replacement of divalent cations such as Zn^{2+} and Ca^{2+} which do not have a naturally occurring spin 1/2 isotope. ^{113}Cd was added as $^{113}CdCl_2$ to the aqueous samples of *E. coli* membranes. Using cross-polarization experiments it was possible to distinguish between free and protein bound $^{113}Cd^{2+}$. Adding Zn^{2+} to the $^{113}Cd^{2+}$-ZitB complex, the intensity of the ^{113}Cd resonance decreased indicating that the signal can indeed be assigned to the protein bound Cd^{2+}. Titration of the $^{113}Cd^{2+}$ and ZitB containing sample with Ni^{2+} and Cu^{2+} revealed, that the method is suitable for identification of other possible metal substrates of the protein. IC_{50} values have been determined for all metal ions studied.

A very elegant way to determine binding parameters without substrate titration was demonstrated by Middleton et al. (26, 62). The authors acquired a series of CP spectra with varying contact

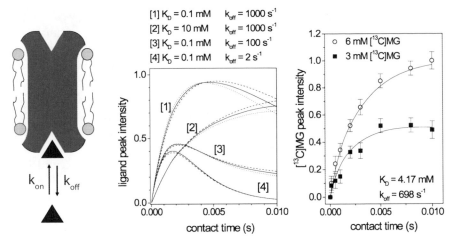

Fig. 1. Cross-polarization has been shown to be a valuable tool to discriminate between bound and nonbound membrane protein ligands. In case of soluble ligands, only the protein bound form will cross-polarize. Middleton and coworkers could show, that the signal buildup depends on K_d and k_{off} (*middle*), which allows to characterize binding kinetics from these buildup curves in a quantitative way. With this approach, the binding of (1-^{13}C)-β-glucuronide to GusB has been analyzed in terms of K_d and k_{off} (*right*). The figures were adapted with permission from (62), copyright 2004 American Chemical Society.

times (0.1–10 ms) on one sample. During the contact time bound substrate molecules build up magnetization, damped by relaxation in the rotating frame. After dissociation of the molecule from the protein the magnetization relaxes and the remaining signal can be detected after the contact time is over. Depending on the k_{off} rate, this can occur multiple times during the contact time (see Fig. 1). Using a Monte Carlo approach to simulate binding trajectories the authors were able to calculate the signal for a specific contact time for any combination of K_d and k_{off}. Principally, one ligand concentration is enough to determine K_d and k_{off} from the cross-polarization dynamics. Practically, more than one concentration is needed, due to contributions of nonspecifically bound substrate (62). The feasibility to analyze the CP buildup curves in order to determine binding parameters was demonstrated for ^{13}C and ^{19}F labeled substrates (26, 62). It was demonstrated that ^{19}F-labeled substrates are suitable to give information about membrane-protein substrate interactions (26). For this study H$^+$/K$^+$-ATPase and its inhibitor trifluoperazine (TFP) (63) were used. The overexpressed protein made up to 70% of the total inner membrane protein. The authors were able to verify that TFP binds to the same site in the protein as the potassium ions and dissociates slowly ($k_{off} < 100$ s^{-1}). In a series of experiments the K_d was determined to be 4 mM. An example data set is shown on Fig. 1.

All examples discussed so far involved soluble substrates. This is a favorable situation which is not always encountered. In fact, many interesting substrates which interact with proteins in the membrane are at least partially lipophilic. These could be, for example, drugs which have to cross the membrane and which are pumped out by multidrug efflux pumps. In such cases, a simple CP experiment to differentiate protein-bound from nonbound ligand populations might not be sufficient. The nonbound form binds to the membrane and could also become visible under CP obscuring the spectrum of the protein-bound form. One solution is offered by difference spectroscopy in which the signal from unspecific interactions are identified and subtracted through suitable control experiments. Alternatively, differences in the molecular dynamics of specifically and unspecifically bound ligands could be utilized, for example by adjusting the temperature in a way that only the desired population cross-polarizes or becomes visible in homonuclear dipolar recoupling experiments. The choice of the particular experiment enables some flexibility with respect to the time range of molecular motions which can be probed. There might be situations in which cross-polarization does not allow to differentiate between both populations while dipolar recoupling experiments effectively filter out the protein-bound species (57, 64, 65). Another alternative has been demonstrated for the nucleoside transporter, NupC, which is a homolog of the mammalian transporter CNT1 (62). The authors used cross-polarization with polarization inversion (CPPI) (66) which allows utilizing differential relaxation behavior of bound and nonbound ligand signals for filtering the protein-bound form only.

4.2. Ligand Structure and Dynamics

There are a number of examples in which it has been demonstrated, that the structure of ligands bound to membrane proteins can be determined by ssNMR. Such structural information in combination with biochemical and mutagenesis data, homology modeling, and computer docking experiments allow for a detailed view into the binding site, even in the absence of a high resolution protein structure (68). One important practical criterion is the formation of a high affinity ligand–protein complex which ensures that most ligands exist in their protein bound form, hence minimizing obscuring NMR signals from nonbound or unspecifically bound ligands. The consideration of substrate hydrophobicity is of practical importance in terms of necessary control experiments as discussed before.

The 3D structure of a bound ligand manifests itself in chemical shifts, which could be analyzed in case of peptidic ligands using empirical approaches such as TALOS+ (68), in terms of J-couplings, which are linked to structure via the Karplus relationship or most directly via through-space dipole couplings between nuclei. TALOS+ correlates secondary structures with chemical shift assignments of 200 proteins, which allows predicting phi and psi dihedral angles from chemical shifts of peptides or proteins with unknown structure. Usually, due to the complexity of the molecular systems, it will not be possible to measure all interactions.

In most cases, peaks will be too broad to resolve J-couplings. In many cases, the signal-to-noise ratio will be too low for complex multidimensional correlation experiments. Therefore, it has to be decided in each particular case which type of experiment offers a maximum of information in the context of available experimental resources. In the following, we demonstrate with some examples, how structural data have been obtained from chemical shifts and from distance measurements.

4.2.1. Conformation Determinations Based on Chemical Shifts

In principle, chemical shifts are sensitive to the local molecular structure (69, 70). A structural analysis can be based on quantum chemical calculations, which is usually restricted to small molecules. For larger systems, such as peptides and proteins, an empirical correlation has been found between observed secondary chemical shifts and those found in proteins with known structure and similar amino acid sequence segments. Secondary chemical shifts are differences between observed chemical shift and standard random coil values. This allows deriving a backbone model based on chemical shifts. A suitable tool for such predictions is the TALOS+ program (68). Alternatively, if structure or chemical shift assignment of the nonbound ligand is known, chemical shift changes upon binding can be mapped onto the structure indicating conformational changes and interaction sites with the protein.

Chemical shift mapping has been used to investigate binding of neurotoxin II (NTII) from the venom of the Asian cobra bound to the nicotinic acetylcholine receptor (nAChR) (71). NTII is a 61 amino acids peptide from the short α-neurotoxin family and is an antagonist of nAChR. With the aid of ^{13}C–^{13}C PDSD and ^{13}C–^{15}N TEDOR experiments more than 90% of the residues of the bound NTII have been identified using extensive or uniformly ^{13}C, ^{15}N labeled samples. Comparison with the solution state NMR spectra (72) of the free NTII showed that 75% of the signals have the same chemical shifts, which indicated that the conformation of NTII was not changing considerably upon binding. Resonances exhibiting larger chemical shift changes allow the identification of NTII regions playing a role in the interaction with the receptor.

Secondary backbone structures based on chemical shifts have been reported for two neuropeptides bound to G-protein coupled receptors (GPCRs) (45, 46). These studies have been especially challenging, since availability of sufficient amounts of functional GPCRs is a major bottleneck. A minimum of approximately 50–100 nmol binding sites, i.e., approximately 1 mg of GPCR is needed, which is a challenge in terms of sample production. Multidimensional heteronuclear experiments for sequential resonance assignment will not be feasible due to signal to noise limitations. In addition, the protein and membrane environment produces a significant ^{13}C background overlapping with the ligand resonances. However, in case of short neuropeptides (<20 residues), experiments which allow residue-specific assignment together with background filtering might be sufficient to obtain chemical shifts of carbons within the bound ligand. One such experiment is the dipolar double-quantum single quantum correlation experiment (73) which has been first demonstrated by Baldus and coworkers to be suitable for the investigation of a neuropeptide-GPCR complex (45). In this experiment, correlations between directly coupled ^{13}C nuclei are observed. The correlation pattern allows identification of the amino acids and suppresses the ^{13}C natural abundance background. The obtained data sets do not allow sequential assignment, but this is less important for the short neuropeptides used for these studies. Such conformational studies have been carried out with neurotensin bound to the rat neurotensin receptor (45) and bradykinin (BK) bound to the human bradykinin B2 (B2R) receptor (46). Both GPCRs are activated upon binding of the agonist neurotensin or bradykinin, respectively.

In case of the neurotensin receptor, a uniformly ^{13}C, ^{15}N-labeled, C-terminal fragment (8–13) of neurotensin, which binds with high affinity, has been analyzed in its receptor bound form (45). The ^{13}C chemical shifts of bound NT were used to predict backbone torsion angle restraints and were used for structure calculations, which returned a linear β-strand structure. Experiments were carried out in frozen samples in both detergent solubilized and lipid reconstituted form. Further MAS-NMR experiments and MD simulations of this NT fragment in different environments did show that NT is unstructured in solution but can adopt a small subset of all possible conformations in a membrane environment (74). There is only little overlap between the receptor-bound and membrane-bound form. The authors concluded that this observation does not support the hypothesis of neuropeptide preorganization within the membrane prior to receptor binding.

Structural studies on bradykinin were carried out in a similar fashion using two differently ^{13}C labeled peptides bound to the human B2R receptor (46). As for NT, assignment walks in the DQSQ spectra allowed the determination of chemical shifts for all backbone ^{13}C atoms of BK. Typical spectra are shown in Fig. 2a. Secondary ^{13}C chemical shifts are shown in Fig. 2b and were used for torsion angle predictions by TALOS, which again were used as input parameters for structure calculation by CYANA (75). An overlap of lowest energy structures is shown in Fig. 2c. BK shows a C-terminal β-turn, and an N-terminal α-helical bend resulting in a twisted S-shaped structure. The observed line width for both neurotensin and bradykinin bound to the receptor was relatively large. It was shown that chemical shift values recalculated from the 500 lowest-energy structures lie within the experimental line width, which is therefore caused by structural flexibility of the ligand trapped during freezing. The solid-state backbone structure of BK was used for docking and MD simulation studies on a homology

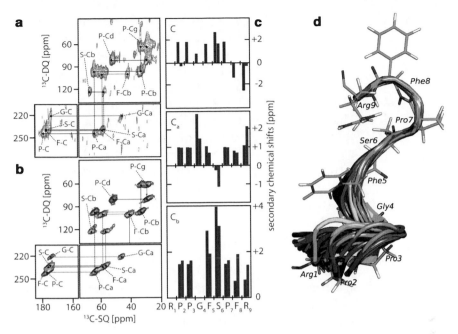

Fig. 2. The structure the neuropeptide bradykinin bound to the human bradykinin 2 receptor has been determined by ssNMR. Two-dimensional ^{13}C double-quantum single-quantum correlation experiments have been recorded for ^{13}C labeled bradykinin bound to unlabeled receptor (**a**) as well as in a frozen solution of DDM micelles (**b**). *Horizontal* and *vertical lines* indicate assignment walks. Secondary backbone chemical shift values (**c**) are pointing to distinct backbone conformations of BK in B2R (*blue*) and in DDM (*red*). These values are used for torsion angle prediction followed by structure calculation. The resulting backbone structure model of bradykinin (**d**) shows a twisted S-shape structure with a N-terminal α-helical bend and a C-terminal β-turn (46).

model of B2R (76). The simulation did provide a first hint towards hydrogen bond formation associated with receptor activation upon BK binding.

4.2.2. Conformation Determinations Based on Distance Measurements

Due to their versatile chemical composition, there are no general empirical data which correlate chemical shift values with three-dimensional structures for small, nonpeptidic organic molecules. Therefore, structural studies solely based on chemical shift analysis would only be possible in the context of quantum mechanical chemical shift calculations, which is not yet a routine tool. A more direct approach is to measure directly internuclear distances between pairs of NMR-active isotopes introduced at specific sites within the ligand. A few but precise distances are often sufficient to restrict the conformational space in such a way, that the bound conformation can be deduced. These constraints are accessible through the determination of internuclear dipole couplings. For homonuclear couplings, rotational resonance and for heteronuclear interactions, REDOR NMR techniques have been used. The most often detected isotopes are ^{13}C and ^{19}F. These approaches will be illustrated in the following for the case of substrate studies involving the Na$^+$/K$^+$-ATPase and for the structure determination of acetylcholine bound to the nicotinic acetylcholine receptor.

^{19}F–^{13}C REDOR NMR was used to determine the conformation of derivatives of the cardiotonic steroid, ouabain, at the digitalis receptor site of the renal Na$^+$/K$^+$-ATPase (77). Ouabain is an inhibitor of the Na$^+$/K$^+$-ATPase ($K_d \approx 5$ nM) and is used in the treatment of cardiac problems. Ouabain consists of a steroid moiety and a rhamnose sugar group linked with an ether bond. The conformation of the inhibitor is determined by the relative orientation of the steroid group with respect to the sugar ring and can be described by two torsion angles. Distances measured between ^{13}C and ^{19}F atoms placed carefully in the steroid and sugar units were used for torsion angle calculations. Ouabain derivatives were shown to adopt a conformation in which the sugar group is almost perpendicular to the plane of the steroid ring system. Analyzing the ^{13}C chemical shift changes upon binding, the authors were able to derive conclusions about the location of the sugar group in the digitalis receptor site, which has to be close to aromatic side chains. Investigating the dynamic properties of the inhibitors lead to additional information about how these molecules interact with the receptor. Dynamic studies have been carried out on deuterated ligands using ^{2}H quadrupole echo NMR. The ^{2}H spectral line shapes were determined by the quadrupolar couplings which in turn are influenced by the local dynamics of the deuterated ligand. The results indicated that the sugar moiety is loosely bound to the receptor site, while the steroid group is much more restrained. All these results together with previous site-directed mutagenesis data let the authors reconcile the structure of the inhibitor and its location at the binding site.

The conformation of acetylcholine (ACh) bound to the nicotinic acetylcholine receptor (nAChR) in its native membrane environment has been determined by Williamson et al. (78) using rotational resonance. This method is usually applied to systems containing isolated spin pairs. In case of two spin systems the spin dynamics allows very accurate distance determination. The disadvantage of this method is that several selectively labeled substrates are usually needed in order to obtain enough distance restraints for conformation analysis. Verhoeven et al. (51) described a new method to extract dipolar couplings and calculate distances between ^{13}C spins from multiple spin systems. The distances determined with this method were less accurate than those from two spins but the errors were still below 10%. Williamson et al. (78) used this method to determine the conformation of uniformly ^{13}C labeled acetylcholine (ACh) bound to the nicotinic acetylcholine receptor (nAChR) (Fig. 3a). Refinement of the torsion angles led to two families of structures related by mirror symmetry (Fig. 3b). The bound acetylcholine was shown to adopt a bent conformation with a distance of 5.1 Å between the carbonyl and the quaternary ammonium group. Using this conformation acetylcholine was successfully docked in the binding pocket of the agonist-bound crystal structure

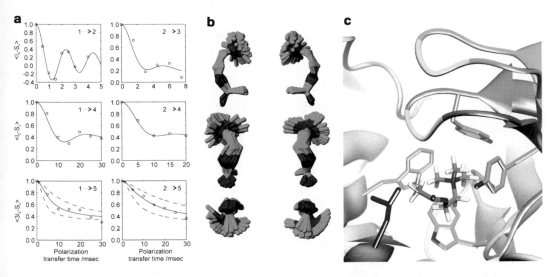

Fig. 3. The conformation of ¹³C-labeled acetylcholine (ACh) in the binding site of the nicotinic acetylcholine receptor has been determined by rotational resonance (RR) NMR. (**a**) The RR polarization transfer curves plotted for the calculated distances (*solid line*) fit well the experimental data (o). (**b**) The two families of structures related by symmetry which fulfill the NMR constraints. The six distance data were not enough to distinguish between these mirror-symmetry conformations. (**c**) A close view of ACh docked in the binding pocket of the homologous soluble Ach-binding protein (AchBP) from *Aplysia californica* with available crystal structure of 2 Å resolution (78). Figures were kindly provided by P.T.F. Williamson, University of Southampton and adapted with the author's permission (Copyright 2007, National Academy of Sciences, U.S.A.).

of a soluble acetylcholine binding protein (Fig. 3c). Protein residues participating in substrate binding were possible to assign.

4.3. Determination of Protonation States of Bound Ligands

ssNMR can also be used to monitor the protonation state of bound substrates. For example, ¹³C chemical shifts in carboxyl groups or ¹⁵N chemical shifts in imidazole rings are sensitive indicators for the charged state of those groups. This has been used for investigating binding of histamine to the human H1 receptor (79). The authors observed two distinct correlation patterns in ¹³C–¹³C PDSD spectra of bound and ¹³C-labeled histamine, which could be assigned to two differently charged species carrying a protonated amino group and either a neutral or a protonated imidazole group. These assignments were supported by DFT chemical shift calculations for both charged states. The authors suggested that histamine binds to the receptor in its dicationic form upon which a protonation switch is triggering receptor activation.

5. Monitoring Protein Response Upon substrate binding

Rearrangements within membrane proteins upon substrate binding can be monitored either by mapping chemical shift changes or through direct distance measurements. In principle, chemical shift mapping provides a global overview of the protein response using uniformly labeled samples. Unfortunately, this is often challenged by limited resolution and difficulties in obtaining a complete resonance assignment. This method relies therefore on favorable samples. Alternatively, detecting chemical differences in highly conserved and essential residues can be achieved by selective labeling and in combination with site-directed mutagenesis. Furthermore, this can be combined with the determination of internuclear distances. In the following section, we highlight examples demonstrating how these approaches can be used to characterize the ligand binding event.

A good example of how chemical shift mapping can be used to gain information about protein–ligand interactions is the study of Lange et al. (80). They monitored chemical shift changes in the uniformly labeled ion channel KcsA-Kv1.3 upon kalitoxin (KTX) binding as well as changes in uniformly labeled KTX bound to unlabeled KcsA. These data allowed for determination of the bound kalitoxin structure and establishing a structural model for the ligand–protein complex. Changes in the KcsA ion selectivity filter upon toxin binding have been detected in ^{13}C–^{13}C PDSD experiments as shown in Fig. 4. Significant chemical shift changes of some resonances indicated the interaction sites of KcsA-Kv1.3 for KTX. Determination of the backbone structure of both the toxin and a segment of KcsA-Kv1.3 indicated that KTX binding to the KcsA-Kv1.3 channel results in significant structural rearrangements in both the toxin and the protein.

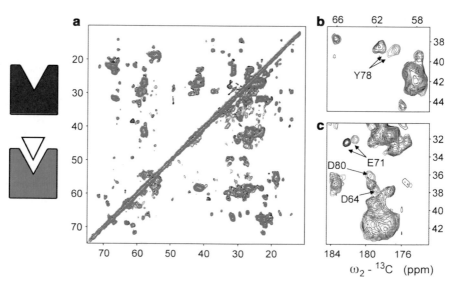

Fig. 4. Example for the use of ssNMR to monitor the protein response upon substrate binding. A comparison of ^{13}C–^{13}C PDSD spectra of the free, uniformly ^{13}C labeled ion channel KcsA-Kv1.3 (*red*) and in complex with unlabeled K$^+$ channel scorpion toxin KTX (*green*) (**a**). These data allow identification of residues in the KTX binding region as well as in the channel's selectivity filter (**b**, **c**) (80). The figure was kindly provided by M. Baldus, University of Utrecht (reprinted by permission from Macmillan Publishers Ltd: Nature (80), copyright 2006).

A less favorable case in terms of spectral resolution was encountered for the *E. coli* multidrug transporter EmrE. Agarwal et al. (81) studied the interaction between the small multidrug resistance protein, EmrE, and its high affinity ($K_d = 2$ nM) substrate tetraphenylphosphonium (TPP$^+$). Despite the small size of EmrE (14 kDa) the resolution of the spectra obtained from 2D crystals of U-^{13}C, ^{15}N EmrE was not sufficient for resonance assignment. The much larger line-broadening compared to KcsA is probably due to equilibrium fluctuations on a micro- to millisecond timescale. Using ^{13}C–^{13}C spin diffusion experiments, the authors tried to observe chemical shift changes upon binding of TPP$^+$. A slight shift and broadening of almost all aliphatic and C′ resonances were observed. This nonspecific protein response has been interpreted as potential helix–helix rearrangements within the protein. In order to obtain more specific data on essential residues, filtering experiments based on REDOR were used which suppress all ^{13}C–^{15}N spin pairs. As a result, carboxyl side chains such as those of the highly conserved E14, essential for binding, could be detected. An alternative approach to simplify the NMR spectra of EmrE was selective labeling of this essential E14 residue using cell free expression. For an unambiguous assignment Lehner et al. (82) used a single glutamate mutant EmrE E25A with ^{13}C labeled E14. Double quantum filtering experiments were performed to detect the ^{13}C resonances of the labeled glutamate with and without the substrate ethidium. Two sets of ^{13}C signals were observed for E14 in the substrate-free protein sample which indicated structural asymmetry of the binding pocket formed through dimerization of EmrE. Ethidium added to the protein induced chemical shift changes and altered line-shapes consistent with substrate binding to both E14 residues in the dimer.

Chemical shift data in combination with empirical tools such as TALOS+ allow valuable conclusions about ligand induced changes on the level of secondary structure. For monitoring tertiary structural changes, distance measurements are needed as demonstrated for the serine bacterial chemoreceptor within intact membranes (83). With a combination of site-directed mutagenesis, site specific labeling and ^{13}C–^{19}F REDOR NMR the authors were able to measure distances between the α1 and α4 helices in the periplasmic domain which contains the ligand-binding site. The protein was ^{13}C-labeled at a single cysteine introduced in helix α1 while all phenylalanine residues which are located in helix α4 carried a ^{19}F-label. The cysteine mutations were introduced in such a way that isolated ^{13}C–^{19}F spin pairs could be created. Upon ligand binding, helix movements on the order of approximately 1 Å were observed by ^{13}C–^{19}F distance measurements. Furthermore, interhelical ^{13}C–^{13}C distances were measured in the transmembrane region between ^{13}C$_\beta$ Cys in TM1 and two different ^{13}C′-Phe residues in TM2 using rotational resonance. The observed interhelical distances of 5.0–5.3 Å did not change considerably in the presence of the ligand (84). These data show that ligand binding induces tertiary structure changes in the periplasmic domain but not in the transmembrane domain. Based on these data, discrimination between previously suggested models for signal transduction has been possible.

6. Outlook

The examples presented here demonstrate that ssNMR is able to cover a wide range of problems regarding protein ligand interactions. Applying different techniques several questions can be addressed, including obtaining structure and affinity as well as dynamic data. Even kinetic data following substrate modification by a protein can be obtained by time-resolved solid-state NMR as shown recently for *E. coli* diacylglycerol kinase, which transfers the gamma-phosphate of ATP to diacylglycerol within the lipid bilayer (85). One of the major limitations for most ssNMR studies is sensitivity. To overcome this problem new methods such as dynamic nuclear polarization (DNP) (86), paramagnetic relaxation enhancement (87), or new pulse sequences that allow shorter recording times (88, 89) will offer new routes for applications.

References

1. Berman HM, Westbrook J, Feng Z, Gilliland G, Bhat TN, Weissig H, Shindyalov IN, Bourne PE (2000) The protein data bank. Nucleic Acids Res 28:235–242

2. Watts A (2005) Solid-state NMR in drug design and discovery for membrane-embedded targets. Nat Rev Drug Discov 4:555–568

3. Laws DD, Bitter H-M, Jerschow A (2002) Solid-state NMR spectroscopic methods in chemistry. Angew Chem Int Ed Engl 41: 3096–3129

4. Renault M, Cukkemane A, Baldus M (2010) Solid-state NMR spectroscopy on complex biomolecules. Angew Chem Int Ed Engl 49:8346–8357

5. Higman V, Flinders J, Hiller M, Jehle S, Markovic S, Fiedler S, van Rossum B-J, Oschkinat H (2009) Assigning large proteins in the solid state: a MAS NMR resonance assignment strategy using selectively and extensively ^{13}C-labelled proteins. J Biomol NMR 44:245–260

6. Castellani F, van Rossum B, Diehl A, Schubert M, Rehbein K, Oschkinat H (2002) Structure of a protein determined by solid-state magic-angle-spinning NMR spectroscopy. Nature 420:98–102

7. Schubert M, Manolikas T, Rogowski M, Meier B (2006) Solid-state NMR spectroscopy of 10% ^{13}C labeled ubiquitin: spectral simplification and stereospecific assignment of isopropyl groups. J Biomol NMR 35: 167–173

8. Hong M, Jakes K (1999) Selective and extensive ^{13}C labeling of a membrane protein for solid-state NMR investigations. J Biomol NMR 14:71–74

9. Lee K, Androphy E, Baleja J (1995) A novel method for selective isotope labeling of bacterially expressed proteins. J Biomol NMR 5:93–96

10. Heise H, Hoyer W, Becker S, Andronesi O, Riedel D, Baldus M (2005) Molecular-level secondary structure, polymorphism, and dynamics of full-length α-synuclein fibrils studied by solid-state NMR. Proc Natl Acad Sci U S A 102:15871–15876

11. Waugh DS (1996) Genetic tools for selective labeling of proteins alpha-^{15}N-amino acids. J Biomol NMR 8:184–192

12. Tong K, Yamamoto M, Tanaka T (2008) A simple method for amino acid selective isotope labeling of recombinant proteins in *E. coli*. J Biomol NMR 42:59–67

13. Sobhanifar S, Reckel S, Junge F, Schwarz D, Kai L, Karbyshev M, Löhr F, Bernhard F, Dötsch V (2010) Cell-free expression and stable isotope labelling strategies for membrane proteins. J Biomol NMR 46:33–43

14. Maslennikov I, Klammt C, Hwang E, Kefala G, Okamura M, Esquivies L, Mörs K, Glaubitz C, Kwiatkowski W, Jeon Y, Choe S (2010) Membrane domain structures of three classes of histidine kinase receptors by cell-free expression and rapid NMR analysis. Proc Natl Acad Sci U S A 107:10902–10907

15. Hefke F, Bagaria A, Reckel S, Ullrich S, Dötsch V, Glaubitz C, Güntert P (2011) Optimization of amino acid type-specific ^{13}C and ^{15}N labeling for the backbone assignment of membrane proteins by solution- and solid-state NMR with the UPLABEL algorithm. J Biomol NMR 49:75–84

16. Patching S, Herbert R, O'Reilly J, Brough A, Henderson P (2004) Low ^{13}C-background for NMR-based studies of ligand binding using ^{13}C-depleted glucose as carbon source for microbial growth: ^{13}C-labeled glucose and ^{13}C-forskolin binding to the galactose-H^+ symport protein GalP in *Escherichia coli*. J Am Chem Soc 126:86–87

17. Perler F, Davis E, Dean G, Gimble F, Jack W, Neff N, Noren C, Thorner J, Belfort M (1994) Protein splicing elements: inteins and exteins— a definition of terms and recommended nomenclature. Nucleic Acids Res 22:1125–1127

18. Dawson PE, Muir TW, Clark-Lewis I, Kent SB (1994) Synthesis of proteins by native chemical ligation. Science 266:776–779

19. Schwarz D, Junge F, Durst F, Frolich N, Schneider B, Reckel S, Sobhanifar S, Dotsch V, Bernhard F (2007) Preparative scale expression of membrane proteins in *Escherichia coli*-based continuous exchange cell-free systems. Nat Protoc 2:2945–2957

20. Noirot C, Habenstein B, Bousset L, Melki R, Meier B, Endo Y, Penin FO, Böckmann A (2011) Wheat-germ cell-free production of prion proteins for solid-state NMR structural studies. N Biotechnol 28:232–238

21. Kigawa T, Muto Y, Yokoyama S (1995) Cell-free synthesis and amino acid-selective stable isotope labeling of proteins for NMR analysis. J Biomol NMR 6:129–134

22. Kigawa T, Yabuki T, Yoshida Y, Tsutsui M, Ito Y, Shibata T, Yokoyama S (1999) Cell-free production and stable-isotope labeling of milligram quantities of proteins. FEBS Lett 442:15–19

23. Abdine A, Verhoeven M, Warschawski D (2011) Cell-free expression and labeling strategies for a new decade in solid-state NMR. N Biotechnol 28:272–276

24. Klammt C, Schwarz D, Fendler K, Haase W, Dötsch V, Bernhard F (2005) Evaluation of detergents for the soluble expression of alpha-helical and beta-barrel-type integral membrane proteins by a preparative scale individual cell-free expression system. FEBS J 272:6024–6038

25. Schwarz D, Klammt C, Koglin A, Löhr F, Schneider B, Dötsch V, Bernhard F (2006) Preparative scale cell-free expression systems: new tools for the large scale preparation of integral membrane proteins for functional and structural studies. Methods 41:355–369

26. Boland M, Middleton D (2004) Insights into the interactions between a drug and a membrane protein target by fluorine cross-polarization magic angle spinning NMR. Magn Reson Chem 42:204–211

27. Privé GG (2009) Lipopeptide detergents for membrane protein studies. Curr Opin Struct Biol 19:379–385

28. Abe R, Caaveiro J, Kudou M, Tsumoto K (2010) Solubilization of membrane proteins with novel n-acylamino acid detergents. Mol Biosyst 6:677–679

29. Landsmann S, Lizandara-Pueyo C, Polarz S (2010) A new class of surfactants with multinuclear, inorganic head groups. J Am Chem Soc 132:5315–5321

30. Popot J (2010) Amphipols, nanodiscs, and fluorinated surfactants: three nonconventional approaches to studying membrane proteins in aqueous solutions. Annu Rev Biochem 79:737–775

31. Privé GG (2007) Detergents for the stabilization and crystallization of membrane proteins. Methods 41:388–397

32. Holloway PW (1973) A simple procedure for removal of triton X-100 from protein samples. Anal Biochem 53:304–308

33. Rigaud JL, Levy D, Mosser G, Lambert O (1998) Detergent removal by non-polar polystyrene beads. Eur Biophys J 27:305–319

34. Seddon AM, Curnow P, Booth PJ (2004) Membrane proteins, lipids and detergents: not just a soap opera. Biochim Biophys Acta 1666:105–117

35. Margolles A, Putman M, van Veen H, Konings W (1999) The purified and functionally reconstituted multidrug transporter LmrA of *Lactococcus lactis* mediates the transbilayer movement of specific fluorescent phospholipids. Biochemistry 38:16298–16306

36. Hellmich U, Haase W, Velamakanni S, van Veen H, Glaubitz C (2008) Caught in the act: ATP hydrolysis of an ABC-multidrug transporter followed by real-time magic angle spinning NMR. FEBS Lett 582:3557–3562

37. Shi L, Ahmed M, Zhang W, Whited G, Brown L, Ladizhansky V (2009) Three-dimensional solid-state NMR study of a seven-helical integral membrane proton pump-structural insights. J Mol Biol 386:1078–1093

38. Levy D, Chami M, Rigaud JL (2001) Two-dimensional crystallization of membrane proteins: the lipid layer strategy. FEBS Lett 504:187–193

39. Shastri S, Vonck J, Pfleger N, Haase W, Kuehlbrandt W, Glaubitz C (2007) Proteorhodopsin: characterisation of 2D crystals by electron microscopy and solid state NMR. Biochim Biophys Acta 1768: 3012–3019

40. Hiller M, Krabben L, Vinothkumar K, Castellani F, van Rossum B-J, Kühlbrandt W, Oschkinat H (2005) Solid-state magic-angle spinning NMR of outer-membrane protein G from *Escherichia coli*. Chembiochem 6: 1679–1684

41. Alvarez F, Orelle C, Davidson A (2010) Functional reconstitution of an ABC transporter in nanodiscs for use in electron paramagnetic resonance spectroscopy. J Am Chem Soc 132:9513–9515

42. Kijac A, Li Y, Sligar S, Rienstra C (2007) Magic-angle spinning solid-state NMR spectroscopy of nanodisc-embedded human CYP3A4. Biochemistry 46:13696–13703

43. Rahman M, Patching S, Ismat F, Henderson P, Herbert R, Baldwin S, McPherson M (2008) Probing metal ion substrate-binding to the *E. coli* ZitB exporter in native membranes by solid state NMR. Mol Membr Biol 25:683–690

44. Middleton DA, Robins R, Feng X, Levitt MH, Spiers ID, Schwalbe CH, Reid DG, Watts A (1997) The conformation of an inhibitor bound to the gastric proton pump. FEBS Lett 410:269–274

45. Luca S, White J, Sohal A, Filippov D, van Boom J, Grisshammer R, Baldus M (2003) The conformation of neurotensin bound to its g protein-coupled receptor. Proc Natl Acad Sci U S A 100:10706–10711

46. Lopez J, Shukla A, Reinhart C, Schwalbe H, Michel H, Glaubitz C (2008) The structure of the neuropeptide bradykinin bound to the human g-protein coupled receptor bradykinin b2 as determined by solid-state NMR spectroscopy. Angew Chem Int Ed Engl 47:1668–1671

47. Williamson PTF, Roth JF, Haddingham T, Watts A (2000) Expression and purification of recombinant neurotensin in *Escherichia coli*. Protein Expr Purif 19:271–275

48. Lee KH, Kwon YC, Yoo SJ, Kim DM (2010) Ribosomal synthesis and in situ isolation of peptide molecules in a cell-free translation system. Protein Expr Purif 71:16–20

49. Watts A, Straus S, Grage S, Kamihira M, Lam Y, Zhao X (2004) Membrane protein structure determination using solid-state NMR. Methods Mol Biol 278:403–473

50. Raleigh D, Levitt M, Griffin R (1988) Rotational resonance in solid state NMR. Chem Phys Lett 146:71–76

51. Verhoeven A, Williamson PTF, Zimmermann H, Ernst M, Meier BH (2004) Rotational-resonance distance measurements in multi-spin systems. J Magn Reson 168:314–326

52. Szeverenyi NM, Sullivan MJ, Maciel GE (1982) Observation of spin exchange by two-dimensional Fourier transform ^{13}C cross polarization-magic-angle spinning. J Magn Reson 47: 462–475

53. Gullion T, Schaefer J (1989) Rotational-echo double-resonance NMR. J Magn Reson 81: 196–200

54. Grage SL, Watts A, Webb GA (2006) Applications of REDOR for distance measurements in biological solids. Annu Rep NMR Spectrosc 60:191–228

55. Hing AW, Vega S, Schaefer J (1992) Transferred-echo double-resonance NMR. J Magn Reson 96:205–209

56. Michal CA, Jelinski LW (1997) REDOR 3D: heteronuclear distance measurements in uniformly labeled and natural abundance solids. J Am Chem Soc 119:9059–9060

57. Hohwy M, Jakobsen HJ, Eden M, Levitt MH, Nielsen NC (1998) Broadband dipolar recoupling in the nuclear magnetic resonance of rotating solids: a compensated C7 pulse sequence. J Chem Phys 108:2686–2694

58. Levitt MH (2002) Symmetry-based pulse sequences in magic-angle spinning solid-state NMR. In: Grant DM, Harris RK (eds) Encyclopedia of nuclear magnetic resonance: supplementary volume. Wiley, England, pp 165–196

59. Spooner PJ, Rutherford NG, Watts A, Henderson PJ (1994) NMR observation of substrate in the binding site of an active sugar-H$^+$

symport protein in native membranes. Proc Natl Acad Sci U S A 91:3877–3881

60. Spooner P, Veenhoff L, Watts A, Poolman B (1999) Structural information on a membrane transport protein from nuclear magnetic resonance spectroscopy using sequence-selective nitroxide labeling. Biochemistry 38: 9634–9639

61. Spooner PJ, O'Reilly WJ, Homans SW, Rutherford NG, Henderson PJ, Watts A (1998) Weak substrate binding to transport proteins studied by NMR. Biophys J 75:2794–2800

62. Patching S, Brough A, Herbert R, Rajakarier A, Henderson P, Middleton D (2004) Substrate affinities for membrane transport proteins determined by ^{13}C cross-polarization magic-angle spinning nuclear magnetic resonance spectroscopy. J Am Chem Soc 126: 3072–3080

63. Im WB, Blakeman DP, Mendlein J, Sachs G (1984) Inhibition of (H^+/K^+)-ATPase and H^+ accumulation in hog gastric membranes by trifluoperazine, verapamil and 8-(n,n-diethylamino)octyl-3,4,5-trimethoxybenzoate. Biochim Biophys Acta 770:65–72

64. Lee YK, Kurur ND, Helmle M, Johannessen OG, Nielsen NC, Levitt MH (1995) Efficient dipolar recoupling in the NMR of rotating solids. A sevenfold symmetric radiofrequency pulse sequence. Chem Phys Lett 242: 304–309

65. Appleyard AN, Herbert RB, Henderson PJ, Watts A, Spooner PJ (2000) Selective NMR observation of inhibitor and sugar binding to the galactose-H^+ symport protein GalP, of *Escherichia coli*. Biochim Biophys Acta 1509:55–64

66. Wu X, Zilm KW (1993) Complete spectral editing in CPMAS NMR. J Magn Reson A 102:205–213

67. Watts A (2002) Direct studies of ligand-receptor interactions and ion channel blocking (review). Mol Membr Biol 19:267–275

68. Shen Y, Delaglio F, Cornilescu G, Bax A (2009) TALOS+: a hybrid method for predicting protein backbone torsion angles from NMR chemical shifts. J Biomol NMR 44:213–223

69. Wishart DS, Sykes BD (1994) The ^{13}C chemical-shift index—a simple method for the identification of protein secondary structure using ^{13}C chemical-shift data. J Biomol NMR 4:171–180

70. Wishart DS, Sykes BD, Richards FM (1992) The chemical-shift index—a fast and simple method for the assignment of protein secondary structure through NMR-spectroscopy. Biochemistry 31:1647–1651

71. Krabben L, van Rossurn BJ, Jehle S, Bocharov E, Lyukmanova EN, Schulga AA, Arseniev A, Hucho F, Oschkinat H (2009) Loop 3 of short neurotoxin II is an additional interaction site with membrane-bound nicotinic acetylcholine receptor as detected by solid-state NMR spectroscopy. J Mol Biol 390:662–671

72. Bocharov EV, Lyukmanova EN, Ermolyuk YS, Schulga AA, Pluzhnikov KA, Dolgikh DA, Kirpichnikov MP, Arseniev AS (2003) Resonance assignment of ^{13}C-^{15}N-labeled snake neurotoxin II from *Naja oxiana*. Appl Magn Reson 24:247–254

73. Hong M (1999) Solid-state dipolar inadequate NMR spectroscopy with a large double-quantum spectral width. J Magn Reson 136:86–91

74. Heise H, Luca S, de Groot B, Grubmüller H, Baldus M (2005) Probing conformational disorder in neurotensin by two-dimensional solid-state NMR and comparison to molecular dynamics simulations. Biophys J 89:2113–2120

75. Güntert P (2004) Automated NMR structure calculation with CYANA. Methods Mol Biol 278:353–378

76. Gieldon A, Lopez JJ, Glaubitz C, Schwalbe H (2008) Theoretical study of the human bradykinin-bradykinin b2 receptor complex. Chembiochem 9:2487–2497

77. Middleton D, Rankin S, Esmann M, Watts A (2000) Structural insights into the binding of cardiac glycosides to the digitalis receptor revealed by solid-state NMR. Proc Natl Acad Sci U S A 97:13602–13607

78. Williamson PT, Verhoeven A, Miller KW, Meier BH, Watts A (2007) The conformation of acetylcholine at its target site in the membrane-embedded nicotinic acetylcholine receptor. Proc Natl Acad Sci U S A 104:18031–18036

79. Ratnala V, Kiihne S, Buda F, Leurs R, de Groot H, DeGrip W (2007) Solid-state NMR evidence for a protonation switch in the binding pocket of the H1 receptor upon binding of the agonist histamine. J Am Chem Soc 129:867–872

80. Lange A, Giller K, Hornig S, Martin-Eauclaire MF, Pongs O, Becker S, Baldus M (2006) Toxin-induced conformational changes in a potassium channel revealed by solid-state NMR. Nature 440:959–962

81. Agarwal V, Fink U, Schuldiner S, Reif B (2007) Mas solid-state NMR studies on the multidrug transporter emre. Biochim Biophys Acta 1768:3036–3043

82. Lehner I, Basting D, Meyer B, Haase W, Manolikas T, Kaiser C, Karas M, Glaubitz C (2008) The key residue for substrate transport Glu(14) in the EmrE dimer is asymmetric. J Biol Chem 283:3281–3288

83. Murphy O, Kovacs F, Sicard E, Thompson L (2001) Site-directed solid-state NMR measurement of a ligand-induced conformational change in the serine bacterial chemoreceptor. Biochemistry 40:1358–1366

84. Isaac B, Gallagher G, Balazs Y, Thompson L (2002) Site-directed rotational resonance solid-state NMR distance measurements probe structure and mechanism in the transmembrane domain of the serine bacterial chemoreceptor. Biochemistry 41:3025–3036

85. Ullrich SJ, Hellmich UA, Ullrich S, Glaubitz C (2011) Interfacial enzyme kinetics of a membrane bound kinase analyzed by real-time MAS-NMR. Nat Chem Biol 7:263–270. doi:10.1038/nchembio.543

86. Maly T, Debelouchina G, Bajaj V, Hu K, Joo C, Mak-Jurkauskas M, Sirigiri J, van der Wel P, Herzfeld J, Temkin R, Griffin R (2008) Dynamic nuclear polarization at high magnetic fields. J Chem Phys 128:052211

87. Ishii Y, Wickramasinghe NP, Chimon S (2003) A new approach in 1D and 2D ^{13}C high-resolution solid-state NMR spectroscopy of paramagnetic organometallic complexes by very fast magic-angle spinning. J Am Chem Soc 125:3438–3439

88. Franks WT, Atreya HS, Szyperski T, Rienstra CM (2010) GFT projection NMR spectroscopy for proteins in the solid state. J Biomol NMR 48:213–223

89. Lopez J, Kaiser C, Asami S, Glaubitz C (2009) Higher sensitivity through selective ^{13}C excitation in solid-state NMR spectroscopy. J Am Chem Soc 131:15970–15971

Chapter 6

Identifying and Measuring Transmembrane Helix–Helix Interactions by FRET

Damien Thévenin and Tzvetana Lazarova

Abstract

Specific interactions between helical transmembrane domains (TMs) play essential roles in the mechanisms governing the folding, stability and assembly of integral membrane proteins. Thus, it is appealing to identify helix–helix contacts and to seek the structural determinants of such interactions at the molecular level. Here, we provide a protocol for detecting and measuring specific helix–helix interactions in liposomes by Förster resonance energy transfer (FRET), using peptides corresponding to the TM domains of an integral membrane protein. We give a detailed procedure and practical guidelines on how to design, prepare, handle, and characterize fluorescently labeled TM peptides reconstituted in large unilamellar lipid vesicles. We also discuss some critical aspects of FRET measurements to ensure the correct analysis and interpretation of spectral data. Our method uses tryptophan/pyrene as the donor–acceptor FRET pair, but it can be easily adapted to other fluorescence pairs and to other membrane mimetic environments. The ability to identify crucial interhelical contacts is a valuable tool for the study of the stability, assembly, and function of the important and experimentally challenging helical membrane proteins.

Key words: MPs—Membrane proteins, GPCRs—G-protein-coupling receptors, FRET—Förster resonance energy transfer, Donor–acceptor pair, Specific helix–helix interactions, Homo- and heterodimerization

Abbreviations

Ala	Alanine
Cys	Cysteine
DMPC	1-2-Dimyristoyl-*sn*-glycero-3-phosphocholine
FRET	Förster Resonance Energy Transfer
LMV	Large multilamellar vesicles
LUV	Large unilamellar liposomes
Lys	Lysine
Phe	Phenylalanine
Pyr	Pyrene
TM	Transmembrane
TMa-W	Donor peptide with tryptophan in its sequence
TMb-Pyr	Acceptor peptide with a pyrene-butyric acid derivative on the third N-terminal Lys

Nagarajan Vaidehi and Judith Klein-Seetharaman (eds.), *Membrane Protein Structure and Dynamics: Methods and Protocols*, Methods in Molecular Biology, vol. 914, DOI 10.1007/978-1-62703-023-6_6, © Springer Science+Business Media, LLC 2012

TMb-Unl   Unlabeled peptide
Trp   Tryptophan
Tyr   Tyrosine

1. Introduction

Integral membrane proteins (MPs) are involved in almost every aspect of cell biology and physiology (1). Consequently, proper functioning of these proteins is vital to health, and specific defects are associated with many known human diseases (2, 3). Most are anchored to the cellular membrane through one or several trans-membrane (TM) domains that predominantly take on an α-helical secondary structure (4). It is the case for G-protein coupled receptors (GPCRs), aquaporins, potassium channels and receptor tyrosine kinases, to just name a few. Recently, the importance of the TM regions has emerged: TM domains, and more specifically, helix–helix interactions, not only are major determinants of the assembly and stability of the native protein structure but also play significant roles in membrane insertion and folding (5). Importantly, TM domains can also direct the assembly of protein complexes through relatively stable lateral contacts between helices (5, 6) and participate in signal transduction through finely tuned changes in oligomeric state and/or conformation (7, 8). Thus, determining the molecular interactions that stabilize (or destabilize) helical MPs and their oligomers, is fundamental to our comprehension of their structure and function (9–12). However, despite recent progress in expression, purification and crystallization, studying helical MPs remains particularly challenging. The difficulties associated with their expression, purification, reconstitution, and characterization arise in part from our limited understanding of the mechanisms that drive their folding, stability, and oligomerization in the membrane.

One way to circumvent these difficulties is to employ a "divide and conquer" strategy, where one studies the fundamental building blocks of helical MPs, the TM domains (13). Studying peptides corresponding to individual TM domains is relevant to understanding folding and assembly of full-length proteins because these processes can be described by a simple, but elegant two-stage model proposed by Engelman and Popot. In this model, the stable insertion and formation of independent TM helices, and the association of these helices in the membrane occurs in two independent, well-separated thermodynamic events (14). Despite its simplicity, this model has proven to adequately guide experimental analysis of full-length MP folding mechanisms (15, 16). This approach also allows obtaining material in quantities sufficient for performing biophysical studies by experimental methodologies not readily accessible to full-length polytopic membrane proteins.

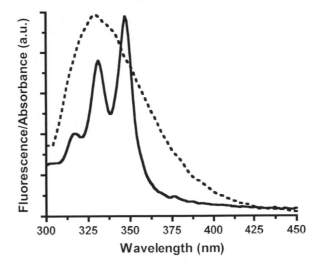

Fig. 1. Schematic spectral overlap of the Trp fluorescence emission spectrum (*dotted line*) of the TMa-W (donor) peptide excited at 295 nm with the UV–VIS absorption spectrum (*solid line*) of the TMb-Pyr (acceptor) peptide in DMPC liposomes.

In this chapter, we provide a detailed protocol and practical guidelines for assessing and measuring homo- and hetero-oligomeric interactions between peptides corresponding to TM domains of helical membrane proteins by Förster Resonance Energy Transfer (FRET). FRET refers to the distance-dependent transfer of energy between an excited donor and a suitable acceptor fluorophore (17). Because the efficiency of energy transfer is inversely proportional to the sixth power of the donor-to-acceptor distance (see Note 1), FRET is a method of choice to detect molecular interactions and measure proximity and conformational changes in solution and membranes (18). In the protocol detailed in this chapter, FRET is used as a means to detect changes in distances due to conformational changes and interactions rather than to measure absolute distances (see Note 2).

In the approach described here, we use the natural amino acid residue tryptophan (Trp) as a donor and pyrene as an acceptor (Fig. 1). This labeling scheme allows (1) minimal disruption of the secondary structure of TM domains and of helix–helix contacts, (2) a sensitive measurement of changes in distances, and (3) a low contribution of nonspecific FRET, arising from random acceptor–donor colocalization in the bilayer (19–23). It also decreases the possibility of nonspecific aggregation that may result from hydrophobic fluorescent probes, and allows the monitoring of homodimerization with pyrene, which forms excited-state dimers upon close encounter with another pyrene molecule (24–26).

The FRET methodology detailed in this chapter, which also applies to other donor–acceptor pairs, represents a valuable tool not only to identify interhelical contacts that are crucial for membrane protein structure and function, but also to gain insights into pathologies arising from mutations (18).

2. Materials

All chemicals can be obtained from Sigma-Aldrich (St. Louis, MO), unless specified otherwise.

2.1. Peptide

- Programs for predicting transmembrane helices in proteins: TMHMM (www.cbs.dtu.dk/services/TMHMM/) (27), MPEx (http://blanco.biomol.uci.edu/mpex/) (28) or TopPred (http://mobyle.pasteur.fr/cgi-bin/portal.py?form=toppred) (29).

- Acetonitrile–water (1:1).

- Trifluoroacetic acid (TFA).

- Indicating Drierite (W.A. Hammond Drierite Company Ltd., Xenia, OH).

- 6 M Guanidine-HCl.

2.2. Spectroscopy

2.2.1. Fluorescence Spectroscopy

- ISS P-1spectrofluorimeter, operating in photon-counting mode.
- 2 × 10 mm QS quartz Ultra-Micro cuvette (Hellma Analytics, Müllheim, Germany).

2.2.2. Circular Dichroism

- Aviv model 215 spectrometer equipped with a Peltier thermal-controlled cuvette holder.
- 0.1-cm path length QS quartz cuvette (350 µL) (Hellma Analytics, Müllheim, Germany).

2.3. Protein Gel Electrophoresis

- 10% NuPAGE Bis-Tris precast polyacrylamide gels (Invitrogen, Carlsbad, CA).

- NuPAGE MES running buffer (Invitrogen, Carlsbad, CA).

- 2× sample loading buffer: 100 mM Tris-Cl, pH 6.8, containing 4% (w/v) SDS, 0.2% (w/v) bromophenol blue, and 20% (v/v) glycerol.

- Mark12 Unstained Standard (Invitrogen, Carlsbad, CA).

- GelCode Blue Safe Protein Stain (Thermo Scientific, Rockford, IL).

2.4. Preparation of Large Unilamellar Liposomes

- Round-bottom glass flask (50 mL).
- 1,2-dimyristoyl-sn-glycero-3-phosphocholine (DMPC) in chloroform (Avanti Polar Lipids, Alabaster, AL).
- Chloroform.
- A source of nitrogen gas.

- Lyophilizer or vacuum chamber.

- Liquid nitrogen or a mix of ethanol–dry ice.

- Avanti Mini-Extruder with heating block (Avanti Polar Lipids, Alabaster, AL).

- Two gas-tight Hamilton syringes (1 mL) (Hamilton Company, Reno, NV).

- Nuclear polycarbonate membrane filter (100 nm pore size, 19 mm diameter) (Whatman—GE Healthcare).

- Drain Disc PE (10 nm diameter) (Whatman—GE Healthcare).

- Buffer: 10 mM HEPES, pH 7, containing 10 mM KCl.

2.5. Phosphate Assay

- Sodium sulfate.

- Ammonium thiocyanate.

- Ferric chloride.

- Chloroform.

- 15 mL plastic tubes.

3. Methods

This section describes how to measure helix–helix interaction between TM peptides in the lipid bilayer by FRET, but also gives a short description of some complementary assays that must be taken to ensure proper experimental procedures and data analysis. For instance, peptides corresponding to the transmembrane domains of interest need to be carefully designed (Subheading 3.1), and accurate determination of peptide and lipid concentrations (Subheadings 3.2 and 3.4, respectively) is essential, as FRET efficiency is determined by the protein-to-lipid ratio (Subheading 3.9). Furthermore, even though FRET is the primary method to identify helix–helix interactions between TM peptides, it is crucial to perform complementary studies to investigate the peptides' propensity to dimerize, to ensure that the attached fluorophore (e.g., pyrene) does not affect secondary structure of the peptide and to confirm whether each peptide variant behaves properly (i.e., inserts and folds as a helical TM domain in membrane bilayers). These properties of the individual peptides can be accessed by SDS-PAGE, circular dichroism, and fluorescence spectroscopies (20–23) (Subheadings 3.5–3.7).

3.1. Designing Peptides Corresponding to Transmembrane Domains

1. Identify the TM segment(s) of the membrane protein of interest. It can be done by inspecting the high-resolution crystal structure, if available, or by using prediction programs such as TMHMM (27), MPEx (28) or TopPred (29) (see Note 3).

2. Because of the high hydrophobicity of such a peptide, one frequent and proven approach to increase solubility and facilitate chemical synthesis and purification (20–23, 30, 31) is to add a poly-lysine (Lys) tag (usually three lysines) at the N- and C-termini of the peptide sequence.

3. To assess interactions between peptides corresponding to two transmembrane domains (TMa and TMb) using FRET, it is necessary to select a proper FRET pair (e.g., tryptophan (Trp)/pyrene—see Fig. 1) and subsequently construct several labeled peptide variants (21–23). In this protocol, tryptophan and pyrene are used as donor and acceptor, respectively (see Note 4):

   • A donor peptide with a Trp residue (TMa-W): If there is no Trp residue in the peptide sequence, the residue considered for substitution should have similar properties than the Trp side chain. Thus, Tyrosine (Tyr) and Phenylalanine (Phe) residues are good choices for substitution. If more than one Tyr or Phe residue is present, substitute the closest one from the N-terminus of the peptide (20).

   • An acceptor peptide with a pyrene-butyric acid derivative on the third N-terminal Lys (TMb-Pyr).

   • An unlabeled peptide (TMb-Unl): substitute Trp residue(s), when present, to Tyr residues.

   If there is more than one cysteine (Cys) residue in the TM sequence, one should consider substituting one of the Cys to an alanine (Ala) to prevent intramolecular disulfide bonds. Representative examples of TM peptides used for monitoring helix–helix interaction by FRET are shown in Table 1. The study of the interaction between the fifth and sixth TM domains of the adenosine A_{2A} receptor (TM5 and TM6, respectively) is used here as a specific example for the general case (TMa + TMb) described in the protocol.

4. These TM peptides are then synthesized using standard Fmoc (N-(9-fluorenyl)methoxycarbonyl) chemistry and purified by reverse phase high-pressure liquid chromatography (see Note 5). For spectroscopic experiments, >95% purity is desired.

5. The peptide should be stored as a powder at –20°C in the presence of a desiccant agent such as Drierite.

3.2. Peptide Stock and Concentration Determination

1. Dissolve the peptide powder in acetonitrile–water (1:1) to obtain a stock solution of about 500 μM (see Note 6). Trifluoroacetic acid (<0.1%) can be added in the event that the peptide is not well dissolved (see Note 7).

Table 1
Representative examples of peptide sequences used to measure specific interactions between TM domains of membrane protein. These peptides correspond to the fifth and sixth TM domains of the adenosine A_{2A} receptor

Amino acid sequence	Name	Role
KS**LAIIVGLFALCWLPLHIINCFTFF**CPD	TM6 native sequence	n/a
*KKK***LAIIVGLFAL<u>A</u>WLPLHIINCFTFF**<u>A</u>PD*KK*	TM6-W	Donor
*KKK***LAIIVGLFAL<u>AY</u>LPLHIINCFTFF**<u>A</u>PD*KK*	TM6-Y	Unlabeled
KKK(pyrene)**LAIIVGLFAL<u>AY</u>LPLHIINCFTFF**<u>A</u>PD*KK*	TM6-Pyr	Acceptor
VVP**MNYMVYFNFFACVLVPLLLMLGVYL**RIF	TM5 native sequence	n/a
*KKK***MNYMVYFNFFACVLVPLLLMLGVYL**R*KKK*	TM-5	Unlabeled
*KKK***MN<u>W</u>MVYFNFFACVLVPLLLMLGVYL**R*KKK*	TM5-W	Donor
KKK(pyrene)**MNYMVYFNFFACVLVPLLLMLGVYL**R*KKK*	TM5-Pyr	Acceptor

The residues predicted to be part of the TM domains are indicated in *bold*. The Lys residues added at the N- and C-termini are indicated in *italics*. The residues that differ from the native sequence are *underlined*. Adapted from (23)

2. Record the UV absorbance spectrum of the peptide in 6 M guanidine-HCl at 280 nm, performing an appropriate correction for a nonzero baseline. Calculate the peptide stock concentration using appropriate molar extinction coefficients, which can be calculated from the peptide sequence using the online tool ProtParam (http://www.expasy.ch/tools/protparam.html) (32) (see Note 8).

3.3. Preparation of Large Unilamellar Liposomes

1. Unilamellar liposomes are prepared using about 10 mg of lipid (DMPC) dissolved in chloroform in a round bottom flask (see Note 9).

2. Chloroform is removed by using a rotary evaporation system or alternatively under a light stream of nitrogen gas while gently rotating the flask until a thin lipid film is formed.

3. Remove the residual solvent by keeping the lipid film under vacuum for several hours or overnight.

4. Hydrate the lipid film with 400 µL of 10 mM HEPES, pH 7 and 10 mM KCl for at least 30 min while keeping the temperature above the gel–liquid crystal transition temperature (T_m) of the lipid (in the case of DMPC, keep above 23°C), with periodic vortexing.

5. Freeze-thaw the resulting multilamellar vesicles (LMV) using liquid nitrogen (or a mix of ethanol–dry ice) and warm water for at least seven cycles. This method, in combination with

vortexing during thawing cycles, helps disrupting the LMV and improves the homogeneity of the size distribution of the final suspension.

Safety concern—Wear protective gear and handle liquid nitrogen, dry ice and ethanol according to institutional chemical hygiene plan.

6. To obtain large unilamellar vesicles (LUV) (see Note 10), the LMV dispersion is forced through a polycarbonate membrane with 100 nm pore diameter (i.e., extrusion process) by using a Mini-Extruder at 30°C, following the supplier's instruction (see Note 11). The total concentration of lipid is ~20 mM (see Subheading 3.4). Liposomes should be used immediately, keeping the temperature above their T_m during the experiment (see Note 12).

3.4. Phospholipid Analysis with Ammonium Ferrothiocyanate

Because the extrusion process results in a loss of lipids, it is necessary to determine exactly the concentration of lipids after extrusion. The colorimetric method presented here is based on complex formation between ammonium ferrothiocyanate and a phospholipid. It is a simpler and more rapid method than acid digestion and colorimetric determination of the inorganic phosphate, but still allows measurements of phospholipids in the range of 0.01–0.1 mg (33).

1. Ammonium ferrothiocyanate is prepared by dissolving 2.7 g of ferric chloride hexahydrate ($FeCl_3 6H_2O$) and 3 g of ammonium thiocyanate (NH_4SCN) in deionized distilled water and made up to 100 mL total volume. This solution is stable for a few months at room temperature.

2. 50 μL of extruded liposome is added to 12.45 mL of chloroform in a 15 mL Falcon tube.

3. Samples for the standard curve are prepared in 15 mL Falcon tubes by adding between 0.1 and 1 mL of a solution of phospholipids (2 mg in 20 mL chloroform), to the representative volumes of chloroform that make the final volume of 2 mL.

4. To each tube, 2 mL of the ammonium ferrothiocyanate solution are added.

5. The solution is then vigorously vortexed for 1 min.

6. After separation, the lower chloroform phase is removed and absorbance of each standard and sample is read at 488 nm.

7. The concentration of phospholipids is calculated by linear regression of the standard curve.

3.5. Protein Gel Electrophoresis

This separation technique is a commonly used method in the evaluation of TM peptide–peptide interactions and it can provide an accurate estimate of the oligomeric state(s) of TM peptides. Migration of TM peptides at apparent molecular weights greater than ~2-fold their formula MWs on SDS-PAGE is often considered to be evidence of high affinity self-association (22).

1. Mix appropriate volumes of peptide stock solution in acetonitrile–water with 10 mM HEPES, pH 7 and 10 mM KCl buffer to obtain solutions of increasing peptide concentrations (2–30 μM).

2. The peptide solutions are then mixed with 2× sample loading buffer and loaded on 10% NuPAGE Bis-Tris polyacrylamide gels (Invitrogen) along with the molecular weight standards.

3. Run the gel at 200 V, at room temperature with MES running buffer.

4. Stain and visualize the gel with the GelCode Plus staining reagent (Thermo Scientific, Rockford, IL).

3.6. Far-UV Circular Dichroism

CD spectroscopy is used as a complementary spectroscopic method to access the secondary structure component of constructed peptides (i.e., their helical fold in lipid bilayer), and also to determine whether residue substitutions and pyrene conjugation affect the secondary structure of the TM peptide (Fig. 2) (20, 22, 23) (see Note 13).

1. Mix an appropriate volume of peptide stock solution in acetonitrile–water with pH 7, 10 mM HEPES, 10 mM KCl buffer to obtain a 20 μM solution (see Note 14). In the case of

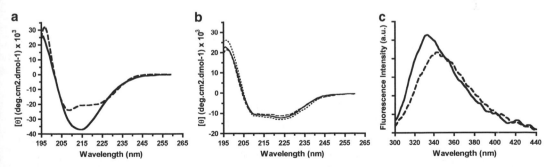

Fig. 2. (**a**) Representative circular dichroism spectra of the TM6-W peptide in buffer 10 mM HEPES and 10 mM KCl, pH 7 (*dashed line*) and in presence of DMPC liposomes (*solid line*). The CD spectrum in buffer shows a broad negative peak at about 215 nm and a positive peak at about 195 nm, indicative for β-strand structure. In the presence of liposomes, the peptide forms α-helical structures as seen from the two minima at 208 and 222 nm and a maximum at 195 nm. (**b**) CD spectra of TM5-W peptide (*dashed line*), TM5-Unl (*dotted line*) and TM5-Pyr (*solid line*) in DMPC liposomes. The CD spectra of the three samples are identical, showing that neither the attached pyrene label, nor the substitution of Tyr for Trp residue perturbs the secondary structure of the peptide. (**c**) Representative Trp emission spectra of the TM6-W peptide in 10 mM HEPES and 10 mM KCl, pH 7 (*dashed line*) and in presence of DMPC liposomes (*solid line*).

measurements with liposomes, the volume of buffer added should take into account the volume of liposome solution that will be added in the subsequent steps to obtain a final peptide concentration of 20 μM.

2. Incubate 15 min at 25°C.

3. Add the appropriate volume of liposome stock solution to obtain a peptide-to-lipid molar ratio of 1:100 (see Note 15).

4. Incubate 30 min at 25°C (see Note 16).

5. Record sample and appropriate reference spectra at 25°C using a 0.1-cm path length quartz cuvette (Hellma Analytics, Müllheim, Germany), from 260 to 190 nm with 1-nm step resolution and an integration time of 3 s (see Note 17). Accumulate at least three to four scans to increase the signal-to-noise ratio.

6. Obtain the corrected CD sample spectrum by subtracting the reference spectrum recorded with peptide-free solutions (i.e., buffer or liposomes alone) from each average peptide spectrum.

7. For comparison between different samples and for verification of the average secondary structure content, convert arbitrary CD units (mdeg) to values of Mean Residue Molar ellipticity (θ), using the equation:

$$[\theta] = \frac{\theta_{obs}}{10 \times lcn} \quad (\text{in degrees } cm^2 d\,mol^{-1}).$$

where, θ_{obs} is the observed ellipticity in millidegrees, l is the optical path length in centimeters, c is the final molar concentration of the peptides, and n is the number of amino acid residues.

8. If the peptide forms α-helical structure in the presence of liposomes, the resulting spectrum should display characteristic features: two minima at 208 and 222 nm and a maximum at 190 nm (Fig. 2a, b) (see Note 18).

3.7. Fluorescence Spectroscopy. Intrinsic Tryptophan Fluorescence Emission

The CD spectrum of a peptide reflects primarily its secondary structural conformation rather than its interaction with the membrane. On the other hand, the high sensitivity of Trp fluorescence emission maximum to the solvent environment makes it a suitable and efficient probe for accessing peptide insertion into lipid bilayer.

1. Prepare samples as described in Subheading 3.6. However, lower peptide concentrations (see Note 19) should be used in fluorescence measurements to prevent inner filter effects while keeping the same peptide-to-lipid ratio as compared to the CD measurements.

2. Turn on the lamp of the instrument and allow it to warm up and stabilize for about 30 min before starting the measurements.

3. Set up the excitation wavelength to 280 nm or 290 nm (see Notes 19 and 20), emission wavelengths from 300 to 600 nm and slits widths. Use the emission and excitation polarizers oriented at 0° and 90°, respectively to minimize the relative contribution of the scattered light.

4. Add samples into a 2 × 10 mm QS quartz Ultra-Micro cuvette and record the emission spectra between 300 and 600 nm.

5. Subtract the blank spectra (i.e., buffer alone or liposomes alone) from the corresponding sample spectra. A fluorescence emission maximum at about 350 nm indicates that the Trp residue is fully exposed to the polar aqueous environment, while a blue-shifted λmax and an intensity increase of Trp emission indicate that the Trp residue is buried in the hydrophobic core of the liposomes (Fig. 2c). Detailed study of the interactions between TM peptides and membrane bilayer by the fluorescence methods can be found in (20).

3.8. Fluorescence Energy Transfer Measurements

In this protocol, FRET is measured as a function of acceptor fraction. In other words, the peptide-to-lipid ratio is kept constant, while the ratio of donor (TMa-W) to acceptor peptides (TMb-Pyr) is varied. The total TMb peptide concentration is kept constant by adjusting the concentration of the TMb acceptor peptide with unlabeled peptide (TMb-Unl). It is desirable to use low peptide concentrations to avoid FRET arising from random proximity of the acceptors and donors.

1. Prepare appropriate samples by mixing stock solutions of the donor, acceptor, and unlabeled peptides as suggested in Table 2.

Table 2
Example of sample preparation to measure helix–helix interactions between TMa and TMb peptides in lipid bilayer by FRET

TMa-W (μM)	10	10	10	10	10	10	0
TMb-Pyr (μM)	0	2	4	6	8	10	10
TMb-Unl (μM)	10	8	6	4	2	0	10

TMa-W and TMb-Pyr are used as donor and acceptor, respectively. In this protocol, the peptide-to-lipid ratio is kept constant (1:100), but the ratio of donor (TMa-W) to acceptor peptides (TMb-Pyr) varies. The total TMb peptide concentration is kept constant by adjusting the concentration of the TMb acceptor peptide with unlabeled peptide (TMb-Unl). The volume of buffer added should take into account the volume of liposome solution that will be added in the final step to get a final concentration of 20 μM of peptides and 2 mM of lipids. Concentration of peptide stock solutions = 500 μM. Concentration of the liposomes stock solution = 20 mM. Final total volume = 150 μL

2. Add 10 mM HEPES buffer, pH 7, containing 10 mM KCl to the peptide mix.

3. Incubate for 15 min at 25°C

4. Add liposomes to maintain a peptide-to-lipid ratio of 1:100 and adjust the final volume to 150 μL with the corresponding buffer.

5. Incubate for 30 min at 25°C.

6. Record the spectra as described in Subheading 3.7 (steps 2–4).

3.9. FRET Data Processing and Analysis

1. Before calculating FRET efficiency, it is necessary to correct all sample spectra (Fig. 3a) for:

   • Light scattering due to the liposome size: subtract the spectrum obtained with liposomes containing unlabeled peptides (at the same concentration as in the sample containing labeled peptides).

   • Contribution of acceptor emission: subtract the respective spectrum of the acceptor peptide recorded at the excitation wavelength of the donor peptide.

2. Plot corrected spectra (Fig. 3b).

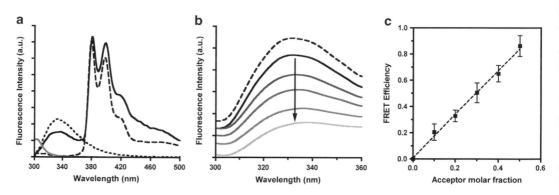

Fig. 3. Measurements of FRET between the TMa-W (donor) peptide and the TMb-Pyr (acceptor) peptide in DMPC liposomes. (a) Representative fluorescence emission spectra of TMa-W peptide (*dotted line*) and TMa-W + TMb-Pyr (donor + acceptor) peptides (*solid black line*) incorporated in DMPC liposomes. These two spectra were obtained after subtracting the background spectrum of liposomes (*solid grey line*) and the acceptor spectrum (TMb-Pyr peptide) (*dashed line*) upon excitation at 290 nm. The spectrum of TMb-Pyr is scaled to (donor + acceptor) spectrum at 380 nm. All spectra were measured in buffered solutions of 10 mM HEPES, pH 7 and were recorded upon excitation at 290 nm. For methodology details *see* Subheading 3.9 and Table 2. (b) Representative FRET spectra of TMa-W peptide mixed with TMb-Pyr peptide in DMPC liposomes upon excitation at 290 nm. 10 μM TMa-W peptide alone (*dashed line*) and in the presence of 2, 4, 6, 8, and 10 μM TMb-Pyr peptide (*solid lines, arrow* indicates increasing concentrations of TMb-Pyr peptide). The samples were prepared as explained in Subheading 3.8 (Table 2). FRET is monitored by decreases of Trp emission at 330 nm at increasing TMb-Pyr molar fraction. (c) Analysis of FRET data. Energy transfer efficiency was calculated as described in Subheading 3.8, step 3 and plotted as a function of acceptor molar fraction. FRET signal arises predominantly from the sequence specific contacts between TMa-W and TMb-Pyr peptides, but not from their random colocalization in the bilayer (see Note 4).

3. Calculate the energy transfer, efficiency E, from the corrected donor emission intensities in the absence (F_0) and presence of acceptor (F) according to:

$$E = 1 - \frac{F(\lambda_{em}^{D})}{F_0(\lambda_{em}^{D})},$$

where λ_{em}^{D} is the wavelength of the donor emission maximum (330 nm in the case of Trp).

4. Plot E as a function of acceptor molar fraction (Fig. 3c). The linear dependent increase of the energy transfer on the acceptor molar fraction indicates the formation of dimers and rules out formation of higher order aggregates (34).

5. Quantitative measurements of dimerization energetics are possible but are beyond the scope of this chapter. In these types of measurements, the peptide–lipid concentration is varied, while the donor–acceptor ratio is held constant. The readers should refer to the work of the Hristova lab for more details on protocol and data analysis (35, 36).

3.10. Detecting Homodimerization by Monitoring Pyrene Excimers Formation

As mentioned in the introduction, the pyrene chromophore conjugated to a Lys at the N-terminus of the peptide allows the monitoring of TM peptide homodimerization. Two pyrene rings can form excited-state dimers (excimers) when they reside within 10 Å of each other (26). This excimer state can be easily monitored, as it has a unique fluorescence emission peak at ~480 nm. Similar to the FRET measurement described in Subheading 3.8, the peptide-to-lipid ratio is kept constant, while the concentration of TMb-Pyr is varied. The total TMb peptide concentration is kept constant by adjusting the concentration of TMb-Unl.

1. Mix appropriate volumes of TMb-Pyr and TMb-Unl peptide stock solutions.

2. Incubate for 15 min at 25°C.

3. Add liposomes to yield a peptide-to-lipid ratio of 1:100 and a final volume of 150 μL.

4. Incubate 30 min at 25°C.

5. Set the excitation wavelength to 345 nm and collect the emission spectra between 360 and 500 nm. Use the emission and excitation polarizers oriented at 0° and 90°, respectively.

6. Subtract from each sample spectrum, the spectrum obtained with liposomes containing unlabeled peptides (at the same concentration as in the sample containing labeled peptides) (Fig. 4a).

7. Plot the fluorescence intensities observed at 480 nm against TMb-Pyr. The linear increase of the excimer intensity with

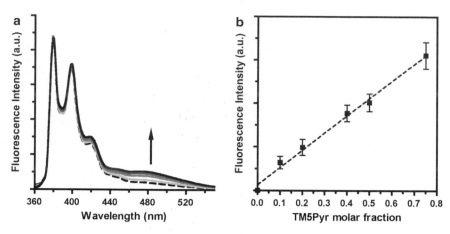

Fig. 4. Homodimerization of TM5 peptide in POPC liposomes upon excitation at 345 nm (**a**) Representative fluorescence emission spectra of TM5-Pyr peptide in the presence of 2 (*dashed line*), 4, 8, 10, and 15 μM of TM5-Unl peptide (*solid lines*, *arrow* indicates increasing concentrations of TM5-Unl). Each sample is prepared as described in Subheading 3.10. (**b**) Fluorescence intensity of the excimer (at 480 nm) represented as a function of the molar fraction of TM5-Pyr in liposomes.

increase in molar fraction of TMb-Pyr in the sample indicates the formation of specific homodimers, but not aggregates (Fig. 4b).

4. Conclusion

The method described in this chapter allows for robust identification of helix–helix interactions between TM peptides in the lipid bilayer by FRET. However, important control assays (e.g., determination of peptide and lipid concentrations, secondary and tertiary structures of labeled and unlabeled peptides in liposomes) need to be systematically performed to assure a correct interpretation of the results. This approach can also be easily adapted to study TM interactions in other membrane mimetic environments such as micelles (21–23), bicelles or surface-supported bilayers.

5. Notes

1. Förster Resonance Energy Transfer (FRET) is a distance-dependent transfer of energy from an excited fluorophore (donor molecule, D) that initially absorbs the energy, to a suitable acceptor fluorophore (acceptor molecule, A). This process occurs whenever the emission spectrum of D overlaps with the absorption spectrum of A (Fig. 1a). The efficiency of

energy transfer for a single donor–acceptor pair at a fixed distance is:

$$E = \frac{1}{1 + (r / R_0)^6},$$

where (r) is the distance between the fluorophores and (R_0) is the distance at which the efficiency of energy transfer is 50% (Förster distance). R_0 depends on the quantum yield of the donor in the absence of acceptor (Q_D), the degree of spectral overlap between the donor emission and the acceptor absorption (J), the refractive index of the solution (n), and the relative orientation in space of the transition dipoles of the donor and acceptor (κ^2), following the equation:

$$R_0 = 9786(Jn^{-4}\kappa^2 Q_D)^{1/6}.$$

2. Even though the evaluation of absolute distances is possible, it is rather challenging because it requires knowing the Förster distance (R_0) with some precision, which, even for a well characterized dye pair with a well known spectral overlap and solvent response, still requires some knowledge of relative dye orientation (37). Therefore, in situations where precise distance information is not vital, the change in FRET efficiency gives very useful information.

3. More information about other important factors (e.g., length of TM segment, flanking residues, membrane insertion capability of the peptide) that need to be considered when designing TM peptides can be found in recent reviews (13, 30).

4. The Trp–pyrene pair is an appropriate FRET donor–acceptor pair, as it satisfies both compatibility and proximity criteria: The fluorescence emission spectrum of Trp (the donor) overlaps with the absorption spectrum of pyrene (the acceptor) (Fig. 1) and the Förster critical distance (R_0) for energy transfer between Trp and pyrene has been calculated as 28 Å (19, 38). The use of Trp–pyrene pair offers other advantages: (1) Use of a naturally occurring residue as a fluorescence chromophore donor without need for chemical modification, (2) Minimal disruption of the TM domains secondary structure and eventually, of helix–helix contacts, as Trp (and Tyr) are often present in the sequence of TM domains, (3) substituting a Tyr for a Trp residue allows to retain the ability to determine peptide concentrations by UV spectroscopy at 280 nm because both Trp and Tyr are chromophores, (4) being able to monitor peptide insertion into lipid bilayer without the need for external fluorescent dyes, due to the solvent-dependent fluorescence properties of Trp, (5) an expected relatively low contribution of FRET arising from random donor–acceptor

colocalization in the bilayer to the measured FRET (~25% FRET) for 1.2% mol acceptor in liposomes, as predicted by simulations and considering $R_0 = 28$ Å for Trp–pyrene (39), and (6) the formation of excited-state dimers (excimers) upon close encounter of two pyrene molecules (25, 26), allows the monitoring of homodimerization of pyrene-labeled peptides.

5. Solid-phase synthesis of peptides has become more amenable to the preparation of TM segments with the availability of optimized amino acid activators, resins and new methodologies (30), but it remains a tedious task. Good quality custom peptides can now be obtained from numerous commercial sources at a reasonable cost (e.g., Anaspec, Fremont, CA, Polypeptides, CA). However, if many peptide variants are needed, it may be worthwhile to synthesize those in-house.

6. Other organic solvents such as hexafluoroisopropanol (HFIP) or trifluoroethanol (TFE) can be used to dissolve TM peptides. However, the concentration of organic solvents should always be kept at a minimum in the final samples. Furthermore, substituted fluoroalcohols (e.g., HFIP and TFE) are known to be strong inducers of helical secondary structure in proteins, and should consequently be used with caution. Moreover, dimethyl sulfoxide and dimethylformamide should be avoided because of their strong absorption. Methanol should be avoided as well, because it does not support TM helical structure and promotes misfolding and aggregation into beta-sheets.

7. Peptide stock solutions can be stored up to 3 months at –20°C in small aliquots to avoid multiple freezing and thawing cycles.

8. 6 M guanidine-HCl is used to completely denature the peptide and to expose Trp and Tyr residues to the polar environment, which ensures an accurate determination of the peptide concentration. If the peptide sequence does not have Trp, Tyr or Phe residues, the only practical option is to do amino acid analysis. It is also crucial to determine the percentage of labeling by UV absorbance, as incomplete labeling of the acceptor greatly affects quantitative FRET measurements.

9. When choosing lipids, two main properties need to be considered:

   (a) The head group charge: It is known that electrostatic interactions between peptide and membrane can play a significant role in the insertion, helix formation and folding of the peptide. In some cases, these interactions are indispensable for the insertion of the peptide into bilayer (13, 20).

   (b) Hydrophobic mismatch: It arises when the length of the hydrophobic region of the TM peptide differs in length from that of the hydrophobic tail of the lipid. This mismatch may cause aggregation, improper orientation or tilting of the peptide inside the bilayer (9, 13, 20). It is then recommended to determine the length of the hydrophobic region of the TM domain of interest from its amino-acid sequence and look for a lipid with a similar hydrophobic tail length.

When possible, it is preferable to work with a mixture of lipids close to the lipid composition of the native membrane where the protein of interest resides physiologically.

10. FRET efficiencies and helix–helix interactions measured in multilamellar vesicles and large unilamellar vesicles are comparable (39), with a negligible (2–5%) decrease in measured FRET efficiencies after extrusion. This decrease may be explained by eliminating the possibility of FRET occurring between donor and acceptor in different bilayers in the multilamellar samples. A potential problem with extrusion is the loss of lipids in the process (13–18%) (39), which emphasizes the need for measuring the lipid concentration after each extrusion (see Subheading 3.4).

11. Attempts to extrude below the T_m will be unsuccessful as the polycarbonate membrane filter has a tendency to get clogged with rigid lipid membranes, which cannot pass through the pores.

12. Store the vesicle solution at 4°C when not in use, but do not freeze. Even though vesicle solutions can remain stable in aqueous solutions for about 3–4 days when stored at 4°C, we encourage preparing fresh liposome solutions before each experiment.

13. The orientation and tilt of the helix can be further investigated by oriented circular dichroism or Fourier transform infrared spectroscopy, but the description of these methods is out of the scope of this chapter.

14. Measuring various peptide concentrations helps to not only identify the minimal peptide concentration with the best signal-to-noise ratio, but also check whether peptides form oligomers. Indeed, the intensity of the CD spectrum (after conversion to mean residue molar ellipticity) increases with peptide concentration if peptides oligomerize (22).

15. Low peptide-to-lipid ratios (e.g., 1:100) are used to minimize FRET arising from random colocalization of the acceptor and donor peptides.

16. From our experience, different TM peptides need different time to equilibrate in the bilayer, which may range from minutes to hours. Therefore, it is essential to ensure that complete equilibrium is reached by recording the Trp fluorescence emission

over time. The equilibrium is reached whenever the emission maximum wavelength and intensity do not change anymore over time. It is also possible by monitoring the formation of helical structure by CD spectroscopy.

17. Quartz cuvettes should be cleaned between samples with concentrated nitric acid and ethanol, followed by extensive washing with nanopure water, as impurities on the cuvette wall can affect measurements.

18. The high absorbance of the buffer may likely prevent collecting spectra below 200 nm. Even though reliable secondary structure calculations (the actual percentages of different structures) are compromised, the validity of spectral comparisons is not.

19. To minimize self-quenching during fluorescence measurements, the UV absorbance of each sample at the excitation wavelength should be in the range of 0.1–0.2.

20. The choice of the excitation wavelength for the donor peptide should be chosen depending on the peptide sequence: Excite at 290 nm when Tyr residues are present in the peptide sequence along with Trp, as exciting at a wavelength shorter than 290 nm would result in the excitation of Tyr residues as well. Having in mind that the emission fluorescent maximum of fluorophores is independent of the wavelength of the excitation, excite the samples at 280 nm if Trp is the only fluorophore present in the peptide to minimize the contribution of the scattering effects on the emission spectra.

Acknowledgments

The authors acknowledge funding from NIH grant R01GM073857 (D.T) and from MCI grant BFU2009-08758/BMC (T.L).

References

1. Wallin E, von Heijne G (1998) Genome-wide analysis of integral membrane proteins from eubacterial, archaean, and eukaryotic organisms. Protein Sci 7(4):1029–1038

2. Overington JP, Al-Lazikani B, Hopkins AL (2006) How many drug targets are there? Nat Rev Drug Discov 5(12):993–996

3. Yildirim MA, Goh K-I, Cusick ME, Barabási A-L, Vidal M (2007) Drug-target network. Nat Biotechnol 25(10):1119–1126

4. Bowie JU (1997) Helix packing in membrane proteins. J Mol Biol 272(5):780–789

5. Curran AR, Engelman DM (2003) Sequence motifs, polar interactions and conformational changes in helical membrane proteins. Curr Opin Struct Biol 13(4):412–417

6. Senes A, Engel DE, DeGrado WF (2004) Folding of helical membrane proteins: the role of polar, GxxxG-like and proline motifs. Curr Opin Struct Biol 14(4):465–479

7. White SH, Wimley WC (1999) Membrane protein folding and stability: physical principles. Annu Rev Biophys Biomol Struct 28: 319–365

8. Moore DT, Berger BW, DeGrado WF (2008) Protein-protein interactions in the membrane: sequence, structural, and biological motifs. Structure 16(7):991–1001

9. Fiedler S, Broecker J, Keller S (2010) Protein folding in membranes. Cell Mol Life Sci 67(11):1779–1798

10. Adamian L, Liang J (2001) Helix-helix packing and interfacial pairwise interactions of residues in membrane proteins. J Mol Biol 311(4): 891–907

11. Faham S, Yang D, Bare E, Yohannan S, Whitelegge JP, Bowie JU (2004) Side-chain contributions to membrane protein structure and stability. J Mol Biol 335(1):297–305

12. Engelman DM, Chen Y, Chin C-N, Curran AR, Dixon AM, Dupuy AD, Lee AS, Lehnert U, Matthews EE, Reshetnyak YK, Senes A, Popot J-L (2003) Membrane protein folding: beyond the two stage model. FEBS Lett 555(1):122–125

13. Bordag N, Keller S (2010) Alpha-helical transmembrane peptides: a "divide and conquer" approach to membrane proteins. Chem Phys Lipids 163(1):1–26

14. Popot JL, Engelman DM (1990) Membrane protein folding and oligomerization: the two-stage model. Biochemistry 29(17): 4031–4037

15. Hunt JF, Earnest TN, Bousché O, Kalghatgi K, Reilly K, Horváth C, Rothschild KJ, Engelman DM (1997) A biophysical study of integral membrane protein folding. Biochemistry 36(49):15156–15176

16. Xie H, Ding FX, Schreiber D, Eng G, Liu SF, Arshava B, Arevalo E, Becker JM, Naider F (2000) Synthesis and biophysical analysis of transmembrane domains of a Saccharomyces cerevisiae G protein-coupled receptor. Biochemistry 39(50):15462–15474

17. Selvin PR (1995) Fluorescence resonance energy transfer. Methods Enzymol 246:300–334

18. Selvin PR (2000) The renaissance of fluorescence resonance energy transfer. Nat Struct Biol 7(9):730–734

19. Wu P, Brand L (1994) Resonance energy transfer: methods and applications. Anal Biochem 218(1):1–13

20. Lazarova T, Brewin KA, Stoeber K, Robinson CR (2004) Characterization of peptides corresponding to the seven transmembrane domains of human adenosine A2a receptor. Biochemistry 43(40):12945–12954

21. Thévenin D, Lazarova T (2008) Stable interactions between the transmembrane domains of the adenosine A_{2A} receptor. Protein Sci 17(7):1188–1199

22. Thévenin D, Lazarova T, Roberts MF, Robinson CR (2005) Oligomerization of the fifth transmembrane domain from the adenosine A_{2A} receptor. Protein Sci 14(8):2177–2186

23. Thévenin D, Roberts MF, Lazarova T, Robinson CR (2005) Identifying interactions between transmembrane helices from the adenosine A_{2A} receptor. Biochemistry 44(49):16239–16245

24. Lakowicz JR (2006) Principles of fluorescence spectroscopy, 3rd edn. Springer, New York, 954

25. Ming M, Chen Y, Katz A (2002) Steady-state fluorescence-based investigation of the interaction between protected thiols and gold nanoparticles. Langmuir 18(6):2413–2420

26. Sahoo D, Weers PMM, Ryan RO, Narayanaswami V (2002) Lipid-triggered conformational switch of apolipophorin III helix bundle to an extended helix organization. J Mol Biol 321(2):201–214

27. Krogh A, Larsson B, von Heijne G, Sonnhammer EL (2001) Predicting transmembrane protein topology with a hidden Markov model: application to complete genomes. J Mol Biol 305(3):567–580

28. Snider C, Jayasinghe S, Hristova K, White SH (2009) MPEx: a tool for exploring membrane proteins. Protein Sci 18(12):2624–2628

29. von Heijne G (1992) Membrane protein structure prediction. Hydrophobicity analysis and the positive-inside rule. J Mol Biol 225(2):487–494

30. Cunningham F, Deber CM (2007) Optimizing synthesis and expression of transmembrane peptides and proteins. Methods 41(4):370–380

31. Liu LP, Deber CM (1998) Guidelines for membrane protein engineering derived from de novo designed model peptides. Biopolymers 47(1):41–62

32. Gasteiger E, Hoogland C, Gattiker A, Duvaud S, Wilkins MR, Appel RD, Bairoch A (2005) Protein identification and analysis tools on the ExPASy server. In: Walker JM (ed) The proteomics protocols handbook. Humana, New Jersey, pp 571–607

33. Stewart JCM (1980) Colorimetric determination of phospholipids with ammonium ferrothiocyanate. Anal Biochem 104(1):10–14

34. Stryer L (1978) Fluorescence energy transfer as a spectroscopic ruler. Annu Rev Biochem 47:819–846

35. Chen L, Merzlyakov M, Cohen T, Shai Y, Hristova K (2009) Energetics of ErbB1 transmembrane domain dimerization in lipid bilayers. Biophys J 96(11):4622–4630

36. Merzlyakov M, Hristova K (2008) Förster resonance energy transfer measurements of

transmembrane helix dimerization energetics. Methods Enzymol 450:107–127

37. Rasnik I, McKinney S, Ha T (2005) Surfaces and orientations: much to FRET about? Acc Chem Res 38(7):542–548

38. Vekshin N, Vincent M, Gallay J (1993) Tyrosine hypochromism and absence of tyrosine-tryptophan energy transfer in phospholipase A2 and ribonuclease T1. Chem Phys 171(1–2): 231–236

39. You M, Li E, Wimley WC, Hristova K (2005) Forster resonance energy transfer in liposomes: measurements of transmembrane helix dimerization in the native bilayer environment. Anal Biochem 340(1): 154–164

Chapter 7

Studying Substrate Binding to Reconstituted Secondary Transporters by Attenuated Total Reflection Infrared Difference Spectroscopy

Víctor A. Lórenz-Fonfría, Xavier León, and Esteve Padrós

Abstract

The determination of protein conformational changes induced by the interaction of substrates with secondary transporters is an important step toward the elucidation of their transport mechanism. Since conformational changes in a protein alter its vibrational patterns, they can be detected with high sensitivity by infrared difference (IR_{diff}) spectroscopy without the need for external probes. We describe a general procedure to obtain substrate-induced IR_{diff} spectra by alternating perfusion of buffers over an attenuated total reflection (ATR) crystal containing an adhered film of a membrane protein reconstituted in lipids. As an example, we provide specific protocols to obtain melibiose and Na^+-induced ATR-IR_{diff} spectra of reconstituted melibiose permease, a sodium/melibiose co-transporter from *E. coli*. The presented methodology is applicable in principle to any membrane protein, provided that it can be purified and reconstituted in functional form, and appropriate substrates are available.

Key words: Secondary transporters, Melibiose permease (MelB), Membrane proteins, Fourier transform infrared spectroscopy (FT-IR), Attenuated total reflection (ATR), Difference spectroscopy, Substrate binding, Ligand binding

1. Introduction

Secondary transporters are involved in the binding and the active transport of substrates, usually of nutrients (e.g., amino acids, sugars, etc.), to the cell interior. This process is fuelled by the electrochemical gradient of the membrane. On the other hand, receptor–ligand interactions trigger conformational changes in the receptor that act as a signal for downstream proteins to generate a cell response. Given their role, it is not surprising that defects in the interaction of membrane proteins with molecules and cations

Nagarajan Vaidehi and Judith Klein-Seetharaman (eds.), *Membrane Protein Structure and Dynamics: Methods and Protocols*, Methods in Molecular Biology, vol. 914, DOI 10.1007/978-1-62703-023-6_7, © Springer Science+Business Media, LLC 2012

are one of the main reasons for human diseases (1–3). As a result, there is ample interest in methods able to both detect and characterize structurally the interaction of different substrates with membrane proteins, preferably when purified and reconstituted in model lipidic environments mimicking their native milieu (4–6).

Among the many spectroscopies, infrared (IR) spectroscopy is particularly well suited for the study of membrane proteins, being simultaneously and naturally sensitive to the protein backbone, amino acid side chains, lipids, water, and any possible protein cofactor, without requiring to add any external probe (7–9). However, even small proteins contain sufficient atoms to give rise to 10^4 vibrational modes, challenging the extraction of structural information from their IR absorbance spectrum (7). This task is simplified by the use of IR difference (IR_{diff}) spectroscopy, which examines the differences in the sample absorbance before and after inducing a perturbation that alters the state of the protein, usually simulating one or several steps of its physiological reaction (10–12).

By alternating buffers over an attenuated total reflection (ATR) crystal it is possible to change in situ the substrate concentration in contact with a reconstituted membrane protein adhered to the surface of the crystal, generating a substrate-induced IR_{diff} spectrum (13–17). This spectrum will not only inform about the presence (or absence) of substrate binding to the protein, but it will also provide details about the resulting changes triggered in the protein conformation. Moreover, the large and widely distributed number of IR active vibrations in a protein (7) practically guarantees that any substrate–protein interaction will be detectable by IR_{diff} spectroscopy. In contrast, spectroscopic methods that rely on localized probes might be silent to substrate-induced changes not affecting the specific environment of the probe.

For several years we have applied ATR-IR_{diff} spectroscopy to characterize substrate binding to the melibiose permease (MelB) from *Escherichia coli* (15, 18–21). MelB belongs to the glycoside-pentoside-hexuronide:cation symporter family, a sub-member of the major facilitator superfamily (22). It is a Na^+-driven secondary transporter that couples the uphill transport of α-galactosides (e.g., melibiose), or less efficiently β-galactosides (e.g., lactose), to the downhill symport of Na^+ (or H^+ or Li^+) in a sugar–cation ratio of 1-to-1 (23, 24).

Here we provide detailed protocols to obtain melibiose and Na^+-induced ATR-IR_{diff} spectra of reconstituted MelB, and to correct them for unspecific spectral contributions. We also describe several suitable controls by using chemically similar but biologically inactive molecules. The assays described here are likely to be applicable with few modifications to study substrate binding to other reconstituted secondary transporters or even other membrane proteins such as primary transporters, channels, or receptors.

2. Materials

2.1. Buffers

1. For experiments studying the effects of melibiose binding in the presence of Na⁺, prepare a sufficiently large volume of fresh reference buffer (buffer A in Fig. 1a, b), containing 100 mM KCl, 10 mM NaCl, and 20 mM MES, and adjust it to pH 6.6 with KOH. To prepare the substrate-containing buffer (buffer B in Fig. 1a, b) take part of the buffer A and add solid melibiose to a final concentration of 10 mM. For control experiments, prepare 50 mM sucrose in the same way, but substituting melibiose by sucrose in buffer B (see Note 1).

2. For experiments studying the effects of Na⁺ binding, prepare a sufficiently large volume of a stock buffer containing 100 mM KCl and 20 mM MES at pH 6.6. Buffer B is prepared by adding

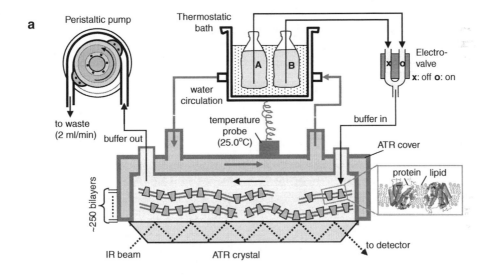

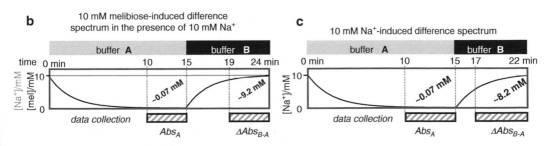

Fig. 1. (a) Schematic illustration of the experimental setup used to obtain substrate-induced ATR-IR$_{diff}$ spectra. Reference buffer A, and a buffer with substrate/s, buffer B, were alternatively and repetitively perfused over the sample deposited on the ATR crystal, using an electro-valve and a peristaltic pump. The sample and buffer temperature was controlled with a thermostatic bath. (b and c) Protocol for buffer exchange and spectra acquisition (for clarity only one cycle is shown, but typically >25 of such cycles are performed for one experiment). The estimated substrate concentration profile during the buffer exchange and during spectra acquisition is indicated.

to the stock buffer solid NaCl to a final concentration of 10 mM. Buffer A is prepared instead by adding to the stock buffer additional solid KCl to a final concentration of 110 mM. This increase in concentration compensates for ionic strength differences between buffers A and B. For control experiments assessing the effect of differences in ionic strength, we use the stock buffer as buffer A (100 mM KCl), while adding additional 2 mM KCl to prepare buffer B (102 mM KCl).

2.2. Sample

We used purified MelB cysteine-less (C-less) reconstituted in commercially available *E. coli* lipids (Avanti Polar Lipids, Inc.). MelB was purified and solubilized in DDM from *E. coli* cells as described (25), with some modifications (21). For lipid reconstitution, the detergent-solubilized MelB sample was incubated for 10 min with small unilamellar liposomes of *E. coli* total lipid extract (Avanti Polar Lipids, Inc.) in a ratio of 1:2 (w/w of protein/lipid). The formation of proteoliposomes was induced overnight by the stepwise addition of 120 mg of prewashed Bio-Beads SM-2 (Biorad Laboratories) per ml of protein solution, under continuous agitation. The next day the Bio-Beads were removed using a polypropylene column (Qiagen), and the proteoliposomes were centrifuged at $300,000 \times g$ for 30 min and washed with the final buffer (three cycles). Finally, they were resuspended at 5–8 mg/ml of protein, and stored at −80°C until use. All the steps were done at 4°C.

2.3. Instrumentation

1. An FT-IR spectrometer. We use an FTS6000 model from Bio-Rad Laboratories (now owned by Agilent Technologies), equipped with a N_2 liquid cooled mercury-cadmium-telluride (MCT) detector. The spectrometer interior is purged with dry air, with a nominal dew-point lower than −50°C.

2. An ATR accessory compatible with the sample compartment of the FT-IR spectrometer. We use a Horizon ATR (Harrick Scientific Products, Inc.).

3. A trapezoidal $50 \times 10 \times 2$ mm ATR crystal made of germanium (Ge), mounted in a steel holder (Harrick Scientific Products, Inc.) (see Note 2).

4. A homemade hollow ATR cover made of steel (see Fig. 1a). Two of the entries lead to the cover interior, where a flow of liquid from a thermostatic bath allows for temperature control. A temperature probe connected to the thermostatic bath is fitted into the cover to control precisely its temperature. The other two entries of the cover lead to the ATR crystal, and are used as inlets and outlets to perfuse the buffers over the sample.

5. A circulation thermostatic bath working in external temperature control mode, connected to the ATR cover and to a temperature probe.

6. An electro-valve. We use a solenoid-operated three-way pinch electro-valve from Cole Parmer. The electro-valve is connected to an electronic device, which is controlled by a desk computer using a driver provided by the supplier. A homemade routine in Visual Basic provides integrated control of the electro-valve and the software of the FT-IR spectrometer.

3. Methods

The idea behind substrate-induced ATR-IR$_{diff}$ spectroscopy is simple: substrate binding is assessed by increasing the substrate concentration in a solution in contact with the lipid-reconstituted protein film, which is adhered to the surface of an ATR crystal (see Fig. 1a). If the tested substrate binds to the protein, the protein structure (and hence its vibrations) will change, and this will generate an IR$_{diff}$ spectrum. Not only substrate–protein interactions can be detected, but information about the changes triggered in the protein conformation and in amino acid side chain environments as well as protonation states can be obtained from the resulting IR$_{diff}$ spectrum.

In practice, some practical points have to be taken into account. As we have discussed previously (15) the experimental IR$_{diff}$ spectrum of the sample contains at least four possible contributions: (1) a protein (and lipid) IR$_{diff}$ spectrum induced by substrate(s) binding to the protein, which is the signal of interest; (2) a buffer IR$_{diff}$ spectrum induced by the substrate effect on the water structure; (3) an IR absorbance of the substrate(s) used to induce the IR$_{diff}$ spectrum (in our case melibiose, since Na$^+$ is not IR active); and (4) a swelling/deflating of the deposited sample film, giving an IR$_{diff}$ spectrum from the gain/loss of water and the concomitant loss/gain of protein–lipid in the volume probed by the evanescent IR beam. To correct for the undesired contributions we need to collect, besides the substrate-induced IR$_{diff}$ of the sample (ΔAbs_{sample}), the IR absorbance of the sample (Abs_{sample}) and the buffer (Abs_{buffer}), and the substrate-induced IR$_{diff}$ of the buffer (ΔAbs_{buffer}). Furthermore, several negative controls will be required to assess the correction procedure, and to discard (or identify) contributions from lipid–substrate interactions to the IR$_{diff}$ spectra of the sample.

It is important to remember that unlike in the more commonly used transmittance mode, in ATR the sample is probed by an evanescent electromagnetic field that penetrates only some few hundreds of nm from the ATR surface into the sample (see Note 3). We recommend to potential users of this technique to acquire some knowledge about the basis of ATR-IR spectroscopy (26–28), and to be aware of some of its peculiar features and distortions (26, 29, 30).

3.1. Melibiose-Induced IR Difference Spectra

3.1.1. Buffer Measurements

1. Fill the MCT detector dewar of the FT-IR spectrometer with liquid nitrogen (see Note 4). Place a clean mounted ATR crystal on the ATR accessory. Control its temperature to $25.0 \pm 0.1°C$ using the thermostatted ATR cover (Fig. 1a). Wait for at least 30 min to let the detector temperature stabilize and the water vapor in the spectrometer decrease to its initial level (see Note 5). Collect a single-beam spectrum by averaging scans for around 10–20 min at 4 cm^{-1} resolution (see Note 6): e.g., 2,000–4,000 symmetric interferograms co-added at a 40 kHz mobile mirror speed.

2. As schematically shown in Fig. 1a, connect (1) the buffer bottles to the two electro-valve *ins*; (2) the electro-valve *out* to the ATR cover *inlet*; and (3) the cover *outlet* to the peristaltic pump (see Note 7). The two buffers used in the experiments, reference buffer (buffer A) and the buffer containing the substrate to be tested (buffer B), should be maintained roughly at the same temperature as the ATR crystal (see Note 8). Start the perfusion of buffers, and adjust the flow to ~2 ml/min. Launch the acquisition program, which alternates perfusion from buffer A to buffer B at user-selected times using the computer-controlled electro-valve (see Note 9), followed by data acquisition (~5 min, corresponding to 1,000 scans). A scheme illustrating one cycle of buffer exchange and data acquisition protocol is provided in Fig. 1b. The output of one cycle is an absorbance spectrum of the buffer A (Abs_{buffer}) and a difference spectrum of the B-minus-A buffers (ΔAbs_{buffer}). Repeat this cycle 10–50 times.

3. On the next day, add liquid nitrogen to the MCT detector and wait for equilibration. Measure a spectrum of water vapor. First collect a spectrum of 100 scans at 4 cm^{-1}. Then, open briefly the sample chamber of the IR spectrometer and let in some air from the room (containing water vapor), collecting immediately a spectrum with 20 scans. Subtract both to obtain a spectrum of water vapor (see Note 10).

4. Browse through the collected spectra, and manually remove those affected by the exhaustion of the liquid nitrogen in the detector dewar, or by the exhaustion of one of the buffers if needed (see Note 11). Erase any other suspicious or outlayer spectra (see Note 12). Following this protocol we obtained 42 spectra for both Abs_{buffer} and ΔAbs_{buffer} (Fig. 2, gray lines), equivalent to 42,000 co-added scans (17 h of useful experimental time). Average those to obtain a final Abs_{buffer} and ΔAbs_{buffer} spectrum (Fig. 2, black lines). Subtract any traces of water vapor absorbance using the collected water vapor spectrum (see Fig. 2a, insert).

3.1.2. Sample Measurements

1. Follow all the steps described in step 1 of Subheading 3.1.1. In addition, collect a background single-beam spectrum at 2 cm^{-1} resolution co-adding 2,000 scans (see Note 6). This background

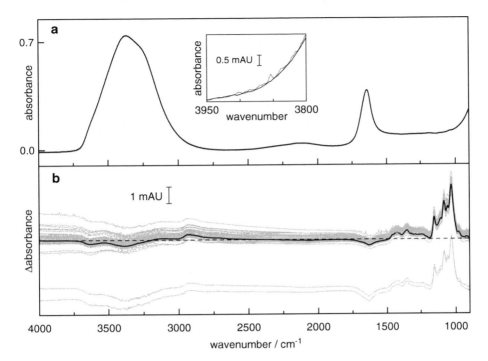

Fig. 2. Raw data obtained for the buffer-alone experiment, displaying both individual (*gray lines*) and averaged (*black lines*) spectra. (**a**) Absorbance of the reference buffer. (**b**) Melibiose-induced difference absorbance of the buffer. The *dashed line* indicates the zero. Note the difference in scale between (**a**) and (**b**). (Insert) Before (*gray line*) and after (*black line*) the subtraction of a water vapor spectrum.

will be later needed to obtain the absorbance spectrum of the dry film of the sample at 2 cm⁻¹ resolution.

2. This step is optional (see Note 13). Collect a fresh Abs_{buffer} spectrum. Add 1 ml of buffer A to the clean ATR crystal and cover it for temperature control. Allow at least 30 min for temperature equilibration, and collect 2,000 scans at 4 cm⁻¹ resolution. Rinse the crystal with deionized water and drain it, but without removing it from the ATR accessory (see Note 14).

3. Prepare a dry film of reconstituted MelB (see Note 15) as follows. Add 20 μl of MelB reconstituted in liposomes (~150 μg), and spread it using a rounded tip until the ATR surface is covered (see Note 16). Let the solution on the ATR dry gently by exposing it to the ambient atmosphere of the laboratory, ~40–50% relative humidity (see Note 17). When the dry film is formed, wait for an additional 30 min and collect 2,000 scans at 2 cm⁻¹ resolution (see Fig. 3a). Correct for water vapor contributions (Fig. 3a, insert), collecting a fresh water vapor spectrum at 2 cm⁻¹ resolution. The absorbance spectrum of the dry film can be used for structural analysis, e.g., to obtain an estimation of the secondary structure of the protein (31), to confirm the absence of structural alterations in a mutant protein with respect to the WT (21), or to reveal signatures of protein aggregation/denaturation (32).

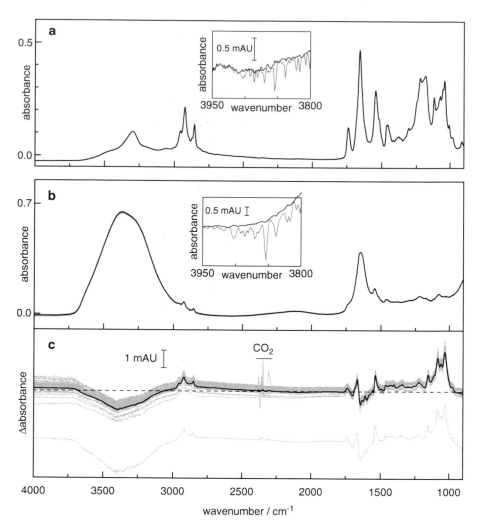

Fig. 3. Raw data obtained for the reconstituted MelB sample, displaying both individual (*gray lines*) and averaged (*black lines*) spectra. (**a**) Absorbance spectrum of the dry protein–lipid film. (**b**) Absorbance spectra of the protein–lipid film hydrated with the reference buffer. (**c**) Melibiose-induced difference spectra of the protein–lipid hydrated film. Note the difference in scale between (**b**) and (**c**). (Inserts) Before (*gray line*) and after (*black line*) the subtraction of a water vapor spectrum.

4. Make the necessary arrangements (see step 2 of Subheading 3.1.1) and start the perfusion of the crystal with buffer A at a flow rate of 2 ml/min. When the buffer arrives at the waste recipient, switch off the peristaltic pump, and wait 30 min to 1 h for the swelling of the protein–lipid film to stabilize. At this point an absorbance spectrum can be optionally collected to confirm that the protein–lipid film has not been washed out and remains close to the ATR surface.

5. Resume the perfusion of the film with buffer A. After ~10 min launch the program and start the measurements. After several hours, add liquid nitrogen to the detector if necessary.

6. On the next day, add liquid nitrogen to the MCT detector, wait for equilibration, and collect a water vapor spectrum.

7. Depending on the stability of the protein the experiment can be continued using the same sample (see Note 18). Simply, change the buffers to measure the effect of a different substrate (see Note 19), perfusing the new buffer A for at least ~30 min before starting the measurements.

8. Browse through the collected data, and manually remove spectra affected by any artifacts (see Note 20). We obtained 34 spectra for both Abs_{sample} and ΔAbs_{sample} (Fig. 3b, c, gray lines), equivalent to 14 h of useful experimental time (see Note 21). Average the individual spectra, obtaining a final Abs_{sample} and ΔAbs_{sample} spectrum made of 34,000 co-added scans each (Fig. 3b, c, black lines). Subtract any traces of water vapor (Fig. 3b, insert).

3.1.3. Corrections

Absorbance Spectrum

The Abs_{sample} spectrum (Fig. 3b) is dominated by the buffer (water) absorbance. The most convenient way to remove this contribution is to subtract digitally the Abs_{buffer} spectrum (Fig. 2a) from the sample spectra. We use the intense band at 3,400 cm^{-1} from the water O–H stretching to guide us in the subtraction process. In the example illustrated in Fig. 4a, the optimal subtraction factor was found to be 0.880. Although the water band at 3,400 cm^{-1} was not perfectly cancelled (Fig. 4b, black line), the arising negative and positive lobules were of low intensity. Figure 4b also shows the results obtained when using instead of 0.88 a subtraction factor of 0.89 and 0.87 (gray lines), giving an idea of the spectral uncertainty associated with the subtraction of the buffer absorbance. The obtained absorbance spectrum of the protein–lipid sample free from the buffer contribution can be subjected—similar to the absorbance spectrum of the dry film—to a structural analysis (see Subheading 3.1.2, step 3), most commonly using band-narrowing and curve-fitting methods (33–35).

The factor used to subtract the buffer absorbance can provide interesting information about the sample deposited over the ATR crystal. Given that the estimated effective penetration depth will be ~2.4% higher for the hydrated sample than for the buffer (see Note 22), a subtraction factor of 0.88 implies that 85.9% of the volume probed by the evanescent IR beam is occupied by the buffer. The rest, 14.1% of the volume, corresponds to the protein and lipids. Incidentally, this relatively high percentage of volume occupied by the lipid–protein suggests that the sample does not form proteoliposomes after hydration (see Note 23), but likely exist as patches of stacked lipid bilayers (see Note 24). This difference is quite relevant in practice, since in the latter case both the cytosolic and periplasmic binding sites of MelB will be directly accessible to the substrates, while in the former case one of them might not be.

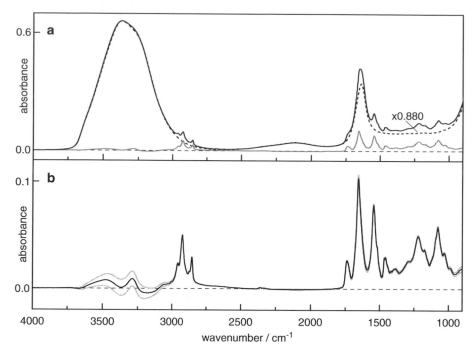

Fig. 4. Correction of the absorbance spectrum of the reconstituted MelB film for the buffer absorbance. (a) Absorbance of the protein–lipid hydrated film (*black line*), shown also in Fig. 3b. Scaled absorbance of the reference buffer (*short dashed line*), shown also in Fig. 2a. Their subtraction with a factor of 0.880 (*gray line*). (b) Expanded view of the resulting buffer-corrected spectrum of the reconstituted MelB (*black line*). The *gray lines* correspond to spectra obtained using subtraction factors of 0.890 and 0.870.

Difference Spectrum

As we have commented in the introduction of Subheading 3, the experimental IR_{diff} spectrum contains several unspecific contributions that need to be corrected before obtaining a substrate-induced IR_{diff} spectrum of a protein. Contributions 2 (water difference spectrum induced by melibiose) and 3 (absorbance of melibiose) can be simultaneously corrected by subtracting ΔAbs_{buffer} (Fig. 2b) from ΔAbs_{sample} (Fig. 3c) with an appropriate scaling factor (Fig. 5a). The factor used in subtraction was 0.635, chosen to remove two well-resolved intense bands characteristic of the melibiose, at 1,154 and 1,030 cm^{-1} (see Fig. 5a). The result of the correction is shown in Fig. 5b (gray line). Taking into account that ~86% of the probed sample volume is occupied by the buffer, and considering the concentration of melibiose in the buffer to be ~9 mM (see Fig. 1b), a subtraction factor of 0.635 implies that the true concentration of melibiose in the sample volume that contributes to the IR signal is ~7 mM (see Note 25).

Contribution 4 (swelling of the sample film) can be corrected in the following way. First compute the difference spectrum between Abs_{sample} and Abs_{buffer} (see Note 26). This spectrum mimics the absorbance changes when the sample film deflates: protein–lipid

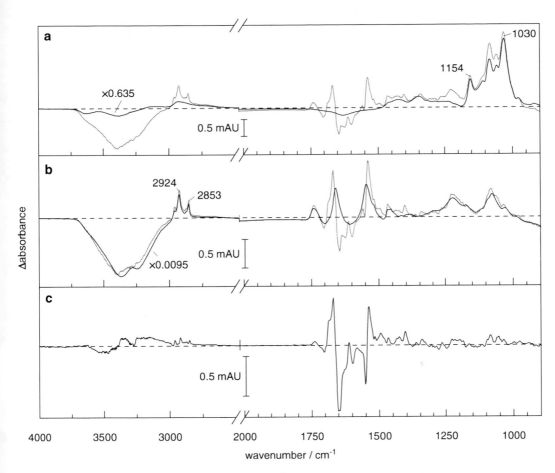

Fig. 5. Correction steps to obtain the melibiose-induced IR$_{diff}$ spectrum of MelB in the presence of Na$^+$. (a) Raw melibiose-induced IR$_{diff}$ spectrum of MelB (*gray line*). Scaled melibiose-induced IR$_{diff}$ spectrum of the buffer (*black line*). (b) Melibiose-induced IR$_{diff}$ spectrum of MelB corrected for the melibiose effect on the buffer absorbance (*gray line*). Scaled IR$_{diff}$ spectrum for the sample film deflation (*black line*), obtained as Abs$_{sample}$ – 1.024 × Abs$_{buffer}$. (c) Final melibiose-induced IR$_{diff}$ spectrum of MelB corrected for all the unspecific spectral contributions.

concentrates, with the concomitant water exclusion from the probed volume. Such spectrum is shown in Fig. 5b (black line), with a scaling factor of 0.0095, used to cancel the lipid CH$_2$ bands at 2,924 and 2,853 cm^{-1}. The result of this correction is shown in Fig. 5c (black line), which should specifically correspond to the vibrational changes on the protein–lipid sample caused by melibiose binding to MelB (see Note 27). Again, the subtraction factor used provides information by itself. We can conclude that in the presence of melibiose the protein–lipid in the volume probed by the IR beam increases by 0.95% at the expense of water, i.e., the separation between the protein–lipid layers is slightly decreased.

3.2. Na$^+$-Induced IR Difference Spectra

The same protocol described for collecting melibiose-induced difference spectra of the buffer and the sample can be used for other substrates. Depending on the type of substrate, its affinity, and the

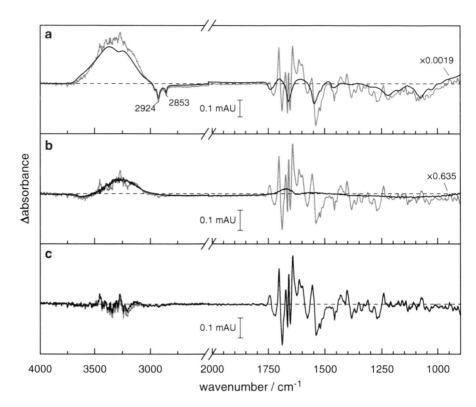

Fig. 6. Correction steps to obtain the IR$_{diff}$ spectrum corresponding to Na$^+$ binding to MelB. (a) Raw Na$^+$-induced IR$_{diff}$ spectrum of MelB (*gray line*). Scaled IR$_{diff}$ spectrum for the sample film swelling (*black line*). (b) Na$^+$-induced IR$_{diff}$ spectrum of MelB corrected for the film swelling (*gray line*). Scaled Na$^+$-induced IR$_{diff}$ spectrum of the buffer (*black line*). (c) Final Na$^+$-induced IR$_{diff}$ spectrum of MelB corrected for all the unspecific spectral contributions, using two different subtraction factors for the Na$^+$ effect on the buffer: 0.635 (*black line*) and 0.859 (*gray line*).

concentration used, the scheme for buffer perfusion times might need to be adapted. Figure 1c shows the scheme we use to obtain a Na$^+$-induced difference spectrum of MelB. The only difference with the previously described experiment with melibiose is a reduced exposure to the substrate before the data acquisition starts (compare Fig. 1b, c). The reason is that MelB shows higher affinity for Na$^+$ than for melibiose in the presence of Na$^+$ (23), and thus less time is needed to accumulate a sufficient local substrate concentration to see substantial binding (see Note 28).

The corrections involved for Na$^+$ are the same as those described in the previous section for melibiose. The main difference is that Na$^+$ does not absorb IR radiation, although it alters slightly the vibration pattern of water and hence its absorbance. Since these alterations are rather small, in this case it is better to start correcting for differences in the film swelling. As can be seen in Fig. 6a (gray line) the raw Na$^+$-induced difference spectra of the sample shows clearly negative lipid–protein bands and positive water bands, which can be corrected using the buffer-minus-sample absorbance spectrum scaled by 0.0019 (Fig. 6a, black line). From the subtraction factor we can deduce that the film becomes 0.19% less compact upon exposure to the buffer containing Na^+.

For the correction of the small unspecific effects of Na^+ in the water absorbance we need to subtract the Na^+-induced difference spectrum of the buffer (Fig. 6b, black line). Due to its broad spectral features, the subtraction factor is hard to obtain using spectroscopic criteria. However, we expect this factor to be at least 0.635 (the factor for melibiose correction), and at most 0.859 (the fraction of buffer volume in the sample). Figure 6c shows the corrected Na^+-induced difference spectrum of MelB using a correction factor of 0.635 (black line) and 0.859 (gray line), giving both almost indistinguishable results, i.e., the actual subtraction factor used is irrelevant for the result (see Note 29). Finally, in our experiments the actual Na^+ concentration in the probed sample volume might range from ~6.1 mM, for melibiose-like delayed diffusion, to ~8.2 mM, a difference that holds no significant consequences for Na^+ binding to C-less MelB (see Note 30).

3.3. Controls

3.3.1. Film Stability

The first control we recommend to perform is to assess the physical stability of the protein–lipid film deposited over the ATR crystal. A stable film is critical to obtain reliable substrate-induced IR_{diff} spectra. For this control we collect IR_{diff} spectra using identical buffers for A and B. Since the buffers are identical, changes in ΔAbs_{sample} will only appear when the film swelling does not stabilize with time, or when the sample is continually washed out from the ATR crystal by the perfusion of buffers (see Note 31).

3.3.2. Inactive Substrates

As negative controls we perform equivalent experiments as those described above for melibiose and Na^+, but using inactive substrates (15). The rationale behind these controls is twofold. First, the approach allows us to confirm that the IR changes induced by melibiose and Na^+ are selective to the MelB substrates, and are not elicited by chemically similar but biologically inactive molecules. Second, we can confirm that the protocol we developed for the correction of unspecific spectral contributions works. We have used 50 mM sucrose instead of 10 mM melibiose, and additional K^+ instead of Na^+ (creating a 2 mM ionic strength difference), confirming that after the corrections for all unspecific spectral contributions the obtained substrate-induced IR_{diff} spectra are flat and devoid of peaks above the noise (Fig. 7, gray lines).

3.3.3. Effect of the Substrates on the Lipids

It is also advisable to perform control experiments preparing a film from liposomes lacking MelB, and conducting the experiments described above to test for melibiose and Na^+ binding (15). From the obtained flat spectra (Fig. 7, black lines) we could confirm that all the bands in the substrate-induced IR_{diff} spectra of MelB arise from substrate-binding to the protein, and not from binding/interaction

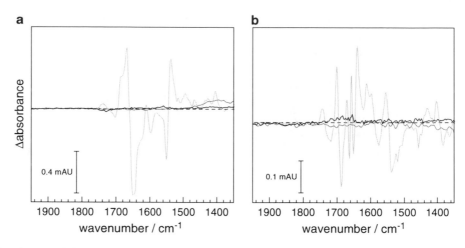

Fig. 7. Negative controls for melibiose (**a**) and Na+ (**b**) binding. Melibiose and Na+-induced IR$_{diff}$ spectra of *E. coli* lipids (*black lines*). Sucrose and K+-induced IR$_{diff}$ spectra of reconstituted MelB (*gray lines*). As a comparison, melibiose and Na+-induced IR$_{diff}$ spectra of reconstituted MelB reproduced from Figs. 5c and 6c (*light gray lines*).

of the substrates with the lipid membrane. Still, we cannot discard that some of the peaks observed in the substrate-induced IR$_{diff}$ spectra of MelB might arise from changes in lipids interacting with the protein, affected by conformational changes in the protein.

4. Notes

1. It is critical for buffers A and B to have the same pH. We have observed that even relatively small differences (0.1 units) in buffer pH largely alter the swelling and stability of the film, compromising the reliability of the obtained substrate-induced IR$_{diff}$ spectra. In our case, we have confirmed that the addition of melibiose does not alter the pH of buffer B.

2. Other IR transparent materials besides Ge can be used in ATR-IR spectroscopy, as for instance Si, ZnSe, ZnS, or diamond. In our case, we have used successfully a ZnSe crystal to obtain IR$_{diff}$ spectra of MelB (18).

3. For a thick film in ATR, 95% of the IR absorbance (or absorbance changes) comes from the sample located $3/2d_p$ from the surface (d_p, penetration depth). The value of d_p under our experimental conditions (refractive index of Ge: $n_1 = 4.0$, and incident angle of the IR beam: $\theta = 45°$) is given by (26) $d_p = (2\pi v \sqrt{8 - n_2^2})^{-1}$, i.e., 0.16, 0.32, and 0.64 μm at 4,000, 2,000, and 1,000 cm^{-1}, respectively, when $n_2 = 1.354$ (corresponding to the refractive index of the hydrated sample (see Note 22)).

4. A standard dewar for an MCT detector filled with liquid nitrogen provides 10–12 h of experimental time. The MCT detector can be refilled in the middle of an experiment if necessary. In this way, longer times of uninterrupted experiments are possible. In this sense, the best timing to add nitrogen is when the perfusion of buffer A starts, which allows for 10 min of detector temperature equilibration before the next spectrum acquisition starts (see Fig. 2b).

5. We recommend the transmittance of the intense water vapor band at 1,653 cm^{-1} to be 98.0% or higher.

6. Increase the IR beam aperture accordingly to the instrumental resolution used, maximizing the IR signal at the detector for better signal-to-noise ratio. Check your spectrometer manual for details.

7. Immerse the buffer bottles in the water reservoir of a thermostatic bath used to control the ATR temperature during the experiments (see Fig. 1a). It is also advisable that the temperature of the room is as close as possible to the temperature used for the experiments to minimize temperature variations in the sample during long experiments.

8. The buffer bottles and the waste recipient should be placed at roughly the same height to avoid the spontaneous flow of buffers from the bottles to the waste (or the other way) when the pump is off.

9. Beware that the change in the composition of the buffers in the ATR sample compartment will be not sharp, but exponential with time (see Fig. 1b, c). Intuitively, the time constant for buffer exchange should be roughly the ATR chamber volume divided by the buffer flow caudal. In agreement with this estimate, in our experimental setup we determined a buffer exchange time constant of 2.4 ± 0.4 min when the flow was ~2 ml/min, i.e., ~7.5 min of buffer perfusion is needed for 95% buffer exchange. Exceptionally longer buffer exchange times might be needed for substrates that bind tightly to the studied transporter.

10. The absorbance spectrum of water vapor should be collected under background absorbance conditions as close as possible to the spectrum we aim to correct for. The reason is that MCT detectors are not perfectly linear, and thus different levels of background absorbance might slightly affect the water vapor spectrum.

11. The first condition can be easily diagnosed by an increasing positive offset in Abs_{buffer} and ΔAbs_{buffer} (or equivalently a reduced intensity in the single-beam spectrum which finally vanishes). The second condition can be recognized by a reduction followed by an absence of bands in ΔAbs_{buffer}.

12. Spectral artifacts might occur for instance when the temperature of the buffers is not well equilibrated when the experiment starts.

13. Ideally, a fresh Abs_{buffer} should be collected for each new experiment to assure the best possible correction of Abs_{sample} for the buffer absorbance. In practice this step can be skipped, and the Abs_{buffer} measured jointly with ΔAbs_{buffer} spectrum can be used with satisfactory results. Note that if this Abs_{buffer} spectrum was measured in a physically different ATR crystal, or in the same crystal but several months before, collecting a new reference spectrum is recommended.

14. If the ATR crystal is accidentally removed from the ATR accessory the background might change. In that case you would need to collect again a single-beam background spectrum.

15. Even though the term "dry film" is customarily used, the film is far from being dry (26). The protein is depleted from the bulk water but not from the bound water. A membrane protein retains 0.15 g of water/g of protein even under a flow of dry nitrogen (36).

16. If 20 μl is a volume too low to be fully spread, add additional 20 μl of deionized water (not buffer) to help.

17. Slightly additional "dryness" can be obtained by placing the cover and gently flowing dry nitrogen (or dry air) through one of the cover inlets. This also allows for strict temperature control of the sample.

18. In our experience with MelB WT and C-less, the same sample can be used without problems for three full days and nights.

19. If you plan to continue using the same sample, take care that buffers do not run out at the end of the experiment.

20. As with the buffer, the last spectra of the sample can show artifacts from the exhaustion of the liquid nitrogen or the buffer (see Note 9). An additional source of potential spectral artifacts is sample loss from the volume probed by the evanescent IR beam during the exchange of buffers. If the lost sample is representative of the average sample (homogenous sample lost) its spectral contribution might be corrected jointly with the film swelling, provided that its intensity is not large enough to overwhelm the underlying substrate-induced IR_{diff} spectra. A more severe situation might occur when inhomogeneities exist in the sample, containing membrane patches with a higher than average lipid–protein ratio. Being less dense, these membrane patches will tend to move upwards in exchange for denser protein-rich membranes, moving away from the probed volume. In this case, the spectral contribution from sample loses will be relatively enriched in lipid, behaving differently from a film swelling (inhomogeneous sample lost). It is important

to remove spectra affected by sample loss before averaging, especially when there are inhomogeneous loss. Sample loss can be detected by subtracting successive Abs_{sample} spectra from each other. After some practice this identification process can be done directly by visual inspection of the ΔAbs_{sample} spectra.

21. In order to obtain a good and reliable substrate-induced IR_{diff} spectrum it is not only necessary to average numerous IR_{diff} spectra with many co-added scans, but also to repeat the same experiment using independent samples.

22. We assume that the sample is a weak absorber and that the film thickness is higher than the evanescent wave penetration ("weak absorber thick film hypothesis"). Then, for unpolarized light with a Ge ATR crystal ($n_1 = 4.0$) and an angle of incidence of $\theta = 45°$, the effective penetration depth (d_e) is (26) $d_e = 3\sqrt{2}n_2 \, / \, (\pi v(16 - n_2^2)\sqrt{8 - n_2^2})$, where n_2 is the real refractive index of the sample and v the wavenumber. The ratio between the d_e of the sample to the buffer will be

$$R = \frac{n_{sample}\sqrt{8 - n_{buffer}^2}\,(n_{buffer}^2 - 16)}{n_{buffer}\sqrt{8 - n_{sample}^2}\,(n_{sample}^2 - 16)}.$$

For the buffer we can approximately take the refractive index as the average refractive index of water: $n_{buffer} = 1.333$. Given that the refractive index of lipids is estimated to be ~1.45 (37) and for proteins 1.55–1.50 (30, 38), we took the refractive index of the proteoliposomes as ~1.5. Then, the average refractive index of the hydrated sample will be approximately: $n_{sample} \approx 1.333 \times 0.88 \times R^{-1} + 1.5 \times (1 - 0.88 \times R^{-1})$. Solving, we obtain $n_{sample} \approx 1.354$, and $R_d = 1.024$, i.e., the effective penetration depth for the sample is 1.024 times that of the buffer.

23. For spherical unilamellar proteoliposomes a 14% of the volume will be occupied by the membrane only if their diameter is ~125 nm or less. Such small liposomes are usually obtained only through pore extrusion, and unlikely to be obtained by simply rehydrating a dry film.

24. A 14.1% of the volume corresponding to the protein–lipid implies that the separation between the (presumably) stacked bilayers would be on average equal to ~6.1 (85.9/14.1) times the bilayer thickness. Considering the protein–lipid membranes to be ~5 nm thick, it implies that the average vertical separation of the layers contributing to the IR signal would be 30 nm, i.e., one membrane every 35 nm. Consequently, given the penetration depth of the evanescent beam between 1,700 and 1,400 cm^{-1}, 0.38–0.46 μm (see Note 3), we can conclude that the 95% of the signal of our sample comes on average from the 16–20 bilayers closer to the ATR surface.

25. This lower than expected concentration probably originates from a delayed diffusion of the melibiose to the ATR surface vicinity, caused by the presence of the stacks of protein–lipid membrane patches. Nevertheless, such lower melibiose concentration is of minor practical concern. For the C-less MelB it might at most reduce the binding occupancy from 69% (at 10 mM melibiose) to 61% (at 7 mM melibiose) (39), decreasing the intensity of the obtaining IR_{diff} spectrum by a modest 12%. Nevertheless, potential users of this technique should be aware that the actual substrate concentration in the probed volume of the sample might be lower than the one expected at the bulk volume.

26. Strictly, we should account for the difference in effective penetration depth between the sample and the buffer. We thus computed $Abs_{sample} - 1.024 \times Abs_{buffer}$ (see Note 22).

27. In some cases the band at 3,400 cm^{-1} from the absorbance of the lost water will not cancel after this procedure, for instance, because the difference in effective penetration depth between the sample and the buffer was not correctly estimated. You can adjust such deviations by subtracting an absorbance spectrum of the buffer alone.

28. Due to the tiny intensity of ΔAbs_{B-A} spectra, the time interval between the acquisition of Abs_A and Abs_B (and so the time for the perfusion of buffer B) should be kept as short as possible. This reduces drifts in the experimental system (e.g., in the sample temperature, in the IR source temperature, etc.), and improves the quality of the experimental ΔAbs_{B-A} spectra. In contrast, the length for the perfusion of buffer A does not affect critically the quality of the Abs_A spectra, and we have used perfusion times of up to 30 min.

29. This also implies that the actual concentration of Na$^+$ in the sample cannot be independently estimated from the subtraction factor used. Therefore, when working with IR silent substrates, as cations, additional care is needed to assure that the actual target concentration is really attained at the ATR surface. This can be done, for instance, by doing parallel experiment with an IR active molecule (as we did with the sugar melibiose).

30. Given the affinity constant for Na$^+$ in the C-less MelB (39), the binding occupancy would vary from 68 to 74%, affecting only modestly the intensity of the IR_{diff} spectrum induced by Na$^+$.

31. Our films of MelB reconstituted in *E. coli* lipids have always shown a good physical stability. Although our experience about strategies for improving film stability is limited, a high ionic strength improves the film stability, leading also to a higher intensity in the obtained IR_{diff} spectra. The lipids might also influence the film stability. We have obtained stable films of

MelB when reconstituted in *E. coli* lipids (15), as well as in egg phospatidyl choline (EPC) (unpublished data). The nicotinic acetylcholine receptor reconstituted in pure EPC, as well as in EPC/phosphatidic acid (PA), EPC/cholesterol (Chol), EPC/PA/Chol, EPC/phosphatidyl serine, and EPC/squalene, has also shown good adherence to a Ge crystal surface (40).

Acknowledgments

This work was supported by a Marie Curie Reintegration Grant PIRG03-6A-2008-231063 (to V.L.-F.), and by Ministerio de Ciencia e Innovación grants BMC2003-04941, BFU2006-04656/BMC, and BFU2009-08758/BMC, the Direcció General de Recerca (DURSI, Generalitat de Catalunya), 1999SGR-0102 grant, and the European Commission Bio4-CT97-2119 grant.

References

1. Shenker A (1995) G protein-coupled receptor structure and function: the impact of disease-causing mutations. Baillieres Clin Endocrinol Metab 9:427–451

2. Abramson J, Wright EM (2009) Structure and function of Na⁺-symporters with inverted repeats. Curr Opin Struct Biol 19:425–432

3. Wright EM, Hirayama BA, Loo DF (2007) Active sugar transport in health and disease. J Intern Med 261:32–43

4. Cooper MA (2002) Optical biosensors in drug discovery. Nat Rev Drug Discov 1:515–528

5. Hovius R, Vallotton P, Wohland T, Vogel H (2000) Fluorescence techniques: shedding light on ligand-receptor interactions. Trends Pharmacol Sci 21:266–273

6. Cooper MA (2004) Advances in membrane receptor screening and analysis. J Mol Recognit 17:286–315

7. Barth A, Zscherp C (2002) What vibrations tell us about proteins. Quart Rev Biophys 35:369–430

8. Jung C (2000) Insight into protein structure and protein-ligand recognition by Fourier transform infrared spectroscopy. J Mol Recognit 13:325–351

9. Berthomieu C, Hienerwadel R (2009) Fourier transform infrared (FTIR) spectroscopy. Photosynth Res 101:157–170

10. Nyquist RM, Ataka K, Heberle J (2004) The molecular mechanism of membrane proteins probed by evanescent infrared waves. Chembiochem 5:431–436

11. Zscherp C, Barth A (2001) Reaction-induced infrared difference spectroscopy for the study of protein reaction mechanisms. Biochemistry 40:1875–1883

12. Rich PR, Iwaki M (2007) Methods to probe protein transitions with ATR infrared spectroscopy. Mol Biosyst 3:398–407

13. Baenziger JE, Miller KW, Rothschild KJ (1992) Incorporation of the nicotinic acetylcholine receptor into planar multilamellar films: characterization by fluorescence and Fourier transform infrared difference spectroscopy. Biophys J 61:983–992

14. Baenziger JE, Miller KW, Rothschild KJ (1993) Fourier transform infrared difference spectroscopy of the nicotinic acetylcholine receptor: evidence for specific protein structural changes upon desensitization. Biochemistry 32:5448–5454

15. León X, Lórenz-Fonfría VA, Lemonnier R, Leblanc G, Padrós E (2005) Substrate-induced conformational changes of melibiose permease from *Escherichia coli* studied by infrared difference spectroscopy. Biochemistry 44:3506–3514

16. Iwaki M, Andrianambinintsoa S, Rich P, Breton J (2002) Attenuated total reflection Fourier transform infrared spectroscopy of redox transitions in photosynthetic reaction centers: comparison of perfusion- and light-induced difference spectra. Spectrochim Acta A 58:1523–1533

17. Nyquist RM, Heitbrink D, Bolwien C, Gennis RB, Heberle J (2003) Direct observation of protonation reactions during the catalytic cycle

of cytochrome *c* oxidase. Proc Natl Acad Sci U S A 100:8715–8720

18. León X, Lemonnier R, Leblanc G, Padrós E (2006) Changes in secondary structures and acidic side chains of melibiose permease upon cosubstrates binding. Biophys J 91: 4440–4449

19. León X, Leblanc G, Padrós E (2009) Alteration of sugar-induced conformational changes of the melibiose permease by mutating Arg[141] in loop 4-5. Biophys J 96:4877–4886

20. Lórenz-Fonfría VA, Granell M, León X, Leblanc G, Padrós E (2009) In-plane and out-of-plane infrared difference spectroscopy unravels tilting of helices and structural changes in a membrane protein upon substrate binding. J Am Chem Soc 131:15094–15095

21. Granell M, León X, Leblanc G, Padrós E, Lórenz-Fonfría V (2010) Structural insights into the activation mechanism of melibiose permease by sodium binding. Proc Natl Acad Sci U S A 107:22078–22083

22. Saier MH Jr (2000) Families of transmembrane sugar transport proteins. Mol Microbiol 35: 699–710

23. Pourcher T, Bassilana M, Sarkar HK, Kaback HR, Leblanc G (1990) The melibiose/Na+symporter of *Escherichia coli*: kinetic and molecular properties. Philos Trans R Soc Lond B: Biol Sci 326:411–423

24. Wilson TH, Ding PZ (2001) Sodium-substrate cotransport in bacteria. Biochim Biophys Acta 1505:121–130

25. Pourcher T, Leclercq S, Brandolin G, Leblanc G (1995) Melibiose permease of *Escherichia coli*: large scale purification and evidence that H^+, Na^+, and Li^+ sugar symport is catalyzed by a single polypeptide. Biochemistry 34: 4412–4420

26. Goormaghtigh E, Raussens V, Ruysschaert JM (1999) Attenuated total reflection infrared spectroscopy of proteins and lipids in biological membranes. Biochim Biophys Acta 1422: 105–185

27. Chittur KK (1998) FTIR/ATR for protein adsorption to biomaterial surfaces. Biomaterials 19:357–369

28. Fringeli UP (2000) ATR and reflectance IR spectroscopy, applications. In: Lindon JC, Tranter GE, Holmes JL (eds) Encyclopedia of spectroscopy and spectrometry. Academic Press. pp 58–75

29. Marsh D (1999) Quantification of secondary structure in ATR infrared spectroscopy. Biophys J 77:2630–2637

30. Boulet-Audet M, Buffeteau T, Boudreault S, Daugey N, Pézolet M (2010) Quantitative determination of band distortions in diamond attenuated total reflectance infrared spectra. J Phys Chem B 114:8255–8261

31. Goormaghtigh E, Cabiaux V, Ruysschaert JM (1994) Determination of soluble and membrane protein structure by Fourier transform infrared spectroscopy. III. Secondary structures. Subcell Biochem 23:405–450

32. Lórenz-Fonfría VA, Villaverde J, Trézéguet V, Lauquin GJ, Brandolin G, Padrós E (2003) Structural and functional implications of the instability of the ADP/ATP transporter purified from mitochondria as revealed by FTIR spectroscopy. Biophys J 85: 255–266

33. Lórenz-Fonfría VA, Villaverde J, Padrós E (2002) Fourier deconvolution in non-self-deconvolving conditions. Effective narrowing, signal-to-noise degradation, and curve fitting. Appl Spectrosc 56:232–242

34. Lórenz-Fonfría VA, Padrós E (2004) Curve-fitting of Fourier manipulated spectra comprising apodization, smoothing, derivation and deconvolution. Spectrochim Acta A 60: 2703–2710

35. Lórenz-Fonfría VA, Padrós E (2004) Curve-fitting overlapped bands: quantification and improvement of curve-fitting robustness in the presence of errors in the model and in the data. Analyst 129:1243–1250

36. Lórenz VA, Villaverde J, Trézéguet V, Lauquin GJ, Brandolin G, Padrós E (2001) The secondary structure of the inhibited mitochondrial ADP/ATP transporter from yeast analyzed by FTIR spectroscopy. Biochemistry 40: 8821–8833

37. Picard F, Buffeteau T, Desbat B, Auger M, Pézolet M (1999) Quantitative orientation measurements in thin lipid films by attenuated total reflection infrared spectroscopy. Biophys J 76:539–551

38. Buffeteau T, Le Calvez E, Desbat B, Pelletier I, Pezolet M (2001) Quantitative orientation of α-helical polypeptides by attenuated total reflection infrared spectroscopy. J Phys Chem B 105:1464–1471

39. Meyer-Lipp K, Séry N, Ganea C, Basquin C, Fendler K, Leblanc G (2006) The inner inter-helix loop 4-5 of the melibiose permease from *Escherichia coli* takes part in conformational changes after sugar binding. J Biol Chem 281: 25882–25892

40. Ryan SE, Demers CN, Chew JP, Baenziger JE (1996) Structural effects of neutral and anionic lipids on the nicotinic acetylcholine receptor. An infrared difference spectroscopy study. J Biol Chem 271:24590–24597

Chapter 8

UV–Visible and Infrared Methods for Investigating Lipid–Rhodopsin Membrane Interactions

Michael F. Brown

Abstract

We describe experimental UV–visible and Fourier transform infrared (FTIR) spectroscopic methods for characterizing lipid–protein interactions for rhodopsin in a membrane bilayer environment. The combination of FTIR and UV–visible difference spectroscopy is used to monitor the structural and functional changes during rhodopsin activation. Investigations of how membrane lipids stabilize various rhodopsin photoproducts are analogous to mutating the protein in terms of gain or loss of function. Interpretation of the results entails a flexible surface model for explaining membrane lipid–protein interactions through material properties relevant to biological activity.

Key words: Flexible surface model, G protein-coupled receptor, Hydrophobic mismatch, Infrared spectroscopy, Lateral pressure, Lipids, Membrane curvature, Protein–lipid interactions, Rhodopsin, Vision

1. Introduction

Rhodopsin is a G protein-coupled receptor (GPCR) that is present in the disk membranes of the retinal rod cells and mediates the process of visual signaling. Experimental studies of rhodopsin allow one to test the role of membrane lipids in governing key biological activities of membrane proteins, including GPCRs, transporters, and ion channels (1). Membrane lipid–protein interactions significantly affect rhodopsin function within the membrane lipid bilayer (2–7). Notably, rhodopsin activation in native and synthetic membranes can be experimentally investigated by the combined use of UV–visible and Fourier transform infrared (FTIR) spectroscopy. Absorption of light by retinal yields 11-*cis* to *trans* isomerization of the ligand bound to rhodopsin within the lipid bilayer. Subsequent deprotonation of the protonated Schiff base

Nagarajan Vaidehi and Judith Klein-Seetharaman (eds.), *Membrane Protein Structure and Dynamics: Methods and Protocols*, Methods in Molecular Biology, vol. 914, DOI 10.1007/978-1-62703-023-6_8, © Springer Science+Business Media, LLC 2012

(PSB) of retinal and activating conformational changes of the protein are probed with UV–visible and FTIR spectroscopy (8–10). The complementary use of these methods with X-ray crystallography is brought out by the crystal structures of rhodopsin in the dark state (11–13) and ligand-free opsin (14) together with molecular visualization software (15). Further analysis of UV–visible and FTIR results for rhodopsin in a native-like bilayer environment entails lipid–membrane protein interactions as described by a flexible surface model (FSM). A chemically nonspecific curvature stress field due to the membrane lipids governs the energetics of membrane proteins. Lipid influences on vision and other key biological functions of GPCRs, transporters, and ion channels are explainable by a simple continuum picture.

2. Materials

Experimental protocols are briefly described below for the preparation and characterization of retinal disk membranes (RDM) and recombinant lipid membranes containing rhodopsin. Native hypotonically washed disk membranes are prepared from bovine retinas following standard procedures (16). Characterization of the photochemical activity of rhodopsin involves bleaching and regeneration of opsin with 11-*cis*-retinal. Membranes with a defined lipid composition are prepared by purifying a detergent extract of rhodopsin by column chromatography on hydroxyapatite. Dialysis in the presence of phospholipids forms membrane bilayers that are harvested by ultracentrifugation.

2.1 Buffers, Detergents, and Lipids

1. Phospholipids: Highly pure phospholipids are readily procured from commercial sources (e.g., Avanti Polar Lipids, Alabaster, AL, USA). Alternatively phospholipids are synthesized as described (17, 18), starting from egg yolk phosphatidylcholine prepared from locally obtained hen eggs and commercially available fatty acids (Nu-Chek Prep, Inc., Elysian MN, USA). For further details see refs. 17, 18. Representative phospholipids include 1-palmitoyl-2-oleoyl-*sn*-glycero-3-phosphocholine (POPC), 1,2-dioleoyl-*sn*-glycero-3-phosphocholine (DOPC), and 1,2-dioleoyl-*sn*-glycero-3-phosphoethanolamine (DOPE).

2. Detergents: Ammonyx LO (Stepan Co., Northfield, IL, USA), dodecyltrimethylammonium bromide (DTAB) (Sigma, St. Louis, MO, USA), lauryldodecylamine oxide (LDAO) (Sigma, St. Louis, MO, USA) (see Note 1).

3. Buffers and solutions: *Homogenizing solution*—30% sucrose (w/w), 5 mM Tris-acetate pH 7.4, 65 mM NaCl, 2 mM $MgCl_2$, and 2 mM EDTA. *Dilution buffer*—10 mM Tris-acetate pH

7.4. *Stock detergent buffer*—30% (v/v) Ammonyx LO or 30% (v/v) LDAO containing 100 mM NH_2OH (hydroxylamine) and 10 mM Na phosphate buffer pH 7.0. *Dialysis buffer*—5 mM HEPES, 1 mM EDTA, and 1 mM DTT (dithiothreitol) (add fresh). *Chromatography buffer*—15 mM Na phosphate pH 6.8 containing 0.02% (w/w) NaN_3 (sodium azide) and 1 mM DTT (add fresh). *Chromatography detergent buffer*—100 mM DTAB plus chromatography buffer.

4. Density gradient solutions: 1.10, 1.11, 1.13, 1.15 g/mL sucrose (23.8, 25.9, 30.3, and 34.0% (w/w), respectively) containing Tris-acetate pH 7.4 and 1 mM EDTA.

5. Column packing materials: Hydroxyapatite (Bio-Gel HTP Gel; Bio-Rad Laboratories, Hercules, CA, USA).

6. Isomerically pure 11-*cis*-retinal can be obtained via the US National Eye Institute (Bethesda, MD, USA).

2.2. Purification of Retinal Disk Membranes Containing Rhodopsin

1. Retinas: Frozen bovine retinas are procured from commercial sources (Lawson, Omaha, NE, USA). Alternatively, fresh bovine eyes are obtained from a local abattoir and are transferred to an ice chest in the dark. Eyes are dissected under dim red light (15-W bulb with Kodak safelight filter 1A) within ≈4 h postmortem by excising the front part of the eye with a sharp scalpel or razor blade, followed by inverting the eye cup, removal of the lens and vitreous humor, and gently separating the retina from the pigment epithelium with a blunt scapula. The retina is cut at the optic nerve with a sharp scissors, and dropped into a glass jar containing homogenizing solution (30% (w/w) sucrose) on ice. Retinas are stored in lightproof glass containers under nitrogen or argon at –80°C until use.

2. Preparation of RDM: All procedures are done in a cold room (4°C) or using an ice bucket under dim red light (see above). Thawed frozen bovine retinas or alternatively freshly obtained retinas are used (see above). Gently separate the retinas and centrifuge at 2,600×*g* for 20 min at 4°C. Transfer the pelleted retinas to a *loose* fitting Teflon homogenizer (machined). Add 30 mL of homogenizing solution (30% sucrose) per 50 retinas under a gentle stream of argon gas. (Do not add excess because that will change the sucrose density for subsequent steps.) Homogenize the retinas by applying about ten strokes slowly under a gentle argon gas stream. Centrifuge at 2,600×*g* for 20 min at 4°C. Remove the supernatant with a spring-loaded syringe or by decanting, and transfer to a *tight* fitting Teflon homogenizer (as obtained from the manufacturer). Add an equal volume of homogenizing solution and apply about 6–8 strokes under a gentle argon gas stream. Centrifuge at 2,600×*g* for 20 min at 4°C and combine the supernatants from the two centrifugation steps.

3. Add two volumes of dilution buffer (10 mM Tris-acetate pH 7.4) to the retinal homogenate and centrifuge at $8,000 \times g$ for 50 min at 4°C. (If necessary increase the time of centrifugation but do not increase the speed above $8,000 \times g$ as this will lead to additional contaminating material in the pellet.) Re-suspend the pellet in a small volume (<25 mL) of 1.10 g/mL sucrose density gradient solution. Prepare discontinuous sucrose step gradients on ice or in the cold room (1.11, 1.13, and 1.15 g/mL; volumes of 10, 10, and 8 mL, respectively) in polyallomer centrifuge tubes (four tubes for 50 retinas or six tubes for 100 retinas). Pull up the pellet with an 18-gauge needle and then expel it through a 21-gauge needle to fragment the crude rod outer segments. Layer onto each of the density gradient tubes and top off with additional 1.10 g/mL sucrose density gradient solution. Centrifuge at $113,000 \times g$ in a swinging bucket rotor for 1 h at 4°C. Collect the carpet (band) at the 1.11/1.13 g/mL interface with a spring-loaded syringe having a cutoff 18-gauge tip (total volume ≈75 mL).

4. Combine the collected bands and dilute twofold with water, add argon gas, and centrifuge at $48,000 \times g$ for 30 min at 4°C. Repeat 1–2 times to ensure the removal of sucrose. Note that water washing (hypotonic) produces membrane fragments due to the osmotic shock. Re-suspend the RDM in either 10 mM or 67 mM Na phosphate buffer pH 7.0 (≈10 mL) and transfer to Eppendorf vials. Overlay with argon gas and store at –80°C until use. For additional details see ref. 16 (see Note 2).

2.3. Characterization of Rhodopsin by Spectrophotometry

1. All procedures are done under dim red light (15-W bulb, Kodak safelight filter 1A). A scanning spectrophotometer (dual beam; see below) is used to acquire UV–visible absorption spectra of rhodopsin by solubilizing a small aliquot of membranes in stock detergent buffer (30% (v/v) Ammonx LO containing 100 mM NH_2OH) at room temperature. First, record a baseline of 10:1 dilution of detergent buffer (3%) (blank sample) versus detergent buffer (3%) (reference) from 700 to 250 nm using 100-μL quartz microcuvettes with a 10-mm path length (see Note 3).

2. Next dissolve 0.1 mL of rhodopsin-containing membrane sample in 0.9 mL of stock detergent buffer at room temperature. Invert the cuvette to mix the contents thoroughly and immediately record an absorption spectrum of the dissolved membranes containing rhodopsin from 700 to 250 nm. Scan several times to ensure reproducibility and to establish that no time-dependent changes are occurring (see Note 3).

3. Fully bleach the sample with ≈6 flashes from a photoflash lamp (Sunpak auto 555 or equivalent) fitted with a high-pass OG515 filter (Schott, Mainz, Germany), and re-scan to record the fully

bleached spectrum with scattered light. Measure both the 500-nm absorption and the 280-nm absorption versus the bleached spectrum and calculate the A_{280}/A_{500} absorption ratio. Note that highly pure samples of native disk membranes typically have A_{280}/A_{500} ratios (unregenerated; see below) of $\approx 2.4 \pm 1$ (see Note 4).

4. Quantify the molar concentration of rhodopsin using $\varepsilon_{500} = 40,600$ M^{-1} cm^{-1} for the molar absorption coefficient. Alternatively use optical density units (OD or A_{500} units). One OD unit is defined as 1 mL of a solution with absorption $A_{500} = 1$ and affords a convenient measure of the equivalents of retinal. The correspondence to the mass of rhodopsin is 1.00 OD = 2.46 nmol = 0.958 mg rhodopsin (assuming $M_r = 39,000$).

2.4. Regeneration Assay for Rhodopsin Photochemical Functionality

1. Conduct regeneration assays of the rhodopsin-containing membranes by adding 11-*cis*-retinal to obtain the percent regenerability, percent of bleached rhodopsin (i.e., opsin) in the original sample, and the regenerated A_{280}/A_{500} ratio. Note that in general the purified RDM may be partially bleached (i.e., containing both rhodopsin and opsin). The percent of bleached rhodopsin as well as the percent of the opsin that can be regenerated from rhodopsin is found by incubating the RDM preparation with added 11-*cis*-retinal.

2. Prepare three 0.2-mL samples of RDM designated as: control, regenerated, and bleached + regenerated. The bleached sample is exposed to light from a photoflash unit as described above. Add at least as many equivalents (ODs) of 11-*cis*-retinal as rhodopsin (e.g., 1.5–2 equivalents, corresponding to 0.5–1 excess) to the regenerated and bleached + regenerated samples. Incubate all three samples for 1.5 h at 37°C. Record the UV–visible absorption spectra as described above.

3. Note that to obtain the percent bleached and percent regenerability it is necessary to correct for partial regeneration of opsin. Assume the fraction of rhodopsin that is unregenerable, when bleached to opsin, is identical to the fraction of unregenerable opsin present in the original RDM sample (partially bleached). In what follows, all amounts refer to the absorbance at 500 nm measured as described above. (It is helpful for the reader to sketch the corresponding absorption spectra.) Let A be the amount of regenerable opsin in the regenerated sample, i.e., before adding 11-*cis*-retinal. B is the total amount of opsin (i.e., regenerable plus unregenerable) in the regenerated sample before adding 11-*cis*-retinal. C is the total absorbance that would be measured if all the opsin in the sample could be regenerated. D is the amount of regenerable opsin in the bleached + regenerated sample. E is the total amount of

rhodopsin in each of the samples (i.e., prior to bleaching or 11-*cis*-retinal addition). F is the unknown absorbance that would be produced if all the unregenerable opsin could be regenerated. Last, G is the amount of rhodopsin in the regenerated sample after incubation with 11-*cis*-retinal. Note the experimentally observed values are D (bleached + regenerated sample), E (control sample), and G (regenerated sample).

4. Assume that the fraction of regenerable opsin is the same before and after bleaching so that $A/B = D/C$. The unknown quantity is then $C = ED/(D - G + E)$ which is the total amount of the visual protein (rhodopsin + opsin) present in the samples. The percent bleached (opsin) in the original sample $= 100B/C$ ($\approx 100A/G = 100(1 - E/G)$ assuming 100% regenerability). The percent regenerability $= 100D/C$ ($\approx 100D/G$). Last, the regenerated A_{280}/A_{500} ratio is A_{280}(control)$/A_{500}$(regenerated). Typical results in our laboratory are: 93% regenerability; 2% bleached; 2.1 post-regeneration A_{280}/A_{500} ratio. For further details see ref. 19.

2.5. Purification of Rhodopsin by Hydroxyapatite Chromatography

1. Rhodopsin is extracted by solubilizing the native disk membranes at 4°C with the cationic detergent DTAB. Centrifuge the purified RDM (see above) at 48,000 × g for 20 min at 4°C. Remove the supernatant and dissolve the pellet in 10–15 mL of chromatography detergent buffer (100 mM DTAB containing 15 mM Na phosphate pH 6.8 with 0.02% NaN$_3$ and 1 mM DTT). Incubate on ice for 1 h and then centrifuge at 163,000 × g for 20 min at 4°C.

2. Purify the detergent-solubilized rhodopsin (50 mg) by column chromatography (column size 2.5 × 6.5 cm) on hydroxyapatite. Equilibrate the column with chromatography detergent buffer (100 mM DTAB, 15 mM Na phosphate pH 6.8, containing 0.02% (w/w) NaN$_3$ (sodium azide) and 1 mM DTT) at 4°C. Apply the sample onto the hydroxyapatite column and elute with a linear gradient of 0–0.5 M NaCl in chromatography detergent buffer with a flow rate of 0.6 mL/min. Identify the rhodopsin-containing fractions by their A_{500} absorption. Note that rhodopsin is typically eluted beginning at ≈4 h in the 100–150 mL fractions, and the yield is typically ≈70%. The A_{280}/A_{500} ratio of the pure rhodopsin fractions is typically 1.6–1.8. For specifics see refs. 20, 21 (see Note 5).

2.6. Preparation of Rhodopsin–Lipid Recombinant Membranes

1. Combine the eluted rhodopsin fractions from hydroxyapatite column chromatography (typically ≈50 mL) and always maintain rhodopsin in DTAB at 4°C to avoid denaturation (see Note 5). Dissolve the phospholipids in 3–4 mL of 300 mM DTAB plus chromatography buffer. Note that rhodopsin is typically recombined with phospholipids (Avanti Polar Lipids,

Alabaster, AL, USA) at a 1:100 molar ratio of rhodopsin–lipid, although ratios from 1:50 to 1:400 are possible. Combine rhodopsin with the solubilized phospholipids at 4°C and adjust the DTAB concentration to 300 mM plus chromatography buffer. Add the rhodopsin plus solubilized phospholipids to a dialysis sack (1.6 cm diameter; ≈2.0 mL solution/centimeter of dialysis tubing) prepared by pre-boiling the dialysis tubing in distilled water (M_r cutoff = 12,000–14,000) (Spectra/Por; Spectrum Laboratories, Inc., Rancho Dominguez, CA, USA).

2. Recombinant proteolipid bilayer membranes are formed by dialysis of the detergent in a cold room at 4°C in darkness with a continuous stream of nitrogen gas (use a gas dispersion tube). Exchange the dialysis buffer eight or more times every 6–8 h. Note that a constant rate of dialysis with regular exchange intervals is desirable for relatively homogeneous vesicle formation. The total volume of dialysis buffer used should be ≈900 mL per milliliter of DTAB solution added to the dialysis sack.

3. Collect the recombinant proteolipid membranes by centrifuging the contents of the dialysis sack at 105,000×g for 1 h at 4°C. Finally, re-suspend the pellet in 67 mM Na phosphate buffer pH 7.0 or 100 mM HEPES buffer pH 7.0 containing 1 mM EDTA, and store in sealed Eppendorf vials under argon at –80°C (stable for several years). For further details see refs. 20, 21.

2.7. Characterization of Rhodopsin–Lipid Membranes by Density Gradient Centrifugation

1. Investigate the homogeneity of proteolipid bilayer membranes with respect to lipid–rhodopsin ratio by isopycnic density gradient centrifugation. Note that proteins are denser than lipids, so the position of the proteolipid band depends on the protein–lipid ratio. Use a gradient maker (commercial or fabricated locally) to pour a linear 0–50% (w/w) sucrose density gradient containing 10 mM Tris-acetate pH 7.4 and 1 mM EDTA in a polyallomer centrifuge tube.

2. Centrifuge at 113,000×g for 6–12 h at 4°C. Note that homogenous samples yield a single tight lipid–rhodopsin band, whereas polydispersity is indicated by diffused multiple bands (not typically observed using DTAB for membrane recombination). If desired, pour a shallower gradient with a smaller range of sucrose density (higher resolution) and repeat the procedure to test for the presence of a single band (see Note 6).

3. Determine the lipid–rhodopsin ratio using phosphorus analysis (22) to quantify the phospholipid content and the A_{500} absorbance for the amount of rhodopsin in the fractions. (The centrifuge tube can be punctured at the bottom to collect the drops corresponding to the fractions.) Use the molar absorptivity of $\varepsilon_{500} = 40,600$ M^{-1} cm^{-1} to quantify the amount of rhodopsin (see Note 6).

3. Methods

Both UV–visible spectrophotometry and FTIR spectroscopy employ either freely suspended membranes or sandwich-type samples of membranes deposited on planar substrates (9, 23, 24). Absorption of light by rhodopsin leads to 11-*cis* to *trans* isomerization of retinal (see above), followed by the reactions: Meta I $\rightleftharpoons$ Meta II$_a$ $\rightleftharpoons$ Meta II$_b$ $\rightleftharpoons$ Meta II$_b$H$^+$, e.g., as described by an activated ensemble mechanism (25, 26) (Fig. 1b). Alternatively, a modified square reaction scheme is applicable, in which a deprotonated early intermediate (Meta I_{380}) is observed prior to the formation of Meta II substates of the extended sequential scheme (27).

3.1. Preparation of Freely Suspended Membrane Samples

1. Membranes are centrifuged, and the pellet is re-suspended in buffer at the desired pH and used immediately. The pH is controlled by equilibration of the sample using various buffers with

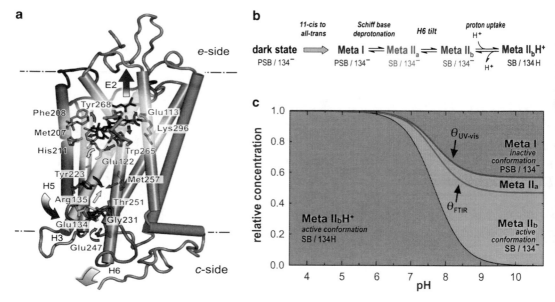

Fig. 1. Structural and thermodynamic model for rhodopsin shows activation of a prototypical G protein-coupled receptor (GPCR) in a membrane environment. (**a**) Rhodopsin in the dark state (Protein Data Bank code 1GZM) has extracellular (*e*) and cytoplasmic (*c*) membrane sides (26). Large-scale conformational changes of rhodopsin involve the E2 extracellular loop and H5 and H6 helices, and initiate the visual process. Volumetric expansion of the receptor within the membrane is described by an activated ensemble mechanism (26). (**b**) Extended reaction scheme for the dynamically activated receptor involves an ensemble of photoproduct substates (PSB≡protonated Schiff base; SB≡Schiff base; 134≡Glu[134]). Light absorption yields 11-*cis* to *trans* isomerization of the retinal ligand, followed by equilibration of Meta I with different isochromic Meta II substates. The Meta I state has a PSB and a deprotonated Glu[134] residue. Deprotonation of the PSB results in the Meta II$_a$ state. In Meta II$_b$ a tilt of helix H6 occurs. Last, in Meta II$_b$H$^+$ proton uptake by Glu[134] occurs giving a pH-dependent driving force for receptor activation. Note that Meta II$_b$ and Meta II$_b$H$^+$ are conformational substates that activate the G protein transducin (G$_t$). (**c**) The extended reaction scheme in panel **b** predicts a complex titration behavior, with the presence in panel of Meta I, Meta II$_a$, and Meta II$_b$ at alkaline pH. A pH-dependent equilibrium with Meta II$_b$H$^+$ occurs at lower pH due to proton uptake by Glu[134] from the aqueous solvent.

overlapping pH ranges, e.g., 200 mM sodium citrate (pH < 6.0), sodium MES (2-(N-morpholino)ethanesulfonic acid) (pH 6.0–7.0), or sodium BTP (1,3-bis(tris(hydroxymethyl) methylamino)propane) (pH > 7.0) buffers.

2. Alternatively the pH is adjusted by titration with 0.1 M HCl or NaOH, e.g., at the extremes of pH where time-dependent changes may be occurring. All pH values are measured at the same temperature as the spectra.

3. To reduce the considerable light scattering, the samples can be lightly sonicated in an ice bath under a gentle argon gas stream (3 min with a 50% duty cycle), using a microtip sonifier (Sonics & Materials, Newtown, CT, USA). Another alternative is to expel the freely suspended membrane suspension through an extruder (Avanti Polar Lipids, Alabaster, AL, USA).

3.2. Preparation of Sandwich-Type Membrane Samples

1. Prepare the membrane films by drying under nitrogen ≈0.4–0.5 nmol of rhodopsin in native disk or recombinant lipid membranes in water or very low ionic strength buffer on 20-mm CaF_2 FTIR windows (note that BaF_2 is more fragile) (28). Gently overlay the films with 80 μL of 200 mM buffer, e.g., citrate, MES, MOPS, or BTP. Let sit for 3 min to ensure adequate hydration and equilibration of the salt concentration and pH of the membrane stacks. Water content in the samples is typically ≈90% (v/v) as estimated from the OH stretch band of water used as an internal marker in FTIR absorption spectra (see Note 7).

2. All pH values should be measured at the same temperature as the spectra. If necessary, correct the nominal pH values for the temperature dependence of the buffer pK_a.

3. Alternatively it may be desirable to avoid drying of the samples (29). Membranes in 50 mM sodium phosphate or sodium BTP buffers are centrifuged at $100,000 \times g$ for 30 min at 4°C. If necessary the pH is adjusted with 0.1 M HCl or NaOH. The pellet is then transferred to CaF_2 or BaF_2 FTIR windows with a 3-μM polytetrafluoroethylene spacer and sealed.

3.3. Experimental Conditions for Actinic Light Exposure

1. Representative conditions for exposure of rhodopsin to actinic light (bleaching) used in UV–visible and/or FITR studies (8–10) are described below. Use either of the following two procedures.

2. Method (1): Samples are photolyzed by exposure for 20 s to a 150-W tungsten-halogen light source fitted with a high-pass filter ($\lambda > 500$ nm) using a fiber optics light guide (Schott-Fostec ACE light source; Fostec, Auburn, NY, USA). This procedure bleaches ≈100% of the rhodopsin.

3. Method (2): Photolyze samples by exposure to an array of six ultra-bright green ($\lambda_{max} = 520$ nm) 5-mm light-emitting diodes (LEDs) (Nichia, Tokushima, Japan) having nominal 16 cd at 20-mA current and 15° emission half angle (9). For rapid-scan experiments (FTIR or UV–visible with diode array) a 100-ms actinic flash is delivered at a LED current of 100 mA, which bleaches ≈70% of the rhodopsin. For conventional FTIR or UV–visible experiments, the sample is illuminated for 2 s at a LED current of 20 mA to completely bleach rhodopsin (see Note 8).

3.4. Experimental Protocols for UV–Visible Spectrophotometry

UV–visible spectrophotometry is used to establish the protonation state of the retinylidene Schiff base (SB) and to quantify the fractional concentrations of the Meta I and Meta II photoproducts of rhodopsin (30). Either a scanning spectrophotometer or a diode-array spectrophotometer can be employed (see Note 9).

1. A scanning spectrophotometer (double beam: e.g., Perkin-Elmer Lambda 19, Waltham, MA, USA; Shimadzu 2500, Columbia, MD, USA; or Cary 4000, Santa Clara, CA, USA; single beam: Cary 50, or equivalent) with thermostatted sample holder is used to record absorption spectra over the wavelength range from 700 to 250 nm (or 650 to 300 nm) (acquisition time of ≈45 s for Shimadzu 2500 or Cary 4000, and ≈6s for Cary 50). (In the case of highly scattering suspended membranes the spectrophotometer can be equipped with an integrating sphere to more efficiently collect the scattered light.) Alternatively, employ a diode-array spectrophotometer (single beam: e.g., Hewlett Packard 8453 or equivalent) (acquisition time of 200 ms). Use either freely suspended or sandwich-type membrane samples with actinic illumination conditions as described above.

2. For experiments with a scanning UV–visible spectrophotometer (double beam), first record a baseline of buffer versus buffer. Either sandwich samples or freely suspended membranes are used. The advantage of sandwich samples prepared by dehydration is the very low light scattering in the visible range (28). Both types of preparation can be employed and checked relative to one another. For sandwich samples, the baseline is subtracted and the absolute spectra are recorded directly (due to the reduced light scattering). Measurements with freely suspended membranes may utilize an identical bleached sample (containing 10 mM hydroxylamine) as the reference to compensate for the high degree of light scattering. In either case, the sample is illuminated by exposure to actinic light as described above. The light and dark absolute spectra are then subtracted to obtain the difference spectra. Multiple dark spectra are recorded to establish the temporal evolution of the spectral drift and to correct the offset of the light minus dark

difference spectra. All spectral corrections involve difference spectra rather than absolute spectra.

3. In UV–visible experiments employing a diode-array spectrophotometer, use a neutral density filter (e.g., 10% transmission; Thorlabs, Newton, NJ, USA, filter ND10B) to minimize sample bleaching caused by the measuring beam. Record multiple dark spectra with a relatively long acquisition time (e.g., 1 s) and average 2–4 spectra to reduce the signal-to-noise ratio. (Note that the dark spectral noise is propagated in the difference spectra.) Photolyze the sample by exposure to actinic light as described above. For light spectra recorded at lower temperatures (10 and 20°C), use a spectral acquisition time of 1 s. At higher temperatures (30 and 37°C) an acquisition time of 200 ms is employed (due to the faster changes). Acquire replicate scans to follow the temporal evolution of the spectra and to check for photoproduct stability. Average multiple light spectra for which the photoproduct is fully stable to increase the signal-to-noise ratio (typically two to four spectra). Subtract the dark reference spectrum, and smooth the resulting difference spectrum using the spectrophotometer software. Experiments are done at least in duplicate and typically the results of four experiments are averaged. Under unstable conditions at least four independent runs are carried out.

4. Establish conditions of (T, pH) needed to stabilize rhodopsin in the Meta II or Meta I states following actinic light absorption, i.e., for the membrane samples to be studied. Results for freely suspended or sandwich-type samples are similar (28). The Meta II photoproduct has a deprotonated SB that absorbs at 380 nm, whereas the Meta I photoproduct has a PSB absorbing at 480 nm. The Meta I state is stabilized at 10°C and pH 9.5 in disk membranes (Fig. 2a). Meta II is typically stabilized at 20°C and pH 5.0 in natural disk membranes and corresponds to the Meta II_bH^+ substate (Fig. 2b).

5. Measure the UV–visible spectra of rhodopsin-containing membranes from 700 to 250 nm as a function of pH at several different temperatures (Fig. 2c, d). Note that at higher temperatures (20–37°C) the 380-nm absorption due to Meta II persists even at very alkaline pH (Fig. 2d); hence the equilibrium is not fully shifted to Meta I. Record the spectra as a function of time to investigate the stability of the photoproducts (see Note 10).

6. Evaluate the Meta II fraction θ_{UV-vis} with a deprotonated SB by decomposing the photoproduct minus dark state difference spectra (Fig. 2c, d) into a linear combination of Meta I and Meta II spectra (8, 10). Alternatively, calculate the fraction of rhodopsin with a deprotonated Schiff base (θ, fractional Meta II) from the absorption difference, e.g., at 380 nm or 480 nm (use Eq. 1 of Subheading 4; see below) (8, 10) (Fig. 2e).

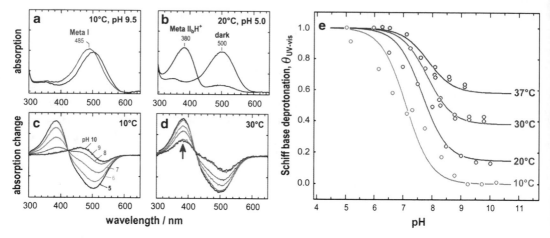

Fig. 2. UV–visible spectroscopy characterizes deprotonation of the retinylidene Schiff base in the Meta I–Meta II equilibrium of rhodopsin in membranes. Results are for hydrated sandwich samples of membranes between CaF_2 windows (used for FTIR; see below) acquired with a diode-array spectrophotometer. (**a**) Light absorption by the dark state ($\lambda_{max} = 500$ nm) forms the Meta I photoproduct ($\lambda_{max} = 485$ nm) at low temperature (10°C) and alkaline pH (9.5) with a protonated Schiff base. (**b**) Illumination of the dark state ($\lambda_{max} = 500$ nm) favors the Meta $II_b H^+$ photoproduct ($\lambda_{max} = 380$ nm) at higher temperature (20°C) and acidic pH (5.0) with a deprotonated Schiff base. (**c**) UV–visible difference spectra (photoproduct minus dark state) at low temperature (10°C) reveal pH dependence of the Meta I–Meta II equilibrium. The pH values are indicated next to the spectra. The equilibrium lies completely to the side of Meta I at alkaline pH (10.0) (evident from lack of 380-nm absorption) and completely to the side of Meta $II_b H^+$ at acidic pH (5.0) (shown by the 380-nm absorption and depletion at 500 nm). (**d**) UV–visible difference spectra showing pH dependence of the Meta I–Meta II equilibrium at higher temperature (30°C). The spectra are at the same pH values as in panel **c**. At higher temperature the equilibrium is not fully shifted to Meta I at alkaline pH. The residual absorption at 380 nm (*arrow*) is due to the Meta II_a and Meta II_b photoproducts and is pH 9.0 (non-zero alkaline endpoint). (**e**) Fraction of rhodopsin with a deprotonated Schiff base (θ_{UV-vis}) plotted as a function of pH. The pH titration curves represent the amount of the Meta II_a, Meta II_b, and Meta $II_b H^+$ substates in equilibrium with Meta I (see Fig. 1b). Significant deviations from classical Henderson–Hasselbalch pH behavior are evident, with a striking temperature-dependent increase of the alkaline endpoint. (Experimental data from ref. 9.)

7. Fit the experimental plots of θ_{UV-vis} versus bulk solution pH to theoretical functions for the binding isotherms for hydronium ions (Subheading 4) (Fig. 2e). If necessary, correct the nominal pH values for the temperature dependence of the buffer pK_a. Typically, at relatively low temperature (10°C) a simple titration equilibrium is seen. However, at higher temperatures (20–37°C) the alkaline endpoint θ_{UV-vis}^{alk} does not reach zero, suggesting a manifold of Meta II substates (Fig. 2e).

3.5. Experimental Protocols for FTIR Spectroscopy

1. Apply FTIR spectroscopy to investigate structural changes occurring in the Meta I–Meta II equilibrium of rhodopsin and the mechanism of the reaction. Prepare sandwich-type samples containing ≈0.4 nmol of rhodopsin (28) in buffers with overlapping pH ranges, e.g., 200 mM citrate for pH < 6.0, MES for pH range from 6.0 to 7.0, and BTP for pH range from 6.0 to 9.5. Measure the buffer pH at the temperature of the FTIR measurements. As an alternative, if necessary correct the nominal pH values for the temperature coefficient of the buffer pK_a values.

2. Employ a FTIR spectrometer (e.g., Bruker Vertex 70 spectrometer, Ettlingen, Germany, or equivalent) with a liquid nitrogen-cooled mercury-cadmium-telluride (MCT) detector and thermostatted sample holder to acquire spectra as a function of pH over the temperature range 0–37°C for sandwich samples of rhodopsin-containing membranes (9, 10).

3. Use the actinic flash illumination protocol as described above. Scale the pre- and post-illumination spectra to the same height of the intense fingerprint mode of retinal near 1,237 cm^{-1} in the dark state. Calculate the difference of the photoproduct minus the pre-illumination dark state using FTIR software (9) (Fig. 3a). Contributions of overlapping buffer bands are

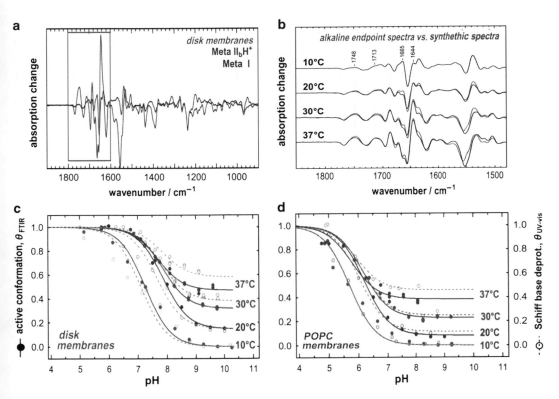

Fig. 3. Fourier transform infrared (FTIR) spectroscopy describes structural changes in rhodopsin occurring upon photoactivation in membranes. Representative data are shown for hydrated sandwich samples between CaF$_2$ windows. (**a**) Reference (basis) FTIR difference spectra (photoproduct minus dark state) obtained for rhodopsin in native disk membranes, corresponding to the Meta II$_b$H$^+$ substate (T = 20°C, pH 5.0) or the Meta I state (T = 10°C, pH 9.5). The conformation-sensitive region is enclosed in the *box*. (**b**) Comparison of alkaline endpoint spectra at different temperatures with synthetic FTIR spectra calculated as a linear combination of the Meta I and Meta II$_b$H$^+$ basis spectra (panel **a**). Note that the Meta II$_b$ and Meta II$_b$H$^+$ photoproducts share similar FTIR signatures. (**c**) pH titration curves for rhodopsin in native disk membranes showing plots of Meta II fraction from FTIR spectroscopy θ_{FTIR} (*filled circles with solid lines*) compared to Meta II fraction θ_{UV-vis} from UV–visible spectrophotometry (*open symbols with dashed lines*, from Fig. 2e). (No distinction is made among Meta II substates.) The alkaline endpoint fractions θ_{FTIR}^{alk} of the FTIR-based pH titration curves at 30 and 37°C are less than the values θ_{UV-vis}^{alk} obtained from the UV–visible titration curves. (**d**) Analogous FTIR and UV–visible pH titration curves for rhodopsin reconstituted into POPC membranes show Meta II fractions θ_{FTIR} and θ_{UV-vis}, respectively. Note the significant influences of the lipid composition versus the native disk membranes in panel **c**, indicating their role in the visual transduction process. (Experimental data from ref. 9.)

corrected as described (9). Average multiple spectra to improve the signal-to-noise ratio (see above). Flatten the baseline of the FTIR spectra using standard FTIR software.

4. Utilize conventional FTIR spectroscopy (24) at lower temperatures (e.g., 10 and 20°C), typically with 4 cm^{-1} resolution and acquisition times of 12 s for pre-illumination spectra and between 1.5 and 12 s for post-illumination spectra, depending on temperature and stability of the photoproduct states. Employ rapid-scan FTIR spectroscopy at the higher temperatures (30 and 37°C) due to faster spectral changes (9, 31). For rapid-scan FTIR the spectra are recorded with a resolution of 4 cm^{-1} and an acquisition time of 6 s for the pre-illumination spectrum, and 120 or 240 ms for single post-illumination spectra (9, 10) (see Note 11).

5. Acquire FTIR spectra for sandwich samples of membranes containing rhodopsin in either the Meta I or Meta II$_b$H$^+$ states; normalize and subtract the pre-illumination spectrum from the post-illumination spectrum as described above (Fig. 3a). Note that for the native disk membranes the Meta II$_b$H$^+$ and Meta I states are stabilized at 20°C and pH 5.0 (acid endpoint) and at 10°C and pH 9.5 (alkaline endpoint), respectively (see UV–visible spectra in Fig. 2a, b). Note also that the Meta II$_b$ and Meta II$_b$H$^+$ substates share a similar active state FTIR signature (Fig. 3b). Use these results as reference (basis) spectra for decomposing the experimental FTIR difference spectra in the conformation-sensitive region from 1,800 to 1,600 cm^{-1} into Meta I and Meta II contributions (see Note 12).

6. For native RDM, record difference FTIR spectra (post-illumination minus pre-illumination) in the pH range from 5 to 10 at a series of temperatures from 10 to 37°C. Use sandwich samples of RDM containing rhodopsin. Note the disk membrane phospholipids include a mixture of phosphocholine (PC), phosphoethanolamine (PE), and phosphoserine (PS) head groups, and various acyl groups including ≈45–50% polyunsaturated docosahexaenoic acid (DHA; C22:6ω3) chains.

7. Fit the experimental FTIR difference spectra in the conformation-sensitive region between 1,800 and 1,600 cm^{-1} as a linear combination (weighted sum) of the Meta I and Meta II$_b$H$^+$ basis (reference) spectra (Fig. 3a). The reference spectra correspond to inactive and active receptor states, respectively. Calculate the Meta II fraction from FTIR spectroscopy (θ_{FTIR}) using the proportions (coefficients) of the weighted sum (linear combination) (Subheading 4). Alternatively use the amide I band at 1,644 cm^{-1} as a marker for Meta II (Subheading 4) (see Note 13).

8. Plot the Meta II fraction θ_{FTIR} as a function of pH and fit the results to simple theoretical functions for the binding

equilibrium of hydronium ions (H_3O^+) (Fig. 3c, d) (Subheading 4). Typically the pH titration curves follow a regular Henderson–Hasselbalch function at 10°C and below. However, at higher temperatures the alkaline endpoint value θ_{FTIR}^{alk} does not fall to zero (Fig. 3b), as also seen for UV–visible data (Fig. 2e).

9. Compare the results for the Meta II fraction θ_{FTIR} (Fig. 3c) to the Meta II fraction θ_{UV-vis} as a function of pH from UV–visible spectrophotometry (Fig. 2e). Typically at 30 and 37°C the alkaline endpoint value θ_{FTIR}^{alk} is less than the θ_{UV-vis}^{alk} value (Fig. 3c). Interpret the difference between θ_{UV-vis}^{alk} and θ_{FTIR}^{alk} by a contribution from the Meta II$_a$ photoproduct to the equilibrium at the alkaline endpoint, amounting to only ≈10% at 37°C and less at lower temperatures (Fig. 3c). Explain the combined results by an extended reaction scheme with an ensemble of Meta II states (e.g., Meta II$_a$, Meta II$_b$, Meta II$_b$H$^+$).

10. For rhodopsin–POPC membranes (1:100 protein–lipid molar ratio), conduct analogous FTIR difference studies over the pH range 5–10 at a series of temperatures between 10 and 37°C (see Note 14). Calculate the Meta II fraction θ_{FTIR} from a linear combination of reference (basis) spectra for Meta I and Meta II$_b$H$^+$ (Fig. 3a), or use the amide I band at 1,644 cm^{-1} as a marker for Meta II (see below).

11. Fit the results versus pH and compare the parameters (pK_a and alkaline endpoint) for the rhodopsin–POPC membranes to the native disk membranes, which have a mix of head groups and polyunsaturated DHA chains. Typically for rhodopsin–POPC membranes, the alkaline endpoint values (Fig. 3d) are less than for the native RDM (Fig. 3c), amounting to shifts of the Meta I–Meta II$_b$ equilibrium constant by <2. However, the apparent pK_a values of the titration curves are decreased by ≈1.5 units, due to a reduction of the equilibrium constant for H$_3$O$^+$ uptake by a factor of ≈30 (9).

12. Conduct more detailed FTIR studies of the individual marker bands in the conformation-sensitive region between 1,800 and 1,600 cm^{-1} for rhodopsin in different membrane lipid environments. The various marker bands are sensitive to conformational changes that occur during receptor activation (9). For example, compare sandwich samples of native RDM (as described above) to recombinant membranes containing rhodopsin with either POPC or DOPC lipids (10).

13. Explore whether the FTIR difference spectra can be represented by a linear combination of just two reference (basis) spectra due to Meta I and Meta II states (9) (Fig. 3a). Note that the individual marker bands (Fig. 4a) encompass the amide I vibrations of the protein backbone at 1,644 cm^{-1} (+) and the C=O stretch due to protonated Glu^{122} at 1,727 cm^{-1} (−) and

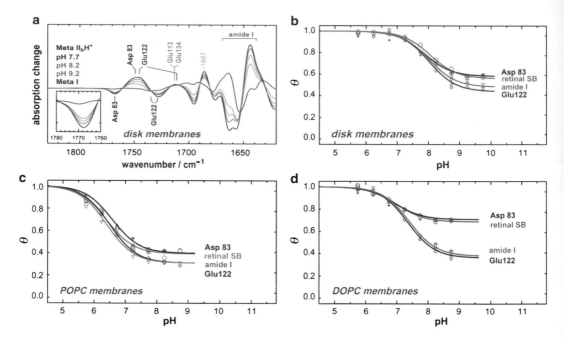

Fig. 4. Combination of FTIR and UV–visible spectroscopy reveals an ensemble of conformational substates in rhodopsin activation. Data are for fully hydrated sandwich samples (10). (**a**) The FTIR marker bands monitor the conformation of the receptor in pH-dependent experiments. The amide I band (due to $C=O$ stretch of peptide groups) is sensitive to the secondary structure; Asp[83] detects the interhelical H1/H2/H7 network; Glu[122] manifests the H3/H5 interhelical network; and Glu[134] of the ERY motif characterizes the H3/H6 network. (**b**) Plots of Meta II fraction θ ($\equiv \theta_{FTIR}$ or θ_{UV-vis}) versus pH for rhodopsin in native retinal disk membranes at 30°C obtained from the various marker bands. (No distinction is made among Meta II substates.) Only small differences are seen in the alkaline endpoint values for the various marker bands. (**c**) Meta II fraction θ for rhodopsin in membranes with the natural phospholipid POPC at 30°C as determined by UV–visible and FTIR spectroscopy. For rhodopsin–POPC membranes minor differences are evident in the alkaline endpoint values for the marker bands. Compared to the disk membranes, the pK_a is strongly shifted to lower values for rhodopsin in the POPC recombinant membranes, and the alkaline endpoint values are lower. (**d**) Meta II fraction θ ($\equiv \theta_{FTIR}$ or θ_{UV-vis}) versus pH for rhodopsin recombined with the synthetic phospholipid DOPC at 30°C obtained from the various UV–visible and FTIR marker bands. For rhodopsin–DOPC membranes the alkaline endpoint values for the retinal SB and Asp[83] bands exceed those for the amide I and Glu[122] marker bands, revealing multiple conformational states. Compared to POPC the pK_a is strongly shifted to higher values and approaches the disk membranes. (Experimental data from ref. 10.)

1,745 cm^{-1} (+) and Asp[83] at 1,768 cm^{-1} (−). Specifically Glu[122] monitors the hydrogen-bonded network between helices H3 and H5, and Asp[83] is a marker for the hydrogen-bonded network among helices H1, H2, and H7 (10).

14. For RDM, plot the Meta II fraction obtained from both the FTIR marker bands (amide I, Glu[122], Asp[83]) and UV–visible spectrophotometry (retinal SB) as a function of bulk solution pH (Fig. 4b) (Subheading 4). If the pH titration curves for the different maker bands are superimposable then just two spectral forms contribute, Meta I and Meta II (i.e., Meta II$_a$, Meta II$_b$, Meta II$_b$H$^+$ share a similar spectral signature). In this case, the structural changes are concerted, and the coefficients are θ ($\equiv \theta_{FTIR}$ or θ_{UV-vis}) for the Meta II fraction and $(1-\theta)$ for

the Meta I fraction (9). However, different photoproducts typically contribute, and the curves for the various marker bands do not superimpose (Fig. 4b). Such nonequivalence implies multiple photoproducts are present, e.g., due to an ensemble of states (Fig. 1b).

15. Fit the intensities of the marker bands versus pH to the modified Henderson–Hasselbalch function with a non-zero alkaline endpoint (use Eq. 3 of Subheading 4). Typically, a non-zero alkaline endpoint is evident for both the FTIR and UV–visible marker bands, indicating the process is not a simple two-state equilibrium (Figs. 3c, 4b). Moreover, the alkaline endpoint obtained from UV–visible spectra (θ_{UV-vis}^{alk}) is usually higher than the endpoint from FTIR spectroscopy (θ_{FTIR}^{alk}) (9) (Figs. 3c and 4b). Although the Meta II$_b$H$^+$ state is predominant at acid pH values, a significant population of Meta II$_a$ and Meta II$_b$ states exists at alkaline pH (9).

16. Interpret the results in structural terms. Note that Asp[83] is a marker for the H1/H2/H7 interhelical network, whereas Glu[122] is a marker for the H3/H5 interhelical network. Data for the retinal SB and Asp[83] marker bands are explained in terms of deprotonation of the PSB, which triggers activating structural changes in the hydrogen-bonded H1/H2/H7 helical network. Use the amide I and Glu[122] marker bands for investigating changes in the H3/H5 helical network. (For example, chromophore movement towards helix H5 occurs with an outward tilt of H6 as seen by spin-label EPR spectroscopy (32).) Note that protonation of Glu[134] of the ERY motif stabilizes the activated receptor due to a second protonation switch and explains the pH dependence of the activation (9).

17. Conduct analogous FTIR studies of rhodopsin in various recombinant membranes, e.g., sandwich samples of rhodopsin recombined with POPC or the synthetic lipid DOPC (see Note 15). Typical data (Fig. 4c) show that for rhodopsin–POPC membranes (1:100 protein–lipid molar ratio) the alkaline endpoint values are generally lower compared to the RDM. There is a striking downshift of the pK_a to lower values, as noted above (Fig. 3d). Additional sample data for rhodopsin–DOPC recombinant membranes (Fig. 4d) show the pH dependence of the retinal SB band follows closely the structural changes monitored by the Asp[83] FTIR difference band. For the Glu[122] or the amide difference bands (Fig. 4d) the alkaline endpoints are substantially lower than for the retinal SB or Asp[83] marker bands.

18. Explain the combined FTIR and UV–visible results by structural changes in hydrogen-bonded networks of rhodopsin (see above). According to the scheme in Fig. 1a, membrane lipids significantly perturb the relative populations of the various Meta II conformational substates (see Note 16).

4. Data Analysis and Interpretation

The results of FTIR and UV–visible spectroscopy are combined to investigate the complex pH-dependent population curves for conformational equilibria among the Meta I, Meta II$_a$, Meta II$_b$, and Meta II$_b$H$^+$ photoproduct states (9, 10). Influences of membrane lipids are analogous to mutating the protein. Further interpretation of the UV–visible and FTIR results in terms of lipid–protein interactions involves an activated ensemble mechanism (Fig. 1b).

4.1. Reduction and Analysis of Experimental UV–Visible and FTIR Spectral Data

1. Use the individual marker bands in UV–visible or FTIR spectroscopy to quantify the fractional Meta II values (θ) as calculated by the formula:

$$\theta = (\Delta A - \Delta A_{\text{Meta I}}) / (\Delta A_{\text{Meta II}_b\text{H}^+} - \Delta A_{\text{Meta I}}), \qquad (1)$$

where $\theta \equiv \theta_{\text{UV–vis}}$ or θ_{FTIR} (10). Note that ΔA is the pH-dependent absorption intensity for a marker band in either UV–visible or FTIR difference spectra of the photoproduct minus dark state. For example, in UV–visible studies the spectral difference ΔA can be monitored at $\lambda = 380$ nm (characteristic of Meta II) or $\lambda = 480$ nm (indicative of Meta I). Equivalently θ can be calculated using the molar absorption coefficients of the photoproducts (8). In FTIR studies, the marker bands are at 1,768 cm^{-1} (Asp^{83}), 1,745 cm^{-1} and 1,727 cm^{-1} (Glu^{122}), and 1,644 cm^{-1} (amide I of Meta II) (Fig. 4a). In the above formula, $\Delta A_{\text{Meta I}}$ and $\Delta A_{\text{Meta II}_b\text{H}^+}$ correspond to the Meta I minus dark state and Meta II$_b$H$^+$ minus dark state reference (basis) spectra, e.g., measured at 10°C and pH 9.5 and 30°C and pH 6.0, respectively (Fig. 2c, d) (see Note 17).

2. In the case of UV–visible spectrophotometry, use Eq. 1 to calculate the fractional Meta II value $\theta_{\text{UV–vis}}$ due to the PSB, using data over the absorption range from 300 to 650 nm (Fig. 2) (9). Typically ΔA at 380 nm is characteristic of Meta II and 480 nm is indicative of Meta I. Alternatively, the experimental difference spectra are fit as a linear combination of Meta I and Meta II$_b$H$^+$ reference (basis) spectra (Fig. 2). Assuming two-component spectra the coefficients are $\theta_{\text{UV–vis}}$ for the Meta II fraction and $(1 - \theta_{\text{UV–vis}})$ for the Meta I fraction (9). Use nonlinear regression fitting software (e.g., Matlab; Mathworks, Natick, MA, USA or Origin; OriginLab, Northampton, MA, USA) to obtain the coefficients (9).

3. In FTIR spectroscopy, the fractional Meta II value θ_{FTIR} is calculated from the difference bands in the conformation-sensitive region between 1,800 and 1,600 cm^{-1} (Fig. 3a). Again there are two possibilities. The Meta II fraction θ_{FTIR} can be

calculated as a linear combination of reference (basis) spectra for the Meta I and Meta II$_b$H$^+$ states (Fig. 3a). (It is assumed that the Meta II spectral signature is nearly the same for the Meta II$_a$, Meta II$_b$, and Meta II$_b$H$^+$ substates as in Fig. 3b.) However, use of only two basis spectra may be an oversimplification, since the changes detected by the various marker bands may differ (10). Alternatively, use Eq. 1 to calculate the fractional Meta II value θ_{FTIR} from the intensities of the individual Asp[83], Glu[122], or the amide I marker bands as a function of pH (Fig. 4).

4. Fit the data obtained for the Meta II fraction as a function of pH from UV–visible or FTIR spectroscopy to various binding isotherms for hydronium ions (use nonlinear regression software). Typically, at relatively low temperature (0–10°C) the spectral results are described by the Henderson–Hasselbalch equation:

$$\theta = 10^{pK_a - pH} / (1 + 10^{pK_a - pH}) \qquad (2)$$

The above formula describes a two-state, acid–base equilibrium between Meta II$_b$H$^+$ (conjugate acid) and Meta I (conjugate base) states, with the acid endpoints at θ_{UV-vis}^{acid} or $\theta_{FTIR}^{acid} = 1$ and the alkaline endpoints at θ_{UV-vis}^{alk} or $\theta_{FTIR}^{alk} = 0$ (Fig. 3c, d) (see Note 18).

5. At higher temperatures (range of 20–37°C), however, the alkaline endpoint values of the pH titration curves typically do not fall to zero (Fig. 3c, d). Fit the fractional Meta II values θ obtained from the UV–visible and/or FTIR marker bands at higher temperatures to a modified Henderson–Hasselbalch function (9) as given by:

$$\theta = (\theta^{alk} + 10^{pK_a - pH}) / (1 + 10^{pK_a - pH}). \qquad (3)$$

Here θ ($\equiv \theta_{UV-vis}$ or θ_{FTIR}) is the Meta II fraction and θ^{alk} ($\equiv \theta_{UV-vis}^{alk}$ or θ_{FTIR}^{alk}) is the alkaline endpoint of the pH titration curve (see Note 18). (Note that Eq. 3 is more general than Eq. 2 which corresponds to the limit that $\theta^{alk} \to 0$.)

4.2. Example of How to Interpret Results in Terms of Lipid–Membrane Protein Interactions

1. The experimental data from UV–visible and FTIR spectroscopy are used to investigate influences of the membrane lipid head group and acyl chain composition on rhodopsin activation (2–5, 7, 33). Lipid substitutions involve varying the head group size and charge as well as the degree of (poly)unsaturation and bulkiness of the acyl chains. Membrane recombinants are selected to distinguish between chemically specific effects and nonspecific lipid–protein interactions due to material properties of the bilayer (1, 5, 9, 10) (see Note 19).

2. Generally the membrane lateral pressure profile (Fig. 5a) originates from a balance of opposing attractive and repulsive forces

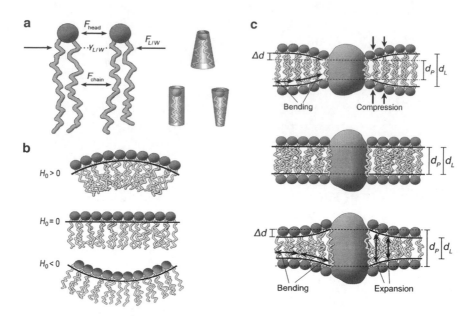

Fig. 5. Example of interpretation of experimental UV–visible and FTIR spectra in terms of lipid–protein interactions. The flexible surface model (FSM) describes how membrane lipids act as agonists or antagonists of rhodopsin function due to nonspecific material properties of the bilayer (1). (**a**) Biophysical properties of membrane lipids involve a balance of attractive and repulsive forces pertinent to average lipid shapes and membrane curvature. The optimum lipid shape is due to the lipid polar head groups and non-polar acyl chains—it is visualized as the frustum of a *right circular cone* or *cylinder*. The *upper base* represents the head groups and the *lower base* is due to the chains. (**b**) Average lipid shape is closely tied to the concept of spontaneous curvature of an individual lipid monolayer (leaflet) (1). The spontaneous curvature (H_0) is due to bending of a neutral (pivotal) plane where the area is constant. It can be towards hydrocarbon ($H_0 > 0$), planar ($H_0 = 0$), or towards water ($H_0 < 0$) as indicated. The local curvature stress is proportional to $k_c|H - H_0|$ where k_c is the curvature elastic modulus (analogous to Hooke's Law). (**c**) The FSM explains functional influences of membrane lipid–protein interactions by elastic deformation of a membrane lipid film ($d_P \equiv$ thickness adjacent to proteolipid interface; $d_L \equiv$ thickness of unperturbed bilayer). Bending of the neutral plane together with expansion/compression competes with the solvation energy of the proteolipid interface. The concept of curvature matching explains membrane lipid influences by nonspecific material properties of the lipid bilayer (frustration). Note the general principles of membrane curvature deformation (1) as revealed by rhodopsin are applicable to other membrane proteins, including GPCRs, transporters, and ion channels.

acting at the level of the polar head groups and the acyl chains of the phospholipids. Head group repulsions and interfacial attraction give an optimal separation for the polar region. Within the hydrocarbon core, van der Waals attraction plus steric repulsion yield a preferred distance between the chains. Relate the lipid influences on the metarhodopsin equilibrium (e.g., Fig. 3) to the head groups of various sizes and to the bulkiness or degree of polyunsaturation of the acyl groups (see Note 19).

3. Note the lateral pressure profile corresponds to the spontaneous (intrinsic) curvature of a monolayer leaflet (designated as H_0) (Fig. 5b) (1, 5). The value of H_0 can be positive (curvature towards hydrocarbon, as in the case of micelles or normal hexagonal phases of amphiphiles), zero (no curvature, as in the textbook bilayer), or negative (curvature towards water, as in the case of microemulsions or reverse hexagonal phases

of amphiphiles) (Fig. 5b). Correlate the shifting of the Meta I–Meta II equilibrium by membrane lipids (see above) with their propensity to form nonlamellar phases, such as the reverse hexagonal (H_{II}) or cubic phases (34). Test whether the influences of bulky unsaturated acyl chains, such as DHA, can be mimicked by relatively small head groups such as PE, and vice versa.

4. Interpret the experimental UV–visible and FTIR results for rhodopsin in terms of models for lipid–membrane protein interactions, such as the FSM (1). The idea behind the FSM is that mismatch of the spontaneous (intrinsic) curvature of a monolayer leaflet to the curvature at the proteolipid interface governs the free energies of proteins such as rhodopsin (1, 21) or ion channels (35–37) (Fig. 5c). The FSM describes a competition between the elastic energy due to deformation of the membrane lipids (within the stress field of the bilayer) and the solvation energy of the proteolipid boundary (formulated in terms of the proteolipid surface tension), e.g., due to hydrophobic mismatch (8). Matching the membrane curvature H to the spontaneous monolayer curvature H_0 provides a driving force for protein conformational transitions that explains many of the lipid influences in membranes (1) (see Note 20).

5. Relate the curvature free energy of the membrane lipid bilayer to the equilibrium constant for the transition between the Meta I and Meta II photoproducts of rhodopsin (21, 33) and further interpret the experimental findings using the FSM (see Note 21).

5. Notes

1. Detergents used for rhodopsin solubilization include Ammonyx LO, sodium cholate, digitonin, n-dodecyl-β-D-maltopyranoside (DDM), DTAB, Emulphogene, LDAO, octaethylene glycol n-dodecyl monoether ($C_{12}E_8$), n-octyl-β-D-glucopyranoside (OG), and Tween 80. Bleached rhodopsin (i.e., opsin) can be regenerated with 11-cis-retinal in the detergents cholate, digitonin, OG, or DDM.

2. Store retinas in lightproof glass containers (wrapped with aluminum foil or electrical tape) under argon. It is recommended to use retinas harvested between the months of October through March due to seasonal variability. Retinas can be stored at −80°C for several years without adverse effects on biophysical and functional parameters.

3. Note that hydroxylamine, NH_2OH, is present in stock detergent buffer, but does not react with rhodopsin in the dark

state. After bleaching, NH_2OH reacts with retinal bound to rhodopsin at the Meta II stage. The reaction is completely forward shifted by le Châtelier's principle to give the final products opsin + retinaloxime.

4. The A_{500} absorption is due to retinal and is a specific marker for rhodopsin. However, the A_{280} absorption includes contributions from aromatic amino acids and is not specific to rhodopsin. Hence the A_{280}/A_{500} ratio provides a simple criterion of the rhodopsin purity.

5. Warning: rhodopsin is unstable in DTAB and thermally bleaches if the sample temperature rises above 4°C. All procedures with DTAB should be done on ice or in the cold room. When in doubt, record a UV–visible absorption spectrum in 3% Ammonyx LO as described above to measure the A_{280}/A_{500} ratio.

6. Because the lipid density is less than protein density, the proteolipid band migrates to a position intermediate between the top and bottom of the centrifuge tube. Typically the lipid–rhodopsin ratio determined in this manner is approximately the same as the stoichiometric ratio before dialysis. However, for careful work this should not be assumed.

7. In control experiments, membranes are washed with a large volume of 1 mM BTP buffer adjusted to the appropriate pH value prior to preparation of the sandwich samples. Typically the same results are obtained as with subsequent pH adjustment, thus verifying the reliability of the pH procedure.

8. The contribution of isorhodopsin to the photoproduct equilibria is found to be small under these conditions by using its FTIR fingerprint bands as spectral markers (<10% of Meta I) (10). Spectra are not typically corrected for the isorhodopsin contribution.

9. A scanning spectrophotometer has the advantage that the sample is exposed to only a narrow band of wavelengths at any one time, and thus exposure to the monitoring light beam is minimal. The disadvantage is that measuring the spectrum is relatively time consuming. A diode-array system has the advantage that all the light is passed through the sample prior to wavelength dispersal and detection, so the spectrum is acquired rapidly. The disadvantage is that the sample is exposed to all of the light during the measurement, which may initiate undesired photochemical processes.

10. Because such experiments involve a combination of elevated temperature and alkaline pH values, photoproduct stability needs to be monitored carefully (9). Thermal stability of the photoproduct states is confirmed by obtaining a series of replicate post-illumination spectra as a function of time. Near room temperature (>20°C) at even mildly alkaline pH values (>8)

the photoproducts change on the time scale of a few seconds, giving a loose bundle state that for an α-helical membrane protein (38) is analogous to the molten globule state for water-soluble proteins.

11. Difference spectra obtained by conventional FTIR spectroscopy and by rapid-scan FTIR spectroscopy utilize pre-illumination spectra acquired with the separate protocols.

12. Notably the reference FTIR spectra for Meta I at the alkaline endpoint become more like Meta II as the temperature is raised. By contrast, the Meta II reference FTIR spectra are less sensitive to temperature. Assuming the PSB photoproduct has a single Meta I difference spectrum, and that the SB photoproduct has a single Meta II difference spectrum, may represent a simplifying approximation (9).

13. Such a decomposition into two basis (reference) spectra assumes the spectral signatures of Meta II_b and Meta II_bH^+ are similar, and that the population of the additional Meta II_a state in the equilibrium mixture is small (9). Differences are seen for the Asp^{83}/retinal SB bands compared to the Glu^{122}/amide I marker bands that moreover depend on the lipid composition (10). Consequently, use of a basis set comprising just two reference spectra (Meta I and Meta II_bH^+) (Fig. 3a) may be an oversimplification. Further investigation may be required, e.g., using singular value decomposition.

14. POPC is a natural phospholipid having a phosphocholine head group at the glycerol sn-3 position, an unsaturated oleoyl (18:1ω9) chain at the glycerol sn-2 position, and a fully saturated palmitoyl (16:0) chain at the glycerol sn-1 position. It is distinct from the natural retinal disk membrane lipids in that a single head group type is present, and that polyunsaturated DHA (22:6ω3) chains are absent.

15. DOPC is a synthetic lipid with a phosphocholine head group at the sn-3 position and two unsaturated oleoyl (18:1ω9) chains at the glycerol sn-2 and sn-1 positions. It differs from the natural lipids in that it has a single type of head group, with two identical unsaturated acyl groups lacking polyunsaturation as in the case of DHA chains.

16. Selective stabilization of photoproducts by membrane lipids affords a way of structurally characterizing the activation mechanism of rhodopsin. For rhodopsin in DOPC membranes the Meta II_a substate is stabilized, in which PSB deprotonation has occurred prior to helix H6 tilt. It provides a missing link between opening the PSB salt bridge in the transition from Meta I and the subsequent tilt of H6 in forming Meta II_b as detected by spin-label EPR spectroscopy (32). A very similar selective stabilization of the Meta II_a substate is also found for recombinants of rhodopsin with DOPE–DOPC lipid mixtures (10).

17. When possible carry out all experiments using rhodopsin from a single batch of retinas to avoid sample variability that may affect the quantitative spectral data reduction and analysis. Alternatively multiple batches can be combined if necessary.

18. Additional data analysis uses the following thermodynamic relations. The standard free energy change ($\Delta G^{o\prime}$) is related to both the standard enthalpy change ($\Delta H^{o\prime}$) and the standard entropy change ($\Delta S^{o\prime}$) through the relation $\Delta G^{o\prime} = \Delta H^{o\prime} - T\Delta S^{o}$. The standard states (symbol o') correspond to a one molar solution of the components (Meta I or Meta II) at pH 7. All symbols have their usual definitions ($R \equiv$ gas constant, $T \equiv$ temperature, $P \equiv$ pressure). The standard free energy change is given by $\Delta G^{o\prime} = -RT \ln K_{eq}{}'$ where $K_{eq}{}'$ is given by $\theta/(1-\theta)$ and $\theta \equiv \theta_{UV-vis}$ or θ_{FTIR} as obtained from the spectral data reduction. The standard enthalpy change is given by $\Delta H^{o\prime} = -R(\partial \ln K_{eq}{}'/\partial 1/T)_P$ and the standard entropy change is given by $\Delta S^{o\prime} = -(\partial \Delta G^{o\prime}/\partial T)_P$. Last, the change in standard heat capacity at constant pressure is given by $\Delta C_P^{o\prime} = (\partial \Delta H^{o\prime}/\partial T)_P$. For further specifics see refs. 5, 39.

19. Lipids with small head groups (DOPE) or increasing acyl chain unsaturation (DOPC versus POPC, Fig. 4c, d) forward shift the metarhodopsin equilibrium towards Meta II. Lipids with larger head groups (e.g., DOPC versus DOPE (21, 33)) and less acyl unsaturation (3, 5) back shift the equilibrium towards Meta I. In general, small head groups such as phosphoethanolamine (PE), bulky polyunsaturated DHA chains, or greater bilayer thickness favor the activated Meta II photoproduct. By contrast larger head groups such as phosphocholine (PC), a decrease in acyl chain unsaturation, or smaller bilayer thickness favor the inactive Meta I state.

20. By connecting the membrane protein shape transitions to changes in membrane lipid curvature forces, a continuum picture explains the influences of membrane lipids on protein functions by the theory of elasticity. The spontaneous (intrinsic) curvature H_0 may be identical with the physical monolayer curvature (denoted by H), or deformation (strain) of the membrane may occur, yielding curvature elastic stress or frustration (5). The competition or balance of the curvature free energy (stress/strain) with the acyl chain stretching energy (due to solvation of the proteolipid interface) explains how membrane lipids affect protein conformational energetics in the case of rhodopsin, and is applicable to (mechanosensitive) ion channels and other membrane proteins.

21. Values of $K_{eq}{}'$ and/or pK_a determined for the extended mechanism (see above) (9, 40) can be further interpreted in terms of standard free energy changes. Note that the lateral pressure profile (41) does not correspond to any measurable quantity—it is calculated theoretically (42). By contrast, the spontaneous

curvature can be found experimentally (43, 44). The membrane lipid curvature free energy can be calculated using experimental estimates of the bending modulus (k_c) and spontaneous curvature (H_0) (33, 45, 46). Use the Helfrich formula $g_c = k_c(H - H_0)^2 + \bar{k}_c K$ as described (1). Here g_c is the curvature free energy per unit area and c_1 and c_2 are the two principal curvatures of the surface. In addition $H \equiv (c_1 + c_2)/2$ is the mean curvature, H_0 is the spontaneous (intrinsic) curvature in the absence of strain, k_c is the bending rigidity (modulus), $K \equiv c_1 c_2$, and $\bar{k}_c$ is the saddle (Gaussian) curvature modulus. Note that the curvature free energy is simply given by $g_c \approx k_c(H - H_0)^2$ assuming that only a single curvature (c_1) contributes. The curvature c_1 can be calculated from the inverse radius of the hexagonal phase cylinders in the presence of hydrocarbon (43).

Acknowledgments

The author is indebted to A. V. Botelho, J. J. Kinnun, J. W. Lewis, K. Martínez-Mayorga, B. Mertz, A. V. Struts, and R. Vogel for discussions. Research in the laboratory of the author is supported by the U.S. National Institutes of Health and is gratefully acknowledged.

References

1. Brown MF (1994) Modulation of rhodopsin function by properties of the membrane bilayer. Chem Phys Lipids 73:159–180

2. Baldwin PA, Hubbell WL (1985) Effects of lipid environment on the light-induced conformational changes of rhodopsin. 2. Roles of lipid chain length, unsaturation, and phase state. Biochemistry 24:2633–2639

3. Wiedmann TS, Pates RD, Beach JM, Salmon A, Brown MF (1988) Lipid-protein interactions mediate the photochemical function of rhodopsin. Biochemistry 27:6469–6474

4. Mitchell DC, Straume M, Miller JL, Litman BJ (1990) Modulation of metarhodopsin formation by cholesterol-induced ordering of bilayer lipids. Biochemistry 29:9143–9149

5. Gibson NJ, Brown MF (1993) Lipid headgroup and acyl chain composition modulate the MI-MII equilibrium of rhodopsin in recombinant membranes. Biochemistry 32:2438–2454

6. Niu S-L, Mitchell DC (2005) Effect of packing density on rhodopsin stability and function in polyunsaturated membranes. Biophys J 89:1833–1840

7. Kusnetzow AK, Altenbach C, Hubbell WL (2006) Conformational states and dynamics of rhodopsin in micelles and bilayers. Biochemistry 45:5538–5550

8. Botelho AV, Huber T, Sakmar TP, Brown MF (2006) Curvature and hydrophobic forces drive oligomerization and modulate activity of rhodopsin in membranes. Biophys J 91:4464–4477

9. Mahalingam M, Martínez-Mayorga K, Brown MF, Vogel R (2008) Two protonation switches control rhodopsin activation in membranes. Proc Natl Acad Sci U S A 105:17795–17800

10. Zaitseva E, Brown MF, Vogel R (2010) Sequential rearrangement of interhelical networks upon rhodopsin activation in membranes: the Meta IIa conformational substate. J Am Chem Soc 132:4815–4821

11. Palczewski K, Kumasaka T, Hori T, Behnke CA, Motoshima H, Fox BA, Le Trong I, Teller DC, Okada T, Stenkamp RE, Yamamoto M, Miyano M (2000) Crystal structure of rhodopsin: a G protein-coupled receptor. Science 289:739–745

12. Li J, Edwards PC, Burghammer M, Villa C, Schertler GFX (2004) Structure of bovine

rhodopsin in a trigonal crystal form. J Mol Biol 343:1409–1438

13. Okada T, Sugihara M, Bondar A-N, Elstner M, Entel P, Buss V (2004) The retinal conformation and its environment in rhodopsin in light of a new 2.2 Å crystal structure. J Mol Biol 342:571–583

14. Park JH, Scheerer P, Hofmann KP, Choe H-W, Ernst OP (2008) Crystal structure of the ligand-free G-protein-coupled receptor opsin. Nature 454:183–188

15. DeLano WL (2004) PyMOL user's guide. DeLano Scientific, San Carlos, California, http://www.pymol.org/

16. Papermaster DS, Dreyer WJ (1974) Rhodopsin content in the outer segment membranes of bovine and frog retinal rods. Biochemistry 13:2438–2444

17. Williams GD, Beach JM, Dodd SW, Brown MF (1985) Dependence of deuterium spin-lattice relaxation rates of multilamellar phospholipid dispersions on orientational order. J Am Chem Soc 107:6868–6873

18. Salmon A, Dodd SW, Williams GD, Beach JM, Brown MF (1987) Configurational statistics of acyl chains in polyunsaturated lipid bilayers from ^{2}H NMR. J Am Chem Soc 109:2600–2609

19. Raubach RA, Franklin LK, Dratz EA (1974) A rapid method for the purification of rod outer segment disk membranes. Vision Res 14:335–337

20. Hong K, Knudsen PJ, Hubbell WL (1982) Purification of rhodopsin on hydroxyapatite columns, detergent exchange, and recombination with phospholipids. Methods Enzymol 81:144–150

21. Botelho AV, Gibson NJ, Wang Y, Thurmond RL, Brown MF (2002) Conformational energetics of rhodopsin modulated by nonlamellar forming lipids. Biochemistry 41:6354–6368

22. Chen PS Jr, Toribara TY, Warner H (1956) Microdetermination of phosphorus. Anal Chem 28:1756–1758

23. Vogel R, Fan G-B, Sheves M, Siebert F (2000) The molecular origin of the inhibition of transducin activation in rhodopsin lacking the 9-methyl group of the retinal chromophore: a UV-Vis and FTIR spectroscopic study. Biochemistry 39:8895–8908

24. Vogel R, Fan GB, Siebert F, Sheves M (2001) Anions stabilize a metarhodopsin II-like photoproduct with a protonated Schiff base. Biochemistry 40:13342–13352

25. Knierim B, Hofmann KP, Ernst OP, Hubbell WL (2007) Sequence of late molecular events in the activation of rhodopsin. Proc Natl Acad Sci U S A 104:20290–20295

26. Struts AV, Salgado GFJ, Martínez-Mayorga K, Brown MF (2011) Retinal dynamics underlie its switch from inverse agonist to agonist during rhodopsin activation. Nat Struct Mol Biol 18(3):392–394

27. Lewis JW, Kliger DS (2000) Absorption spectroscopy in studies of visual pigments: spectral and kinetic characterization of intermediates. Methods Enzymol 315:164–178

28. Vogel R, Siebert F (2003) Fourier transform IR spectroscopy study for new insights into molecular properties and activation mechanisms of visual pigment rhodopsin. Biopolymers 72:133–148

29. Ritter E, Zimmermann K, Heck M, Hofmann KP, Bartl FJ (2004) Transition of rhodopsin into the active metarhodopsin II state opens a new light-induced pathway linked to Schiff base isomerization. J Biol Chem 279:48102–48111

30. Parkes JH, Liebman PA (1984) Temperature and pH dependence of the metarhodopsin I-metarhodopsin II kinetics and equilibria in bovine rod disk membrane suspensions. Biochemistry 23:5054–5061

31. Ritter E, Elgeti M, Hofmann KP, Bartl FJ (2007) Deactivation and proton transfer in light-induced metarhodopsin II/metarhodopsin III conversion. A time-resolved Fourier transform infrared spectroscopic study. J Biol Chem 282:10720–10730

32. Altenbach C, Kusnetzow AK, Ernst OP, Hofmann KP, Hubbell WL (2008) High-resolution distance mapping in rhodopsin reveals the pattern of helix movement due to activation. Proc Natl Acad Sci U S A 105:7439–7444

33. Soubias O, Teague WE Jr, Hines KG, Mitchell DC, Gawrisch K (2010) Contribution of membrane elastic energy to rhodopsin function. Biophys J 99:817–824

34. Deese AJ, Dratz EA, Brown MF (1981) Retinal rod outer segment lipids form bilayers in the presence and absence of rhodopsin: a ^{31}P NMR study. FEBS Lett 124:93–99

35. Nielsen C, Andersen OS (2000) Inclusion-induced bilayer deformations: effects of monolayer equililbrium curvature. Biophys J 79:2583–2604

36. Perozo E, Kloda A, Cortes DM, Martinac B (2002) Physical principles underlying the transduction of bilayer deformation forces during mechanosensitive channel gating. Nat Struct Biol 9:696–703

37. Phillips R, Ursell T, Wiggins P, Sens P (2009) Emerging roles for lipids in shaping membrane-protein function. Nature 459:379–385
38. Vogel R, Siebert F (2002) Conformation and stability of alpha-helical membrane proteins. 2. Influence of pH and salts on stability and unfolding of rhodopsin. Biochemistry 41: 3536–3545
39. Brown MF (1997) Influence of non-lamellar forming lipids on rhodopsin. Curr Top Membr 44:285–356
40. Lewis JW, Martínez-Mayorga K, Szundi I, Kliger DS, Brown MF (2010) Protonation switches in GPCR activation: physiologically significant rhodopsin photointermediates. Biophys J 98:288a
41. Cantor RS (1999) Lipid composition and the lateral pressure profile in bilayers. Biophys J 76:2625–2639
42. Marsh D (1996) Intrinsic curvature in normal and inverted lipid structures and in membranes. Biophys J 70:2248–2255
43. Gruner SM (1989) Stability of lyotropic phases with curved interfaces. J Phys Chem 93: 7562–7570
44. Seddon JM (1990) Structure of the inverted hexagonal (H_{II}) phase, and non-lamellar phase transitions of lipids. Biochim Biophys Acta 1031:1–69
45. Keller SL, Bezrukov SM, Gruner SM, Tate MW, Vodyanoy I, Parsegian VA (1993) Probability of alamethicin conductance states varies with nonlamellar tendency of bilayer phospholipids. Biophys J 65:23–27
46. Rawicz W, Olbrich KC, McIntosh T, Needham D, Evans E (2000) Effect of chain length and unsaturation on elasticity of lipid bilayers. Biophys J 79:328–339

Chapter 9

Proteomic Characterization of Integral Membrane Proteins Using Thermostatted Liquid Chromatography Coupled with Tandem Mass Spectrometry

Sarah M. Moore and Christine C. Wu

Abstract

Due to the hydrophobicity and localization of integral membrane proteins, they are difficult to study using conventional biochemical methods that are compatible with proteomic analyses. This chapter describes the coupling of multiple crucial steps that lead to the optimized shotgun proteomic analysis of integral membrane proteins while maintaining empirical topology information. Namely, a membrane shaving method is utilized to separate protease accessible peptides from membrane embedded peptides and elevated temperatures during chromatographic separation is utilized to augment the recovery of hydrophobic peptides for in-line analysis using tandem mass spectrometry. This combination of steps facilitates increased identification of membrane proteins while also maintaining information regarding protein topology.

Key words: Membrane proteomics, Membrane protein topology, Shotgun proteomic analysis, Hydrophobic peptides, Mass spectrometry

1. Introduction

Integral membrane proteins (IMPs) are a class of proteins that are essential in a variety of cellular functions and constitute the majority of known pharmaceutical drug targets, approximately 70% (1). This class of proteins also forms a critical class of cellular proteins as it is encoded for by 20–30% of the human genome. However, due to their hydrophobic nature, IMPs have thus far been difficult to study and consequently they are under-represented in most proteomic studies (2). Similarly, structural techniques, such as X-ray crystallography and NMR spectroscopy, have generally been low-throughput for the characterization of membrane proteins (1).

Nagarajan Vaidehi and Judith Klein-Seetharaman (eds.), *Membrane Protein Structure and Dynamics: Methods and Protocols*, Methods in Molecular Biology, vol. 914, DOI 10.1007/978-1-62703-023-6_9, © Springer Science+Business Media, LLC 2012

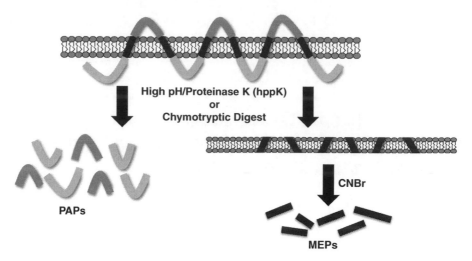

Fig. 1. Membrane Shaving Workflow. Membrane proteins are digested using to high pH/Proteinase K (hppK) method or chymotrypsin, in order to shave off the protease accessible peptides (PAPs). These processes leave the membrane embedded peptides (MEPs) in the membrane. The membrane is solubilized and CNBr is used to digest the MEPs.

The methods described in this chapter facilitate the profiling of IMPs from enriched membrane fractions. Membranes are first shaved using proteases and proteins are separated into protease-accessible peptides (PAPs), generated from soluble domains of IMPs, and membrane-embedded peptides (MEPs), representing the transmembrane domains of IMPs remaining in the lipid bilayer (Fig. 1). This shaving procedure can be performed using two complementary methods. PAPs can be digested from the membrane using either proteinase K or chymotrypsin. Proteinase K is a soluble, non-sequence specific protease that, under physiological conditions, cleaves proteins indiscriminately to the point that the peptides generated are too short for unique sequence identification by mass spectrometry. However, at high pH, the activity of proteinase K is attenuated and the resulting peptides are optimal for analysis by tandem mass spectrometry averaging 5–20 residues in length. Furthermore, the high pH maintains open membrane structures (with exposed "membrane sheet" edges) (3). PAPs can also be prepared using chymotrypsin, a sequence specific protease, allowing for orthogonal sequence coverage of these soluble domains. Because the lipid bilayer is not solubilized and denatured, these methods maintain protein structure information in the form of overall protein topology.

To enhance the recovery of peptides during chromatographic separation, elevated temperatures are maintained to ensure for better LC separations and enhanced recovery which result in increased observed sensitivity during the mass spectrometric measurement. Column heaters (built in-house or commercially purchased,

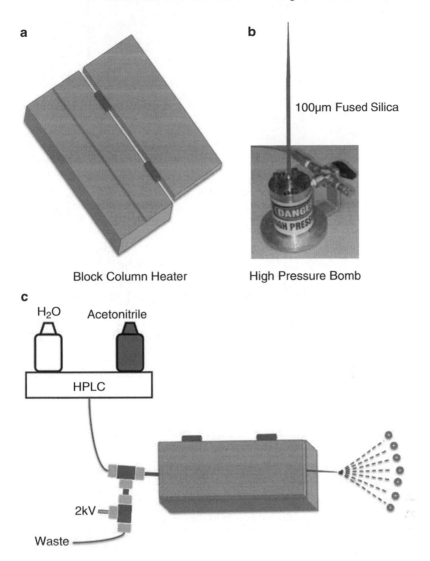

a

Block Column Heater

b

100µm Fused Silica

High Pressure Bomb

c

H₂O   Acetonitrile

HPLC

2kV

Waste

Fig. 2. LC/MS/MS Workflow and Schematic. (**a**) Block column heater, also built in-house is used to heat column. (**b**) 100 µm fused silica column pulled to a 5 µm tip is packed and loaded via a high pressure bomb, built in-house. (**c**) Column is inserted into column heater and connected to HPLC. A 2 kV charge is applied to induce ionization into the MS.

Fig. 2a) can be utilized to maintain constant elevated temperatures of µLC columns. Elevated temperatures during the analysis of enriched plasma membrane samples have been shown to increase protein identifications by 400% and peptide identifications by 500% (4–6).

The methods described in this chapter detail (1) the enrichment of plasma membranes from HeLa cells, (2) two digestion strategies to prepare PAPs and MEPs, and (3) subsequent µLC-MS/MS analysis.

2. Materials

2.1. Plasma Membrane Enrichment

1. HeLa cells (typically overexpressing IMP of interest for better sequence coverage).

2. Rubber policeman for cell harvest.

3. 400 mM Phosphate buffer (400 mM KH_2PO_4/K_2HPO_4, pH 6.7). Make 1 L of 400 mM dibasic potassium buffer (K_2HPO_4) and 1 L of 400 mM monobasic potassium buffer (KH_2PO_4). Add the monobasic buffer to the dibasic buffer slowly while monitoring the pH. Stop adding buffer when the pH reaches 6.7. Store at 4°C.

4. 100 mM phosphate buffer (100 mM KH_2PO_4/K_2HPO_4, pH 6.7). Dilute the 400 mM phostphate buffer in ultrapure water at a 1:4 ratio. This buffer can be stored at room temperature, but needs to be cooled to 4°C before use.

5. Magnesium chloride ($MgCl_2$). Dissolve at 1.0 M in ultrapure water. Store at room temperature.

6. Protease Inhibitor Cocktail, for mammalian cell culture, in DMSO (Sigma, St. Louis, MO).

7. Sucrose. Dissolve at 2.0 M in ultrapure water. Mix over low heat (40°C) to help dissolve. Store at 4°C.

8. Buffers for sucrose gradient: prepare according to instructions in Table 1. Keep on ice or at 4°C.

9. Glass homogenizer and Teflon pestle.

10. Refrigerated tabletop centrifuge.

11. Swinging bucket rotor capable of $135,000 \times g$.

12. Narrow bore transfer pipettes.

13. 1 cc insulin syringes.

14. Sodium carbonate buffer: dissolve Na_2CO_3 at 0.2 M in ultrapure water and adjust to pH 11 with concentrated HCl. Store at room temperature.

2.2. Proteinase K Sample Digestion

1. Solid urea (ultra pure molecular grade).

2. Dithiotheitol (DTT). Dissolve at 0.5 M in ultrapure water and aliquot into single-use aliquots (recommended: 200 μL). Store aliquots at –20°C (stable for 1 year). Do not freeze-thaw aliquots.

3. Iodoacetimide (IAA). Dissolve at 0.5 M in ultrapure water and aliquot into single-use aliquots (recommended: 200 μL). Store aliquots at –20°C (stable for 1 year). Do not freeze-thaw aliquots.

4. Recombinant proteinase K (Roche, Indianapolis, IN): resuspend the proteinase K at 1 mg/mL in ultrapure water, aliquot (recommended: 200 μL) and store at –20°C (stable up to 1 year). Do not freeze-thaw aliquots.

Table 1
Sucrose gradient solutions

Reagents	Sucrose solutions				Final concentration
	0.25 M	0.5 M	0.86 M	1.3 M	
400 mM KH_2PO_4/ K_2HPO_4 buffer	12.5 mL	12.5 mL	12.5 mL	12.5 mL	100 mM
1.0 M $MgCl_2$	0.25 mL	0.25 mL	0.25 mL	0.25 mL	5 mM
2.0 M Sucrose	6.25 mL	12.5 mL	21.5 mL	32.5 mL	Variable
Ultrapure water	31 mL	24.75 mL	15.75 mL	4.75 mL	–
Protease Inhibitor Cocktail	50 µL	50 µL	50 µL	50 µL	–
Final volume	50 mL	50 mL	50 mL	50 mL	–

2.3. Chymotrypsin Sample Digestion

1. 100 mM Tris–HCl (pH 8.0). Add 100 mmol of Tris to 800 mL of ultrapure water. Add concentrated HCl to pH 8.0 then dilute to 1 L with ultrapure water.

2. 8 M Urea/100 mM Tris–HCl (pH 8.0). Dissolve urea in 100 mM Tris–HCl to a final concentration of 8 M.

3. Dithiotheitol (DTT). Dissolve at 0.5 M in ultrapure water and aliquot into single-use aliquots (recommended: 200 µL). Store aliquots at –20°C (stable for 1 year). Do not freeze-thaw aliquots.

4. Iodoacetimide (IAA). Dissolve at 0.5 M in ultrapure water and aliquot into single-use aliquots (recommended: 200 µL). Store aliquots at –20°C (stable for 1 year). Do not freeze-thaw aliquots.

5. Chymotrypsin (Roche, Indianapolis, IN): resuspend the chymotrypsin at 0.5 mg/mL in ultrapure water. Store at –20°C.

2.4. Liquid Chromatography and Mass Spectrometry

1. 90% formic acid (J.T. Baker) (used for samples and chromatography buffers).

2. Buffers for reverse-phase chromatography—mass spectrometry. Buffer A: acetonitrile/formic acid/water, 5/0.1/95 (v/v/v). Use mass spectrometry grade acetonitrile and water (J.T. Baker, Phillipsburg, NJ). Buffer B: water/formic acid/ acetonitrile, 5/0.1/95 (v/v/v).

3. Mass spectrometry grade methanol.

4. Aqua C18 reverse-phase material (Phenomenex, Torrance, CA). Add a small aliquot (~10 mg) to 0.5 mL of mass spectrometry grade methanol in a micro-centrifuge tube just before

use and spin briefly so the solid phase forms a loose pellet at the bottom of the tube. Do not store long-term in methanol.

5. 100 μm internal diameter fused silica tubing (Polymicro Technologies, Phoenix, AZ) for use as chromatography columns.

6. Ceramic scribes.

7. Laser puller (Sutter, Novato, CA) for pulling chromatography column tips.

8. High-pressure bomb (built in-house, plans available upon request) for packing columns capable of handling 2,000 psi (Fig. 2b).

9. Helium for use with high-pressure bomb.

10. LTQ Linear Ion Trap mass spectrometer (ThermoFinnigan, San Jose, CA).

11. Autosampler and HPLC (Agilent, Santa Clara, CA).

12. Microspray source (built in-house, plans available upon request).

3. Methods

3.1. Plasma Membrane Enrichment

1. Harvest HeLa cells expressing the membrane protein of interest at 95% confluency on 500 cm^2 plates using a rubber policeman.

2. Pellet cells by centrifugation ($700 \times g$, 5 min, 4°C), remove liquid and store at –80°C.

3. Thaw cell pellets on ice and resuspend in 5 mL of ice-cold 0.5 M sucrose buffer.

4. Homogenize with a loose fitting, cold Teflon pestle in a glass homogenizer for 25 strokes (see Notes 1–3).

5. Pellet unbroken cells and nuclei by low-speed centrifugation ($700 \times g$, 5 min, 4°C) and collect the postnuclear supernatant (PNS).

6. Layer the PNS onto a 4-step discontinuous sucrose gradient, using the following sucrose concentrations (in order from top to bottom) in a thin-walled 12.5 mL ultracentrifugation tube: 3 mL of 1.3 M sucrose, 3.5 mL of 0.86 M sucrose, 5 mL of PNS in 0.5 M sucrose, 2.5 mL of 0.25 M sucrose (see Note 4).

7. Centrifuge in a swinging bucket rotor ($80,000 \times g$, 60 min, 4°C) (see Note 5).

8. Collect the interface from between the 1.3 M sucrose and 0.86 M sucrose layers using a wide-bore transfer pipet.

9. Double the volume of the interface with cold phosphate buffer (see Note 6).

10. Vortex and centrifuge the sample ($20,000 \times g$, 30 min, 4°C).

11. Discard the supernatant and resuspend the membrane pellet in 0.5 mL of Na_2CO_3 (pH 11) buffer and agitate on ice with five strokes of an insulin syringe every 15 min for 1 h (see Note 7).

12. Pellet the sample by ultracentrifugation ($135,000 \times g$, 45 min, 4°C) and discard the supernatant.

13. Resuspend the enriched membrane pellet in phosphate buffer for protein assay.

14. Store the membrane pellets at –20°C until needed.

15. Similar methods can be preformed using bacterial cells to produce enriched inner membranes that are compatible with the following procedures.

3.2. Protease-Accessible Peptides Preparation: High pH/Proteinase K Method

1. Pellet enriched membrane sample (500 μg protein) by centrifugation ($20,000 \times g$, 4°C, 30 min).

2. Resuspend membrane pellet at 1 mg/mL in 200 mM Na_2CO_3 (pH 11) and agitate on ice using an insulin syringe for five strokes every 15 min for 1 h (see Note 8).

3. Add solid urea to a final concentration of 8 M, vortexing until urea is completely dissolved.

4. Reduce the sample by adding dithiothreitol (DTT) to 5 mM and incubate at 60°C for 20 min with gentle shaking (see Note 9).

5. Alkylate the sample by adding iodoacetamide (IAA) to 15 mM and incubate at room temperature in the dark for 20 min (see Note 10).

6. Add recombinant Proteinase K to an enzyme to substrate ratio of 1:50 (w/w) and incubate the sample overnight at 37°C with shaking.

7. Double the sample volume with aqueous-organic buffer (10% acetonitrile, 2% formic acid) and incubate on ice for 30 min.

8. Pellet membrane-embedded peptides (MEPs) by centrifugation ($20,000 \times g$, 4°C, 30 min).

9. Collect the supernatant (protease-accessible peptides (PAPs)).

10. Store both MEPs and PAPs at –20°C until further preparation or analysis, respectively.

3.3. Protease-Accessible Peptides Preparation: Chymotrypsin Method

1. Pellet enriched membrane sample (500 μg protein) by centrifugation ($20,000 \times g$, 4°C, 30 min).

2. Resuspend pellet in 8 M urea/100 mM Tris–HCl (4 mg/mL).

3. Fully resuspend vesicles by passing through an insulin syringe.

4. Add 100 mM Tris–HCl for a final concentration of 2 M Urea/100 mM Tris–HCl (1 mg/mL).

5. Reduce the sample with 5 mM DTT and incubate at 60°C for 20 min with gentle shaking.

6. Alkylate with 15 mM IAA and incubate for 20 min in the dark at room temperature.

7. Add chymotrypsin at an enzyme to protein ratio of 1:50 (w/w).

8. Pass through an insulin syringe again to again resuspend vesicles and incubate overnight at 25°C.

9. Double the sample volume with aqueous-organic buffer (10% acetonitrile).

10. Pellet membrane-embedded peptides (MEPs) by centrifugation (20,000×g, 4°C, 30 min).

11. Collect the supernatant (protease-accessible peptides (PAPs)) and adjust to 5% formic acid.

12. Store both MEPs and PAPs at –20°C until further preparation or analysis, respectively.

3.4. Membrane Embedded Protein Preparation

1. In a fume hood, solubilize the membrane pellet in 10 μl of cyanogen bromide/90% formic acid (1:2 w/v) and incubate overnight (20°C, in dark).

2. Add 1 volume methanol and 18 volumes 100 mM Tris–HCl (pH 8.0) for a final 5% formic acid, 5% methanol, 90% 100 mM Tris–HCl.

3. Centrifuge (20,000×g, 4°C, 15 min) to pellet lipids.

4. Collect supernatant (MEPs) and store at –20°C until analysis.

3.5. Liquid Chromatography and Mass Spectrometry

Various LC and MS methods can be used with this protocol, depending what the sample is, and what the user is trying to achieve. However, there is one attribute that should be kept constant despite which MS method is used. Samples should be run at elevated temperature to alleviate the difficulties associated with their hydrophobicity. Here is an example of a LC/MS method.

1. Pull a 100-μm inner diameter fused silica column to a 5-μm tip using a laser puller.

2. Place the microcentrifuge tube containing a slurry of methanol and Aqua C18 reverse-phase resin in the high-pressure bomb and seal the lid.

3. Put the chromatography column in the top of the bomb with the tip of the column facing upward. Make sure the bottom of the column is not touching the bottom of the microcentrifuge tube containing the reverse-phase material. Apply 600 psi of pressure to the bomb (see Fig. 2a).

4. Verify that methanol is coming out of the tip of the column. It may be necessary to gently open up the column with a ceramic scribe.

5. While holding onto the column, gently loosen the ferrule holding the column by a quarter of a turn. This allows for the column to be manipulated without venting the bomb. Gently tap the column to the bottom of the microcentrifuge tube a few times. Packing the column a few millimeters at a time allows for more control and reproducible packing dimensions. When the column is not touching the base of the microcentrifuge tube, retighten the ferrule so that the column will stay in the bomb. The resin can be viewed traveling up the column (see Note 11).

6. Pack the column to 15 cm of reverse-phase material.

7. Equilibrate the column on a HPLC pump for a minimum of 30 min by running buffer A through it at a rate of 0.1 mL/min.

8. Thaw the sample and pressure load a 150-μg aliquot at 800 psi onto the loading column after equilibration.

9. After the sample is loaded, desalt it by washing for 10 min with buffer A (0.1 mL/min).

10. Place the column on a microspray source in-line with an ion trap mass spectrometer and align so the tip of the column is in-line with the heated source of the mass spectrometer (Fig. 2c).

11. Start running buffer A through the column at 0.1 mL/min. Adjust the waste line on the mass spectrometer to keep the pressure to 25–35 bar. Analyze the sample using a 2–4 h-long reverse-phase gradient. The acetonitrile gradient for each step of the MudPIT starts at 100% buffer A for 5 min and gradually decreases to 35% buffer A, 65% buffer B over 80 min. The gradient then returns to 100% buffer A over 10 min and remains at 100% buffer A for the rest of the 120 min.

12. Analyze the resulting tandem spectra.

4. Notes

1. Homogenization conditions reported here are optimized for HeLa cells. Homogenization conditions need to be optimized separately for each cell type or tissue.

2. It is important during cell fractionation to keep everything on ice to reduce protein degradation.

3. Over-homogenization of cells and tissue results in DNA contamination from broken nuclei. This will result in poor separation of membrane fractions.

4. Make sure that all interfaces are crisp and not diffuse. Handle the sucrose gradients gently to avoid disturbing the interfaces.

5. It is necessary to stop the centrifuge slowly (or leave the brake off when the centrifuge is stopping) to avoid disturbing the sucrose interfaces.

6. This step dilutes the sucrose enough so that the plasma membranes will pellet to the bottom instead of remaining suspended in the sucrose gradient.

7. This high pH treatment opens membrane structures and releases the soluble luminal proteins. This extraction method serves as an enrichment step for membranes.

8. High pH extraction with mechanical agitation causes membrane vesicles to open and form stable membrane sheets with free edges. This step is important to allow the protease access to both sides of the membrane.

9. The addition of DTT reduces disulfide bonds.

10. IAA is light sensitive, so thaw the aliquot in the dark. IAA alkylates sulfhydryl groups to inhibit the reformation of disulfide bonds.

11. A black backboard should be placed on the wall next to the bomb to facilitate viewing of the resin in the column.

References

1. Wu C, Yates JR (2003) The application of mass spectrometry to membrane proteomics. Nat Biotechnol 21:262–267

2. Speers A, Wu C (2007) Proteomics of integral membrane proteins theory and application. Chem Rev 107:3687–3714

3. Howell KE, Palade GE (1982) Hepatic Golgi fractions resolved into membrane and content subfractions. J Cell Biol 92(3): 822–832

4. Speers A, Blackler A, Wu C (2007) Shotgun analysis of integral membrane proteins facilitated by elevated temperature. Anal Chem 79(12):4613–4620

5. Snyder LR, Kirkland JJ, Glajch JL (1997) Practical HPLC method development, 2nd edn. Wiley, New York, pp 497–509

6. Dolan JW (2002) Temperature selectivity in reversed-phase high performance liquid chromatography. J Chromatogr A 965:195–205

Part II

Computational Methods for Prediction of Membrane Protein Structure and Dynamics

LITiCon: A Discrete Conformational Sampling Computational Method for Mapping Various Functionally Selective Conformational States of Transmembrane Helical Proteins

Supriyo Bhattacharya and Nagarajan Vaidehi

Abstract

G-Protein-coupled receptors (GPCRs) are seven helical transmembrane proteins that mediate cell signaling thereby controlling many important physiological and pathological functions. GPCRs get activated upon ligand binding and trigger the signal transduction process. GPCRs exist in multiple inactive and active conformations, and there is a finite population of the active and inactive states even in the ligand-free condition. An understanding of the nature of the conformational ensemble sampled by GPCRs and the atomic level mechanism of the conformational transitions require a combination of computational methods and experimental techniques. We have developed a coarse grained discrete conformational sampling computational method called "LITiCon" to map the conformational ensemble sampled by GPCRs in the presence and absence of ligands. The LITiCon method can also be used to predict functional selective conformational states starting from the inactive state of the receptor.

LITiCon has been applied to map the conformational ensemble of β2-adrenergic receptor, a class A GPCR. We have shown that β2-adrenergic receptor samples a larger conformational space in the ligand-free state and that different ligands select and stabilize conformations from this ensemble. In this review we describe the LITiCon method in detail and elucidate the uses and pitfalls of this method.

Key words: LITiCon, Agonist-induced conformational changes, GPCR, Coarse grain method

1. Introduction

G-Protein-coupled receptors (GPCRs) are seven helical transmembrane (TM) proteins that transduce signals from outside to the inside of the cell. GPCRs form one of the largest family of drug targets. Ligands, ranging from small molecules to large proteins, bind and activate the receptors. GPCR structures are dynamic, and various active and inactive state conformations exist in a dynamic equilibrium even in a ligand-free state. Each of these conformational states has a functional role in coupling to a variety of extracellular ligands

Nagarajan Vaidehi and Judith Klein-Seetharaman (eds.), *Membrane Protein Structure and Dynamics: Methods and Protocols*,
Methods in Molecular Biology, vol. 914, DOI 10.1007/978-1-62703-023-6_10, © Springer Science+Business Media, LLC 2012

and intracellular proteins, important for functional selectivity (1–3). Agonists are ligands that bind and activate the receptors. It has been shown that agonists selectively stabilize one or more of the conformational states that are sparsely populated in the receptor conformational equilibrium without the agonist (4).

In the last several years, experiments such as fluorescence lifetime analysis have been used to characterize certain features of the agonist-stabilized states, but the activation mechanism that emerges from these observations is still fragmented. Recently, crystal structures of several agonist-bound GPCRs have been reported (5–7) which give valuable insights into agonist selective conformations and the conformational changes that occur upon activation by an agonist. The agonist-bound states are highly dynamic, and the crystal structures represent just one densely populated low energy conformational ensemble in the whole range of ensembles sampled by the receptor. The efficacy of the agonist to activate a receptor not only depends on the lowest energy most populated state but also on the range of conformations that it samples (8). Thus crystal structures are a good starting point to understand the dynamics of a receptor. More insight is needed to understand the activation mechanism than what is provided by the static crystal structures.

Classical all-atom molecular dynamics (MD) methods can explore only a limited conformational space and are inadequate in sampling the larger backbone motions associated with receptor activation, possibly due to energy barriers that exist between active and inactive states. Biased MD methods require a prior knowledge of the active state structure, and coarse-grained MD methods that coarsen the force field are not accurate enough to map the conformational transitions that emerge from atomic details. Thus there is a vital need for advanced computational methods, which can bridge the gaps among experimental observations to give a more complete picture of the activation process in GPCRs.

The LITiCon computational method is a discrete conformational sampling method in the rigid body degrees of freedom for the seven TM helices in GPCRs. In the LITiCon method each TM helix is treated as a rigid body, with seven rigid helices connected by flexible loops. The method involves generating conformations of the receptor with or without ligand bound, by systematically and simultaneously sampling the six rigid body degrees of freedom for all the seven transmembrane helices. Since most of the conformational differences observed between the crystal structures of the active and inactive state of a GPCR can be projected on to the rigid body rotation and tilt degrees of freedom, one can limit the conformational search to just the helical rotation and tilt degrees of freedom. The results from LITiCon provide a mapping of the ligand selective conformational states and a minimum energy pathway for going from the inactive to the ligand-stabilized conformational state (3, 9–15). Here we describe various steps involved in applying the method to understand the activation mechanisms of GPCRs. Starting from a ligand docked receptor model or a ligand bound crystal structure, the LITiCon method generates multiple receptor conformations by sampling the axial rotations of the TM helices that are likely to undergo conformational changes upon activation (to be determined from experimental or mechanistic evidences). Each resulting conformation is then energy minimized, and properties such as total potential energy, ligand binding energy, and number of interhelical hydrogen bonds are calculated. The final ligand-stabilized state is selected from these ensemble of conformations using metrics such as low energy, high number of hydrogen bonds, and experimental information regarding the active state (e.g., mutagenesis). The details of the LITiCon method are discussed in Subheading 3. The LITiCon method thus generates the potential energy surface in the rigid body degrees of freedom that would provide insights into the possible kinetic intermediate states in the minimum energy pathway.

Application and Success of the LITiCon Method: The method has been applied for studying activation mechanisms of rhodopsin (12) and human β2-adrenergic receptor (β2-AR) (11). Here we briefly describe the type of results we got using LITiCon for predicting the active state structure and studying the mechanism of β2-AR activation. Starting from the crystal structure of the inactive state of β2-AR, we docked the agonist norepinephrine, partial agonist salbutamol, and weak partial agonist dopamine using ligand docking methods (11). We then performed the LITiCon procedure as described in this review for TM helices 3, 5, and 6. The global minimum in binding energy was chosen as the ligand-stabilized active state for each ligand studied here. Figure 1a shows the computed binding potential energy landscape for β2-AR bound to the full agonist norepinephrine. Starting from the inactive (carazolol bound crystal, shown as red dot in Fig. 1a) state, we select the receptor conformation that shows the lowest agonist binding energy, as the norepinephrine stabilized β2-AR conformation (green dot in Fig. 1a). This receptor conformation is characterized by an inward movement of three serine residues on TM5, which improve the contact with the agonist, in agreement with experiments (16). The activation pathway, which is the minimum energy pathway going from the inactive to the active state structure, was calculated using a coarse-grained Monte Carlo method (11). Figure 1a shows the sampled conformations along the minimum energy pathway or the activation pathway (shown by black dots), and the average pathway (red dotted line). Figure 1b shows the percent conformational change (extent of movement of the helices) (11) as a function of Monte Carlo step. It is seen that the activation pathway traverses through a stable intermediate shown by the yellow dot in Fig. 1a, b. In the intermediate conformation, the β-OH group of norepinephrine faces D113 on TM3 (Fig. 1c)

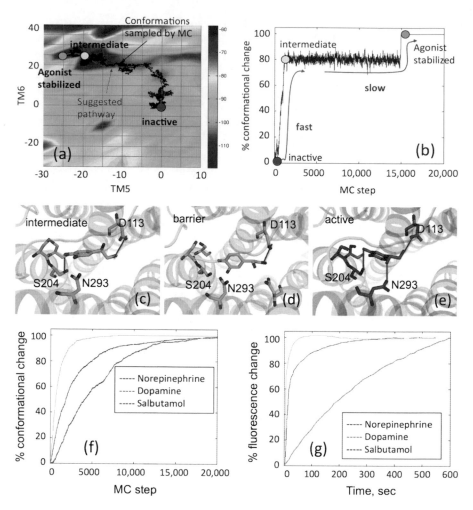

Fig. 1. Analysis of β2-AR activation using LITiCon. (**a**) Binding potential energy landscape of norepinephrine bound β2-AR. *Blue* and *red* regions represent conformations with favorable and unfavorable energy, respectively. The *black points* are the conformations sampled using Monte Carlo while computing the activation pathway. *Red dotted line* represents the average activation pathway determined from the sampled conformations. (**b**) Conformational change as a function of Monte Carlo step; (**c–e**) Binding pocket of norepinephrine in intermediate, barrier, and active β2-AR conformations; (**f**) Percent conformational change as a function of Monte Carlo step for β2-AR agonists computed using LITiCon; (**g**) Experimental fluorescence change as a function of time measured for agonist-bound β2-AR.

and in the final agonist-stabilized state, the β-OH group faces N293 on TM6 (Fig. 1e). The LITiCon computations suggest that activation by norepinephrine consists of a fast step (inactive → intermediate) followed by a slow step (intermediate →agonist-stabilized state) (Fig. 1b). The slow transition from intermediate to the agonist stabilized state is characterized by an energy barrier, where the hydrogen bond between β-OH and D113 is disrupted and new hydrogen bond formed with N293 (Fig. 1d). We performed these analyses for other β2-AR agonists such as dopamine (weak partial agonist) and salbutamol (strong partial agonist). Unlike norepinephrine, the activation pathway of dopamine showed only a fast step, while that of salbutamol showed only a slow step. We validated these findings by estimating the average conformational change as a function of time for multiple populations of receptors, as shown in Fig. 1f, where time is equivalent to the number of Monte Carlo steps. The longer lifetime of a conformational state in the activation pathway will be reflected in the population of that state sampled by the Monte Carlo method. These findings showed qualitative agreement with fluorescence intensity measurements on agonist-bound β2-AR (17, 18) (Fig. 1g). The percentage fluorescence change is a measure of the conformational change predicted by LITiCon in Fig. 1e. Thus LITiCon can be a valuable tool for analyzing the conformational change and activation pathways in GPCRs, and for prediction of active state structure of a GPCR from the inactive state structure.

2. Materials

2.1. Hardware

Computer system: Pentium IV or faster CPU with 1+ GB RAM, Ethernet card (see Note 1), high resolution color monitor for viewing protein conformations.

2.2. Software

1. Operating system: Linux/Unix (LITiCon has been tested on Redhat Enterprise Linux and Sun OS).
2. LITiCon software can be obtained free of cost for academic use by contacting the corresponding author.
3. LITiCon dependencies: CHARMM/NAMD for structure minimization and calculating energy, scwrl for side chain optimization, Perl, VMD/PyMol for preparing/viewing protein structures, Matlab for analysis of results.

3. Methods

The starting conformation for the LITiCon procedure can be a crystal structure of a ligand bound receptor or a structural model of the receptor–ligand complex generated by any computational method. We first identify the TM helices that are in direct contact with the ligand (by visually analyzing the receptor structure) and are likely to undergo conformational changes due to ligand binding. The TM helices can be assigned from the crystal structure or any method of prediction of the transmembrane regions. The kinematic model of the receptor conformations are generated through simultaneous systematic spanning of the rotational orientation of TM helices through a grid of ±5° with respect to the initial state. The range of rotation angle chosen for each TM helix is left to the user. In most cases, starting from the inactive structure, a rotation range of ±30° is sufficient to cover the active-like receptor conformations in the sampled ensemble (11–13). For each conformation thus generated, the following steps are performed:

- Optimization of all side chain conformations using SCWRL 3.0 (19).

- Conjugate gradient minimization of the potential energy of the ligand in the field of the rest of protein fixed until convergence of 0.3 kcal/mol Å RMS deviation in force per atom is achieved.

- Calculation of the ligand binding energy defined as the difference of the potential energy of the ligand with protein fixed and the potential energy of the free ligand calculated in water using Generalized Born solvation method (20).

- Interhelical and ligand–receptor hydrogen bonds using HBPLUS 3.0 (21).

This generates a multidimensional binding energy landscape in the rotational degrees of freedom of the TM helices. Next we identify all the local minima in this landscape and sort them by total number of interhelical hydrogen bonds (HB) and ligand–receptor HB and then by ligand binding energy. The final ligand-stabilized receptor structural model is selected based on low binding energy and high number of HBs.

3.1. Installation

1. Designate a folder in the home directory where you will install LITiCon, known henceforth as the liticon root directory. Copy all files and directories in the installation package to the liticon root directory.

2. Set *LITICON_NAMD_PATH* environmental variable to the "*bin*" folder inside liticon root directory. Way to set environment variable depends on the shell environment. Please consult your system administrator on how to set environment variable.

3. Create a softlink to scwrl4 inside *<liticon-root>/bin/pdb-scwrl/scwrl4*.

4. Copy or create a softlink to namd2 inside folder *<liticon-root>/bin/namd*.

3.2. Input Structure Preparation

1. The input to LITiCon is the initial receptor conformation along with the docked ligand or the crystal structure wherever available. The ligand can be docked using any docking program of users' choice. Once the receptor–ligand complex is generated and optimized, the topology file for the ligand should be generated. The topology file needs to be in the

starting residue no. for TM1

```
NT        1 MGQPGNGSAFLLAPNRSHAPDHDVTQQRDEV 31 (31)

TM 1     32 WVVGMGIVMSLIVLAIVFGNVLVITAIAK 60 (29)          length of TM1

LP 1     61 FERLQ 65 (5)

TM 2     66 TVTNYFITSLACADLVMGLAVVPFGAAHI 94 (29)    end residue no. for TM1

LP 2     95 LMKMWTFG 102 (8)

TM 3    103 NFWCEFWTSIDVLCVTASIETLCVIAVDRYFAI 135 (33)

LP 3    136 TSPFKYQSLLTKNK 149 (14)

TM 4    150 ARVIILMVWIVSGLTSFLPIQMHW 173 (24)

LP 4    174 YRATHQEAINCYANETCCDFFT 195 (22)

TM 5    196 NQAYAIASSIVSFYVPLVIMVFVYSR 221 (26)

LP 5    222 VFQEAKRQLQKIDKSEGRFHVQNLSQVEQDGRTGHGLRRSSK 263 (42)

TM 6    264 FCLKEHKALKTLGIIMGTFTLCWLPFFIVNIVHVI 298 (35)

LP 6    299 QDNLIR 304 (6)

TM 7    305 KEVYILLNWIGYVNSGFNPLIYC 327 (23)

CT      328 RSPDFRIAFQELLCLRRSSLKAYGNGYSSNGNTGEQSGYHVEQEKENKLLCEDLPGTEDFVGHQGTVPSDNIDSQGRNCSTNDSLL
```

Fig. 2. Format of text file defining the TM domains. NT and CT represent the N and the C termini, respectively.

standard CHARMM22 format. Next generate a set of pdb and psf files for the protein–ligand complex using the CHARMM topology file. The psf file can be generated using the psfgen module in VMD.

2. Create a text file containing the definitions for the TM domains as shown in Fig. 2. The definition of the TM domains can come from the crystal structure, or any other source defined by the user.

3. Create a text file containing the hydrophobic center for each TM helix as shown in Fig. 3. The hydrophobic center is the residue numbers at the maximum hydrophobicity along each helix, counting from the N-terminus of the helix, the first N-terminal residue being 1. The default is to enter the number "0" for each helix and this would use the geometric center of the helix. See Note 1.

3.3. Running LITiCon

1. Copy the following input files to the current working directory.

   (a) receptor-ligand pdb and psf

   (b) CHARMM parameter file

   (c) File containing the definitions of TM domains

2. Copy the perl scripts "*liticon-namd.pl*", "*join-scwrl.tcl*", and "*join.pl*" to the working directory.

3. LITiCon is usually run from inside a cluster environment using the command: *./liticon-namd.pl −f protein.psf −b protein.pdb −d TMs.doc −c HPMCenter.txt −p <no. of cpus> −j join-scwrl.tcl −l UNK −r "0 0 0 0 -30 30 0 0 -30 30 -30 30 0 0" −i "0 0 10 0 10 10 0" −s setup* (see Note 2).

HELIX1 21.000

HELIX2 14.000

HELIX3 13.000

HELIX4 21.000

HELIX5 19.000

HELIX6 19.000

HELIX7 22.000

Fig. 3. Format of the text file HPMCenters.txt, the file that is used to calculate the axes of rotation for each helix.

Here, "*TMs.doc*" is a text file containing the TM definitions (Fig. 2), "*HPMCenter.txt*" contains the helical axes (Fig. 3), "*no. of cpus*" denotes the number of processors to be used for the calculations. Since these algorithms are simply parallel, the computational time scales linearly with the number of processors. The recommended number of processors is 32–48 cpus for a typical LITiCon job. "UNK" represents the residue name of the ligand. You can set it to any desired name. The numbers following "*–r*" represents the starting and ending rotation angles for each helix. This is the range of rotation angles chosen for each TM helix. Each adjacent pair of numbers represents the range of rotation angles for helical domains from TM1 to TM7. Having "*0 0*" for a particular helix means that helix will not be rotated. For example, in the above sample command, TM1, TM2, TM4, TM7 are not rotated, TM3, TM5, TM6 are rotated from –30° to +30°. The numbers following "*–i*" represent the increments in rotation for each helix. In the sample command, TMs 3, 5, 6, are rotated by 10°. This allows each helix to be rotated to different extent and over different size rotation angle grids.

4. Once the liticon job is finished, the results are stored in the directories *dir1*, *dir2*, etc. within the working directory. These results must be copied or moved from the individual directories to a final location for later analysis. This task is performed using the script "*join.pl*". This script is run as: "*./join.pl 1 50*", which combines the data from *dir1* to *dir50* (for example) and stores them in a new directory called "*final*".

5. Copy the script <*liticon-root*>*/tools/bind-energy.pl* to the "*final*" directory. Run the script as "*./bind-energy.pl ./*" to calculate the binding energy from the existing energy data.

3.4. Analysis of Results

1. The "final" directory contains the following files:

   (a) "*pref-*-*-*-*-*-*-*.bz2*"; These are bzip2 files containing the protein–ligand complex structures obtained after helical rotations and subsequent side chain optimization and energy minimization. There will be one bzip2 file for every conformation representing a rotational combination, where the name of the file indicates the rotation angles for each helix. For example, "*protein-0-0--10-0-30--20-0.bz2*" represents the structure with no rotations for TM1, TM2, TM4, TM7, –10° for TM3, 30° for TM5, and –20° for TM6. Here we assume that the starting structures are called "*protein.pdb*" and "*protein.psf*". The number of bzip2 files should be equal to the number of rotation combinations calculated by LITiCon.

   (b) "*protenergy.dat*"; This file contains the individual components of the protein energy listed for each rotated conformation. The data columns are:

   Rotations Frame Time Bond Angle Dihed Impr Elec vdW Conf Nonbond Total

   (c) "*protliginterenergy.dat*"; This file contains the protein–ligand interaction energy components. The format of data columns is:

   Rotations Frame Time Elec vdW Nonbond Total

   (d) "*mid-HB-final.dat*" and "*total-HB-final.dat*"; These files contain the number of interhelical and protein-ligand HBs for each structure. "*mid-HB-final.dat*" contains the number of HBs formed in the middle parts of the transmembrane domains.

2. The analysis of the results is performed within Matlab environment, using two main scripts: "*minima.m*" and "*plotany.m*". Create a new "*analysis*" directory and copy the data files from "final" directory and the two Matlab scripts to this folder.

3. Edit the "*minima.m*" script to set the proper parameters for analysis. *n1, n2, ... n7* are the number of rotations for each TM helix. This number can be determined as

$$n = \frac{end\_rotation\_angle - start\_rotation\_angle}{increment} + 1$$

starta, startb, ... startg represent the starting rotation angles of TM1, TM2,...,TM7, respectively. The values of *starta, startb, ... startg* have to be modified according to your system and set the value of "*div*", which is the increment in rotation.

4. Run the "*minima.m*" script in Matlab. This will take couple of minutes depending on your system size. The "*minima.m*" script will analyze the binding energy landscape and identify all the local minima. It will then sort these minima by binding

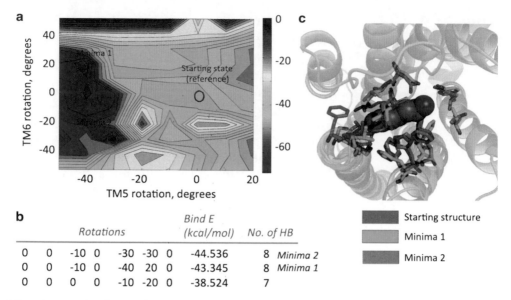

		Rotations					Bind E (kcal/mol)	No. of HB	
0	0	-10	0	-30	-30	0	-44.536	8	*Minima 2*
0	0	-10	0	-40	20	0	-43.345	8	*Minima 1*
0	0	0	0	-10	-20	0	-38.524	7	

Fig. 4. Sample output data from LITiCon. (**a**) Binding energy landscape. The starting structure (0, 0) and the two top ranking minima are *highlighted*; (**b**) List of minima with binding energy and number of HBs; (**c**) starting receptor structure with the conformations corresponding to the two minima superimposed.

energy and number of HBs, as explained in Subheading 1. The list of minima will be stored in a text file called "*results. dat*", along with the energy and number of HBs for each minima.

5. The energy landscape can be visualized using the script "*plotany.m*". Since the energy landscape is multidimensional, this script displays 2D sections of the landscape, where two of the helices are rotated keeping the rest of the helices fixed. "*plotany.m*" can be edited to specify the helices to be rotated and to set the rotation values for the fixed helices. A sample binding energy landscape is shown in Fig. 4a.

6. Figure 4a shows the 2D section of the binding energy landscape for rotations of TM5 and TM6. In this landscape, rotation of TM3 is fixed at $-10°$. This landscape shows two major minima highlighted in Fig. 4a. Figure 4b shows the list of minima computed by "*minima.m*". Figure 4c shows the starting structure of the receptor with the conformations corresponding to the two minima overlaid on one another. In order to select the final agonist-stabilized state, first select the top ranking minima from the list (Fig. 4b) and then visualize the protein conformations for changes in ligand orientation and inter-residue and ligand-residue contacts with respect to the starting structure. The final agonist-stabilized state is then selected based on available mutation data or other experimental information regarding the active state, such as inter-residue distance constraints (see Note 3).

4. Notes

1. The choice of the hydrophobic centers of each helix is used to calculate the axis of rotation for each helix. The axis is calculated for the total of 14 residues, 7 above and below the hydrophobic center.

2. Ideally, LITiCon can be run on a single CPU. In practice, however, running on a single CPU is only feasible for very limited number of helical rotations. It is recommended to invest in a cluster of 32+ CPUs. LITiCon does not use inter-processor communication. So regular Ethernet cards in conjunction with a network switch is sufficient.

3. LITiCon is a computational tool that allows one to explore the conformational landscape of a GPCR and study how ligands modulate this landscape. The conformational states obtained from LITiCon need to be analyzed in conjunction with experimental data in order to achieve meaningful results. In this context, it is important to remember that the most stable minimum by energy may not always correspond to the correct global minimum of the ligand bound receptor. This can happen due to imperfections in the classical force field, limitations of receptor minimization and side chain reassignment, and approximations regarding solvent and membrane environment. In that case, the other minima in the list should also be carefully considered. Often, experimental data such as specific mutations that affect ligand interaction need to be used to eliminate some of the minima that do not satisfy the experimental constraints. A thorough knowledge of the receptor structure is also important for correctly identifying the ligand-stabilized state. LITiCon method should be used as a toolbox to identify ligand stabilized states.

References

1. Kenakin TP (2007) Collateral efficacy in drug discovery: taking advantage of the good (allosteric) nature of 7TM receptors. Trends Pharmacol Sci 28:407–415

2. Mailman RB, Murthy V (2010) Ligand functional selectivity advances our understanding of drug mechanisms and drug discovery. Neuropsychopharmacology 35:345–346

3. Vaidehi N, Kenakin T (2010) Conformational ensembles of seven transmembrane receptors and their relevance to functional selectivity. Curr Opin Pharmacol 10:775–781

4. Niesen MJM, Bhattacharya S, Vaidehi N (2011) The role of conformational ensembles in ligand recognition in G-protein coupled receptors. J Am Chem Soc 133:13197–13204

5. Rasmussen SG, Choi HJ, Fung JJ, Pardon E, Casarosa P, Chae PS et al (2011) Structure of a nanobody-stabilized active state of the β_2 adrenoceptor. Nature 469:175–180

6. Warne T, Moukhametzianov R, Baker JG, Nehmé R, Edwards PC, Leslie AG et al (2011) The structural basis for agonist and partial agonist action on a β_1-adrenergic receptor. Nature 469:241–244

7. Xu F, Wu H, Katritch V, Han GW, Jacobson KA, Gao Z, Cherezov V, Stevens RC (2011) Structure of an agonist bound human A2A adenosine receptor. Science 332:322–327

8. Landes CF, Ramabhadran A, Taylor JN, Salatan F, Jayaraman V (2011) Structural landscape of isolated agonist-binding domains

from single AMPA receptors. Nat Chem Biol 7:168–173

9. Balaraman G, Bhattacharya S, Vaidehi N (2010) Structural insights into conformational stability of wild type and mutant β_1-adrenergic receptor. Biophys J 99:568–577

10. Bhattacharya S, Subramanian G, Hall S, Lin J, Laoui A, Vaidehi N (2010) Allosteric antagonist binding sites in class B GPCRs: corticotropin receptor 1. J Comput Aided Mol Des 8: 659–674

11. Bhattacharya S, Vaidehi N (2010) Computational mapping of the conformational transitions in agonist selective pathways of a G-protein coupled receptor. J Am Chem Soc 132:5205–5214

12. Bhattacharya S, Hall SE, Vaidehi N (2008) Agonist-induced conformational changes in bovine rhodopsin: insight into activation of G-protein-coupled receptors. J Mol Biol 382: 539–555

13. Bhattacharya S, Hall SE, Li H, Vaidehi N (2008) Ligand-stabilized conformational states of human β2 adrenergic receptor: insight into G-protein-coupled receptor activation. Biophys J 94:2027–2042

14. Lam AR, Bhattacharya S, Patel K, Hall SE, Mao A, Vaidehi N (2011) Importance of receptor flexibility in binding of cyclam compounds to the chemokine receptor CXCR4. J Chem Inf Model 51:139–147

15. Vaidehi N (2010) Dynamics and flexibility of G-protein coupled receptor conformations and its relevance to drug design. Drug Discov Today 15:951–957

16. Kikkawa H, Kurose H, Isogaya M, Sato Y, Nagao T (1997) Differential contribution of two serine residues of wild type and constitutively active β2-adrenoceptors to the interaction with β2-selective agonists. Br J Pharmacol 121:1059–1064

17. Swaminath G, Xiang Y, Lee TW, Steenhuis J, Parnot C, Kobilka BK (2004) Sequential binding of agonists to the β2 adrenoceptor: kinetic evidence for intermediate conformational states. J Biol Chem 279:686–691

18. Swaminath G, Deupi X, Lee TW, Zhu W, Thian FS, Kobilka TS, Kobilka BK (2005) Probing the β2 adrenoceptor binding site with catechol reveals differences in binding and activation by agonists and partial agonists. J Biol Chem 280:22165–22171

19. Canutescu AA, Shelenkov AA, Dunbrack RL Jr (2003) A graph-theory algorithm for rapid protein side-chain prediction. Prot Sci 12: 2001–2014

20. Zamanakos G (2001) A fast and accurate analytical method for the computation of solvent effects in molecular simulations. Ph. D. thesis. Caltech, Pasadena

21. McDonald IK, Thornton JM (1994) Satisfying hydrogen bonding potential in proteins. J Mol Biol 238:777–793

Chapter 11

Homology Model-Assisted Elucidation of Binding Sites in GPCRs

Anat Levit, Dov Barak, Maik Behrens, Wolfgang Meyerhof, and Masha Y. Niv

Abstract

G protein-coupled receptors (GPCRs) are important mediators of cell signaling and a major family of drug targets. Despite recent breakthroughs, experimental elucidation of GPCR structures remains a formidable challenge. Homology modeling of 3D structures of GPCRs provides a practical tool for elucidating the structural determinants governing the interactions of these important receptors with their ligands. The working model of the binding site can then be used for virtual screening of additional ligands that may fit this site, for determining and comparing specificity profiles of related receptors, and for structure-based design of agonists and antagonists. The current review presents the protocol and enumerates the steps for modeling and validating the residues involved in ligand binding. The main stages include (a) modeling the receptor structure using an automated fragment-based approach, (b) predicting potential binding pockets, (c) docking known binders, (d) analyzing predicted interactions and comparing with positions that have been shown to bind ligands in other receptors, (e) validating the structural model by mutagenesis.

Key words: Docking, GPCR model, Binding site, Agonist, Antagonist, Taste receptors, Broad tuning, Receptor range, Molecular recognition

1. Introduction

G protein-coupled receptors (GPCRs) are the largest family of membrane proteins serving as key signal-transduction proteins and representing a major class of drug targets (1). Recent breakthroughs in GPCR crystallography (2–5) provide exciting opportunities for structure-based drug design methods that can now use increasingly reliable homology models of GPCR targets (6, 7). Successful computational models of GPCRs have been used for virtual screening, enriching the rate of ligand hits relative to a random collection of compounds, with hit rates ranging from 3 to 21% (8, 9), comparable to virtual screening success rates with X-ray

Nagarajan Vaidehi and Judith Klein-Seetharaman (eds.), *Membrane Protein Structure and Dynamics: Methods and Protocols*, Methods in Molecular Biology, vol. 914, DOI 10.1007/978-1-62703-023-6_11, © Springer Science+Business Media, LLC 2012

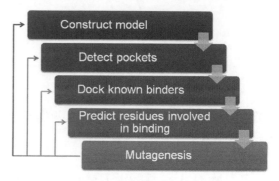

Fig. 1. Scheme of binding-site elucidation procedure.

structures (8). Furthermore, research aimed at elucidating the underlying principles determining the molecular responsiveness range of GPCRs that mediate senses, such as odor (10) and taste (11, 12) receptors, depends on the ability to build reliable models of the interaction sites. A crucial step in understanding specificity and promiscuity in molecular recognition and structure-based design is to identify the residues that are important for ligand binding.

Sequence-based classification systems have been developed to facilitate the analysis of GPCRs, the two most widely used being the *GRAFS* (13) and the UIPHAR (14). The *GRAFS* classifies GPCRs into families: *R*hodopsin (which corresponds to UIPHAR class A), *S*ecretin (UIPHAR class B), *A*dhesion (UIPHAR class B), *G*lutamate (UIPHAR class C), and *F*rizzled/Taste2. While researchers have successfully applied homology-based GPCR structure modeling approaches to ligand-binding elucidation similar to the one described below to understanding the role of modulators in class C (15) and class B GPCRs (9, 16), this review is focused on application of homology-based models primarily to class A (17–19) and Taste2 (T2R) GPCRs (12, 20, 21).

Several approaches to modeling GPCRs have been described (7), including *ab initio* (8, 22) and template based (6, 19, 23). Following previous work (7, 12), here we illustrate the use of a fragment-based approach, using the I-TASSER server (24) for homology modeling of the human bitter-taste receptor hTAS2R46, which belongs to the Frizzled/Taste2 family. The steps toward homology model-assisted elucidation of the binding-site residues discussed in this protocol are described in the flowchart in Fig. 1.

2. Materials

2.1. Sequence of the GPCR Being Studied

The protein sequence of interest may be obtained via NCBI (http://www.ncbi.nlm.nih.gov/protein) or Uniprot (http://www.uniprot.org/) databases, or from the GPCRDB database (http://www.gpcr.org/7tm/).

2.2. Ballesteros–
Weinstein Numbering

To facilitate comparisons between different GPCRs, we use Ballesteros–Weinstein (BW) numbering (25), in which the most conserved residue in a given transmembrane (TM) domain X is assigned the index X.50, and the remaining TM residues are numbered relative to this position. For example, the most conserved position in TM6 is designated 6.50.

2.3. A List of Known
Binders

To dock ligands into the predicted binding site, a list of ligands that are known to bind or activate (known binders or activators) the receptor is needed. These data may be obtained from the GLIDA database, which provides interaction data between GPCRs and their ligands along with chemical information on the ligands (26), directly from the literature, or experimentally.

2.4. Servers Used
for Modeling
in This Protocol

Structure modeling: I-TASSER (http://zhanglab.ccmb.med.umich.edu/I-TASSER).

Cavity detection: QSiteFinder (http://www.modelling.leeds.ac.uk/qsitefinder/).

PocketFinder (http://www.modelling.leeds.ac.uk/pocketfinder/).

Cavity detection allows for the identification of nonorthodox and allosteric binding sites (see also (27)).

2.5. Mutation
Databases

- TinyGRAP database (28) (http://www.cmbi.ru.nl/tinygrap/search/).
- Some mutations can be extracted using GPCRDB (http://www.gpcr.org/7tm/) or the MuteXt repository of mutations (29).

2.6. Structure-Based
Sequence Alignment

TCoffee (http://tcoffee.vital-it.ch/cgi-bin/Tcoffee/tcoffee_cgi/indcx.cgi).

2.7. Ligand
Preparation
and Docking

Software

The Discovery Studio 2.5 software package (Accelrys, Inc.) is used in this protocol as follows:

- Generation of Ramachandran plots of the modeled protein to assess its quality by checking the predicted torsion angles.
- Ligand preparation prior to docking experiments, using the "Prepare Ligands" protocol, which removes duplicate structures, enumerates isomers and tautomers, sets standard formal charges on common functional groups, sets ionization states at a given pH range, and generates 3D conformations.
- Docking of small molecule ligands to the receptor. In this protocol we use the "Flexible Docking" option, which allows for some receptor flexibility during docking of flexible ligands. A typical docking simulation of one compound using the "Flexible Docking" module requires 10 min on a Pentium 4,

2 GB RAM, 2.8 GHz dual core computer (for precomputed protein conformations, and program ran with parallel processing).

- Docking pose analysis is performed using the "Analyze ligand poses" protocol. The protocol enables calculation of the RMSD of the poses to each other or to a reference pose, identification of ligand–receptor hydrogen bonds at varying degrees of detail, and analysis of contacts between the ligand and the receptor (including clashes).

- Plotting a ligand–receptor interaction diagram in 2D using the "draw ligand interaction diagram" tool in the "analyze binding site" module.

Hardware

For best performance, the Discovery Studio software should be installed on a server with an Intel-compatible ≥2 GHz processor with x86 or x86_64 architecture, and an SGE 6.1 grid engine. A minimum of 2 GB of memory for Discovery Studio Client and 2 GB for the Pipeline Pilot Server is required. Ideally, a total of 4 GB should be available if the client and server are installed on the same machine.

2.8. Site-Directed Mutagenesis and Functional Assays

The computational process described in this protocol generates a structural model for the GPCR and the ligand binding site that is next validated using site-directed mutagenesis work. The cDNAs spanning the coding region of the receptor of interest should be cloned into a vector that allows expression in eukaryotic cell lines. Oligonucleotides for mutagenesis and vector-specific primers can be ordered online. Thermostable DNA polymerase with proofreading activity such as *Pfu*-DNA polymerase is preferred. Deoxyribonucleotides (dGTP, dATP, dTTP, and dCTP): prepare a stock solution of, e.g., 2.5 mM each (10 mM total) and store in aliquots at –20°C. There are no specific requirements for the thermocyclers used for the polymerase chain reaction (PCR) amplification.

For agarose gel electrophoresis, use high-quality agarose suitable for DNA-fragment recovery. After gel electrophoresis, the DNA bands of interest are excised from the ethidium bromide-stained gels on a UV-transilluminator (eye protection is necessary!) and purified using a commercially available spin column purification kit. Appropriate restriction endonucleases, T4-DNA ligase, and chemically competent bacterial cells are required for subcloning. For the transfection of eukaryotic cells, use highly purified plasmid DNA of mutated constructs which have been analyzed by DNA sequencing to confirm their integrity.

2.9. Functional Heterologous Expression

We use cells of the human embryonic kidney cell line HEK 293 T stably expressing the G protein chimera Gα16gust44 for functional expression. High-quality fetal bovine serum, Dulbecco's Modified Eagle's Medium (DMEM), and a sterile workbench, as well as incubators providing constant temperature (37°C), humidity, and CO_2 levels (5%), are necessary for cultivation of cells. For materials used in calcium-imaging assays, see Subheading 3.

3. Methods

We illustrate the flowchart in Fig. 1 using the hTAS2R46 example, generally following the steps performed in our recent paper (12):

1. Obtain the sequence of the hTAS2R46 receptor from the Uniprot database (accession no. P59540).

2. Submit to I-TASSER Web site in order to generate a homology model of the protein: go to the I-TASSER Web page (http://zhanglab.ccmb.med.umich.edu/I-TASSER/) and follow these steps:

   (a) Copy-paste the protein sequence onto the provided form or directly upload the sequence from a file.

   (b) Provide an e-mail address where results will be sent upon job completion.

   (c) Provide a name for the protein (optional).

   (d) Specify additional restraints to guide I-TASSER modeling (optional): a file containing contact/distance restraints may be optionally uploaded by the user, as well as specification of the template to be used during the modeling process (this can be achieved by either PDB code specification or by uploading a PDB file, with or without an alignment file; see below).

   (e) The user may also exclude some homologous templates from the I-TASSER template library (optional). This new option is most useful for validation studies, such as testing homology model performance when an experimental structure is available in the database, or when a particular structure is to be excluded from the templates due to its quality, conformational state, or other reasons.

   (f) To submit the sequence for modeling, click the "Run I-TASSER" button. The browser will be directed to an acknowledgment page that will display confirmation of the submitted sequence, a job identification number, restraint information, and a link to the page that will contain the detailed results when the job is complete.

An example of a model generated by I-TASSER for hTAS2R46 is shown in Fig. 2a. This model is based on the (automatically selected) X-ray structures of human β2-adrenergic receptor (β2ADR; PDB code 2RH1), turkey β1-adrenergic receptor (2VT4), human A2A adenosine receptor (3EML), squid rhodopsin (2Z73), and bovine rhodopsin (1l9h).

3. The generated model is validated by examining the Ramachandran plot. Such plots may be generated using a stand-alone modeling software package, such as Discovery Studio 2.5 (Accelrys, Inc.), or by use of different Web servers, such as PROCHECK (http://www.ebi.ac.uk/thornton-srv/software/PROCHECK/). Figure 2b shows a Ramachandran plot generated for the hTAS2R46 model using PROCHECK, where dihedral angles for most residues appear in the core ("allowed") regions of the plot, as expected.

4. Validate the model by comparing the most conserved residue positions in GPCRs named as BW positions with template structures. A structure-based multiple-sequence alignment of the Class A X-ray structures may be obtained by the structure-based sequence alignment option on the TCoffee Web server (http://tcoffee.vital-it.ch/cgi-bin/Tcoffee/tcoffee_cgi/index.cgi); Fig. 3. As expected, the most conserved positions in each TM helix (BW X.50 positions) are in alignment. Next, the model can be aligned to this multiple sequence alignment using the Combine Option on the TCoffee server. All (within

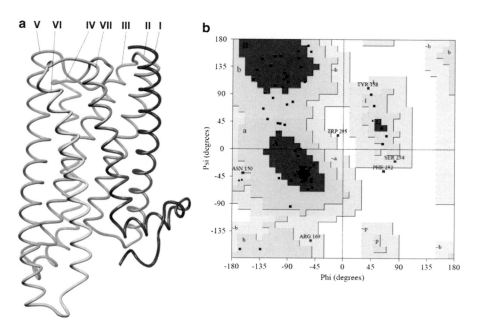

Fig. 2. Model and model quality assessment.

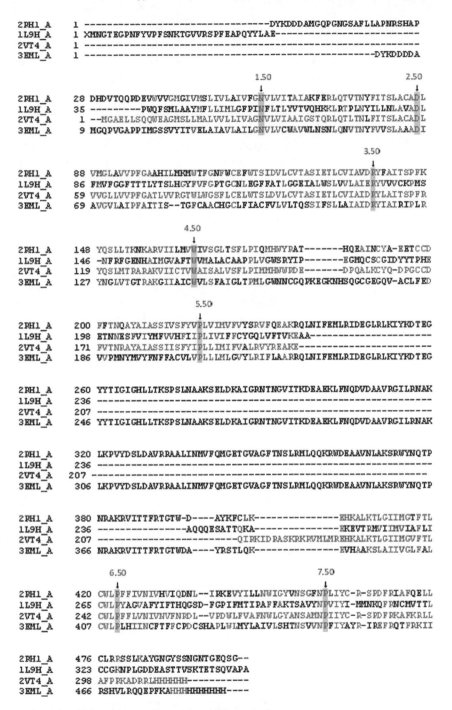

Fig. 3. Structure-based multiple-sequence alignment and BW numbering.

Class A) or several (in other classes) BW positions are expected to align in a correct model. For the Frizzled/Taste 2 receptor hTAS2R46 model the aligned positions include $N21^{1.50}$, $P187^{5.50}$, and $P276^{7.50}$.

5. Loop deletion: Modeling of loops remains one of the greatest challenges in modeling GPCRs due to their length and receptor-to-receptor variability. Several studies have shown that unless the models are based on a closely related template (e.g., β2ADR and β1ADR) (6, 30, 31), models without loops are preferable for use in docking studies. In particular, deletion of the second extracellular loop (ECL2) ((7) and see Note 1) is recommended. ECL2 is defined as the loop connecting TM4 and TM5, residues N150 to R169 in the case of hTAS2R46. The model may be structurally aligned to its X-ray templates to facilitate identification of loop boundaries.

6. Predict the binding pockets: Prediction of the putative ligand-binding pockets within the TM bundle is performed using the QSiteFinder Web server (32), which uses the interaction energy between the protein and a van der Waals methyl probe ($-CH_3$) to locate energetically favorable binding sites, and is especially useful for detection of relatively small sites. Protein residues within 5 Å of each predicted binding site are identified as potentially contacting residues. To validate QSiteFinder performance for GPCRs, we submitted the β2ADR structure (with antagonist deleted; PDB code 2RH1) and found that the predicted antagonist and cholesterol-binding pockets were in good agreement with the published experimentally determined structures (PDB codes 2RH1, 3D4S; data not shown) (33, 34). For hTAS2R46, the server predicted several different sites, of which we chose to further analyze the top binding-site clusters. As shown in Fig. 4, the largest and most energetically favorable binding site is predicted to be located between TM3 and TM7, in agreement with previous work on a related bitter-taste receptor (35, 36): most data for GPCRs in general point toward the same pocket as being the binding site for multiple ligands of many GPCRs (see Note 2). The importance of some positions (such as 3.32, 7.39—BW numbering) was illustrated for a wide range of receptors (see Note 2), but differences between related receptors in the roles of the residues within the main binding site were also shown (see Note 3). For some receptors only subtle changes occur in agonist bound vs. antagonist conformations (4, 37) but may be more pronounced in other receptors, as the A_{2A} adenosine receptor agonist-bound structure indicates (5). In the latter case, models based on an inactive state crystal structure may not be optimal as pockets for docking of agonists (see Note 4).

7. Check for known mutations using mutational databases such as GPCRDB (http://www.gpcr.org/7tm/) and tinyGRAP (http://www.cmbi.ru.nl/tinygrap/). Such mutations may provide supporting evidence for residues predicted in step 6 to be involved in ligand binding (28, 38). On the GPCRDB

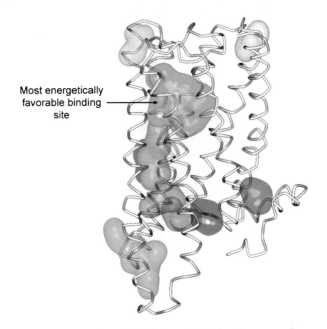

Most energetically favorable binding site

Fig. 4. QSiteFinder mapping of the hTAS2R46 model binding sites.

home page, choose the browse option (by families). Choose the appropriate GPCR family from the hierarchy view, and follow the "open alignment" link in the "alignment" field that appears following selection of the GPCR family. Select the JalView option for visualization of mutational data on the alignment. The known mutated positions are indicated in white font. Each such position is linked to an annotation of the mutation, including the amino acid to which the position was mutated, BW notation, TM domain, and links to the relevant literature. For the case of hTAS2R46, we did not find mutational data in these databases.

8. Dock a known ligand into the binding pocket of choice. In our working example, following Brockhoff et al. (12), we docked strychnine (a known ligand of the receptor) into the most energetically favorable binding site. We use the "Flexible Docking" algorithm as implemented in Discovery Studio 2.5 (Accelrys, Inc.). Additional docking algorithms, such as GOLD (39), GLIDE (40, 41), LigandFit (42), and CDocker (43), have also been successfully used, as reported in the literature. For a comprehensive recent review see Senderowitz et al. (8). Prior to docking, the ligands should be carefully prepared so that they are good representations of the actual ligand structures that would appear in a protein–ligand complex. The following is a "checklist" for ligand preparation:

(a) A 3D structure of the ligand is needed, without any accompanying fragments such as counterions and solvent molecules.

The most stable, or preferably a number of, 3D conformations should be used for docking.

(b) The bond lengths and bond angles are appropriate.

(c) The ligand has all the hydrogens added (filled valances).

(d) The ligand is in its appropriate protonation state for physiological pH values (~7). Inappropriate protonation states may, for example, result in docking of a polar molecule into a hydrophobic region, or cause it to serve as a hydrogen bond acceptor.

(e) Generate tautomers, alternative chiralities, and low-energy ring conformations, where applicable.
Steps (a–e) may be facilitated by using "LigPrep" (Schrödinger, Inc.) software, a recommended option for ligand preparation.

9. After docking, and according to the obtained poses, identify residues that are involved in ligand binding and suggest mutagenesis. Poses generated in the docking experiments are clustered, e.g., by means of root mean square deviation (RMSD) of all poses from the top scoring pose (using "analyze ligand poses" protocol, Discovery Studio 2.5, Accelrys, Inc.). Several distinct poses are chosen for further examination, which includes identification of all residues within 5 Å of the docked ligand, and determination of specific ligand–receptor contacts, such as hydrogen bonds and π–π and π–cation interactions. The latter may also be achieved by generation of ligand plots—2D representations of all ligand-contacting residues, e.g., using LigPlot (44). Residues are then chosen for experimental validation by site-directed mutagenesis, followed by binding/activity assays. The candidate residues are chosen based on their chemical nature, distance, and orientation from the ligand. In addition, it is important to test those positions which discriminate between two distinct ligand poses (for example, occupying two separate regions in the same cavity). For hTAS2R46, such residues include W88$^{3.32}$, A89$^{3.33}$, N92$^{3.36}$, H93$^{3.37}$, N96$^{3.40}$, I245$^{6.55}$, E265$^{7.39}$, and A268$^{7.42}$ (see Fig. 5 for one example of a chosen pose and surrounding residues). Variations in binding residues for different ligands (even of the same pharmacology) are not uncommon (see Note 5).

3.1. In Vitro Mutagenesis

Site-directed mutagenesis can be performed by various methods which rely on the exchange of nucleotides within the coding sequence of the target receptor by way of synthetic oligonucleotides. We typically perform site-directed mutagenesis by PCR-mediated recombination, a method originally described by Fang and colleagues (45). This versatile method is not restricted to the

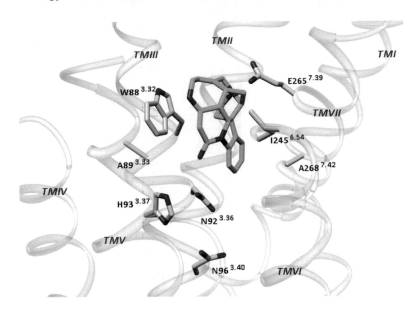

Fig. 5. Strychnine docked into the proposed binding pocket of hTAS2R46.

mutagenesis of one to several nucleotides in the context of an existing double-stranded DNA template (46); it is also useful, as originally intended, to generate chimeric receptor constructs (12) or other major modifications, such as the introduction of epitope tags (47). For successful point mutagenesis, design of synthetic oligonucleotides is crucial. Numerous companies offer rather inexpensive custom oligonucleotide syntheses with rapid turnaround via the Internet. The nucleotide sequence to be modified should be located in the center of a complementary primer pair and flanked by a sufficient number of nucleotides that are an exact match to the sequence of the DNA template to allow efficient annealing during the subsequent PCR. Depending upon the context of the codon(s) being subjected to mutagenesis, oligonucleotides of 21–30 bases are usually sufficient. For a schematic overview of the PCR steps involved in this procedure see Fig. 6.

In the first step, each of the two complementary oligonucleotides is used in combination with a primer specific for upstream and downstream vector sequences, respectively, to produce two subfragments corresponding to the 5′ and 3′ parts of the mutated cDNA (Fig. 6b). When calculating the annealing temperature, it is important to take the number of mismatches between primer and template sequences into account. A reduction of the annealing temperature by 1–1.5°C per percent mismatch is usually sufficient for efficient amplification. The annealing temperature prior to mismatch correction is 3–5°C below the calculated melting temperature, which is generally provided by the company on the

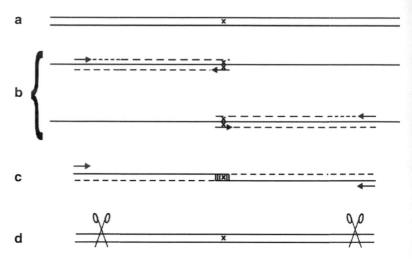

Fig. 6. *In vitro* mutagenesis flowchart.

accompanying data sheet assigned for oligonucleotide synthesis. An example procedure is described below.

PCR sample:

> X µL Template DNA (~10 ng)
> 0.5 µL Forward/reverse mutagenesis primer (10 µM)
> 0.5 µL Reverse/forward vector primer (10 µM)
> 2.5 µL 10× *Pfu*-DNA-polymerase buffer (including magnesium)
> 2.0 µL dNTP-mixture (2.5 mM each)
> 1.0 µL *Pfu*-DNA polymerase (3 U/µL)
> Add 25 µL deionized H₂O

PCR conditions:

1. 5 min 95°C Denaturation (perform "hot-start" during this time)
2. 1 min X°C annealing
3. 2 min 72°C Polymerization (~2 min/1 kb; check with supplier of DNA polymerase for optimal temperature)
4. 0.5 min 95°C
5. 5 min X°C Annealing
6. 10 min 72°C Polymerization

Repeat steps (2–5) 15–20 times (alternatively, after 3–5 initial PCR cycles, you may want to increase the annealing temperature for the remaining cycles as the mutagenesis primers are already incorporated).

After purification of the subfragments from non-incorporated mutagenesis primers and traces of the original template DNA by, e.g., isolating the PCR products from agarose gels, the subfragments are mixed in approximately equimolar amounts and further amplified by PCR using the pair of vector-specific primers (Fig. 6c). If the annealing temperature of the vector-specific primers exceeds the annealing temperature of the mutagenesis primers, which now form the overlap between the subfragments, at least 3–5 PCR cycles at the beginning of the amplification must be performed at a lower annealing temperature. However, a mismatch between the original template DNA and the mutated sequence does not need to be taken into account, as the mutation is already incorporated. A typical recombinant PCR step is shown below:

PCR sample:

    0.5 µL Subfragment A (~5 ng)
    0.5 µL Subfragment B (~5 ng)
    0.5 µL Forward vector primer (10 µM)
    0.5 µL Reverse vector primer (10 µM)
    2.5 µL 10× *Pfu*-DNA-polymerase buffer (including magnesium)
    2.0 µL dNTP mixture (2.5 mM each)
    1.0 µL *Pfu*-DNA polymerase (3 U/µL)
    Add 25 µL Deionized H_2O

PCR conditions:

1. 5 min 95°C Denaturation (perform "hot-start" during this time)

2. 1 min X°C Annealing

3. 2 min 72°C Polymerization (~2 min/1 kb; check with supplier of DNA polymerase for optimal temperature)

4. 0.5 min 95°C

5. 5 min X°C Annealing

6. 10 min 72°C Polymerization

Repeat steps (2–5) 15–20 times (if optimal annealing temperature of overlapping sequence between subfragments A and B is lower than the annealing temperature of the vector primers, perform 3–5 initial PCR cycles at an annealing temperature specific for the overlap and increase the annealing temperature for the remaining cycles).

The use of thermostable DNA polymerases with proofreading function is advisable for the described PCRs to avoid the accumulation of PCR products containing additional, unwanted mutations. Since vector-specific primers are used for the generation of mutated cDNAs, the PCR products contain the multiple-cloning site that enables, after purification and restriction-endonuclease treatment, reintroduction into the same vector (Fig. 6d). After the integrity of the constructs is confirmed by sequencing, the mutated construct can be subjected to functional characterization.

If no particular type of interaction involving the amino acid in question has been predicted by computer modeling, an exchange to alanine ("alanine-scanning mutagenesis" (48)) can be performed as an initial step.

3.2. Functional Heterologous Expression Assays

In our lab, functional characterization of receptor mutants is performed using HEK 293T cells stably expressing the G-protein chimera Gα16gust44. The G-protein chimera consists of Gα16, which couples, upon activation, to the IP$_3$/calcium second messenger-signaling pathway, and the last 44 amino acids of α-gustducin for effective interaction with T2R proteins (49). For the characterization of receptor mutants, it is important to include the parental receptor(s) and an empty cloning vector (mock control) for transient transfection. The cells are seeded onto black 96-well plates with clear bottom and transfected using Lipofectamine 2000 reagent (Invitrogen) at a cell density of ~60–70%. After ~24 h, the cells are loaded with the membrane-permeable calcium-sensitive dye, Fluo4-am, for 1 h in the presence of probenicid, an agent that inhibits the organic anion-transporter type 1, thus preventing rapid extrusion of the dye (50). While working with this or related fluorescent dyes, prolonged exposure of samples to bright light should be avoided. Now the cells are washed three times with buffer (130 mM NaCl, 5 mM KCl, 2 mM CaCl$_2$, 10 mM glucose, 10 mM HEPES, pH 7.4) using an automated microtiter plate washer to remove excess dye. After each washing step, the cells are incubated for ~15 min in 100 μL buffer. Next, the plates are transferred to a fluorometric imaging plate reader (FLIPR, Molecular Devices) and the baseline fluorescence is monitored. After application of an appropriate agonist concentration series, e.g., 0.003–100 μM strychnine for the hTAS2R46 (threefold concentrated in 50 μL applied to 100 μL present in each well), changes in fluorescence are recorded until the peaks of agonist-induced fluorescence are evident. After the signal has returned to baseline, a second application of an agonist stimulating transient calcium release from intracellular stores via an endogenous receptor is recommended to check cell viability (e.g., 100 nM somatostatin-14 stimulating endogenous somatostatin receptor, or 1 μM isoproterenol to activate β-adrenergic receptors. Note that these agonists have to be applied at fourfold concentration, if 50 μL are applied: a 100-μL volume is placed initially in each well, and then 50 μL of agonist solution is added; the total is 150 μL, so a threefold-concentrated agonist solution is needed. Addition of another 50 μL leads to 200 μL total volume, so the agonist concentration should be fourfold). For calculation of dose–response relations, at least two independent experiments with triplicate measurements of each construct and agonist concentration should be performed. Fluorescence changes of mock-transfected cells are subtracted and signals are normalized to background fluorescence. Calculations of EC$_{50}$ concentrations by nonlinear regression of the plots to the function $f(x) = 100/[(EC_{50}/x)^{nH}]$ and generation of the corresponding graphics are performed using Sigma Plot (for more details on the functional assay procedure see, e.g., (51–53)).

**3.3. Interpretation
of the Results**

This can be a rather challenging task. The mutated receptor may deviate from the parental wild-type receptor by its EC_{50} value, threshold concentration, maximal amplitude of the fluorescence signal, or a combination of these parameters. Furthermore, mutants of receptors with multiple agonists, such as many human bitter-taste receptors, may show the aforementioned changes in their dose–response curves for single, multiple, or all agonists. An idealized example of dose–response relationships is depicted in Fig. 7.

Clearly, a pure shift in the EC_{50} value indicates a change in agonist interaction that can be fully compensated for by the application of different agonist concentrations (Fig. 7, curve b). On the contrary, if the EC_{50} concentration remains unaffected but the maximal amplitude has changed, the amino-acid exchange apparently affected receptor activation capability rather than agonist interaction (Fig. 7, curve c). The most difficult outcome, however, is loss of function of the mutant receptor (see Note 6).

Note also that not in all cases where mutation of a residue affects agonist-induced response, a direct contact exists (see Note 7). The effect of mutation on ligand may be due to an allosteric effect as well. Nevertheless, in the few cases in which validation using X-ray structure was possible, the results were encouraging (see Note 8). Furthermore, many successful virtual-screening campaigns have provided confirmation of binding-site models derived as described above (9, 54).

Using the above procedure, residues located close to the docked strychnine molecule (see Fig. 5) were mutated. We show the results for some positions in Table 1, in which the mutations were designed to exchange residues of the strychnine-activated hTAS2R46 for residues of hTAS2R31, a bitter-taste receptor which is not activated by strychnine (12). Indeed the mutations led to a decrease in receptor responsiveness upon stimulation with the

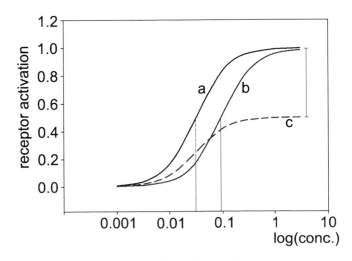

Fig. 7. Idealized dose–response curves of receptor mutants.

Table 1
hTAS2R46 binding-site residues

Mutation	Effect
hTAS2R46 wild type	Threshold ~ 0.1 μM[a]
N92G[3.36]	Nd. Strong decrease in responsiveness[b]
E265K[7.39]	Threshold ~ 30 μM[b]
A268R[7.42]	Threshold ~ 30 μM[b]

Mutagenesis of hTAS2R46 amino-acid residues located close to the docked strychnine molecule. All mutations were designed to cause an exchange of hTAS2R46 residues with hTAS2R31 residues. Threshold values were taken from (55) ([a]) and (12) ([b]). Note that all mutations led to severe decreases in receptor responses upon stimulation with the hTAS2R46 agonist strychnine and hence determination of EC_{50} values was not possible. *Nd.* not determined.

hTAS2R46 agonist strychnine. Interestingly, the double mutation of E265K and A268R exhibited loss of function for strychnine and gain of function for the hTAS2R31-specific agonist aristolochic acid (12).

4. Notes

1. Using GPCR models without loops

   The role of GPCR extracellular loops in binding high-molecular-weight peptidic ligands is well established (56). Studies have demonstrated that the extracellular loops, specifically ECL2, also interact with low-molecular-weight ligands, such as biogenic amines or adenosines (57). The recently reported X-ray structures of various GPCRs demonstrate that these extracellular loops also differ greatly in their structural features (3, 58). Furthermore, nuclear magnetic resonance (NMR) studies have revealed the dynamic and ligand-dependent characteristics of ECL2 in β2ADR (59), and the activation-induced and TM5-coupled changes in the rhodopsin ECL2 (60). The accurate modeling of these loops is therefore far from trivial.

   The available approaches to loop treatment, namely, homology modeling, use of loopless models, and *de novo* prediction, are summarized in our recent review (7). The emerging consensus is that models perform better with no modeled loops at all than with badly modeled loops, and that *de novo* modeling approaches warrant further development.

2. Comparison of known binding-site residues in different GPCRs

Experimental data from Shi and Javitch (61) and de Graaf and Rognan (9) reviews and from recent papers (12, 19–21, 62–69) on residues corresponding to the upper inside part of the TM bundle in solved GPCR structures ((70), and current work) show that they are involved in ligand binding. The residues are shown in Table 2 and listed in Fig. 8.

3. Variation in binding orientation between different receptors

The adenosine receptor 2A X-ray structure (3EML) revealed that the antagonist is positioned higher in the binding cavity than in other structures (74). Recent studies of taste receptors have also revealed different effects of mutations in conserved binding-site residues of different receptors. For example, mutation $W^{3.32}A$ in the hTAS2R38 receptor barely affected binding of the agonist (21), while the same mutation in hTAS2R43 and hTAS2R47 (63) had a pronounced effect on agonist-induced activation. This effect depends on the compound that was used to test binding sensitivity (see Note 4). Overall, while particular residues are shown to be involved in the binding of several ligands in several receptors (see Table 2), effects of mutations may also be ligand specific, indicating binding of different ligands in different subareas and orientations within the same pocket.

4. Variation between agonists and antagonists and use of inactive structures for docking agonists

Virtual screening studies show that *agonists* can be retrieved by using receptor models based on inactive crystal structure templates and, conversely, *antagonists* have been found using *agonist*-based receptor models (see (9) and references therein). However, active structure properties inferred from indirect measurements, as summarized, e.g., in (17), and directly from X-ray structures (4, 5, 75) indicate significant differences from the inactive form. The recent structural data indicates that the degree of variation in the agonist-bound vs. antagonist-bound pockets is receptor dependent.

Strategies for accommodating changes in the binding site include *in silico* activation (76) and agonist-induced modeling, as recently reviewed (7, 77). Other methods use advanced free-energy mapping methods to study activation dynamics and intermediate-state stabilization by ligands (78, 79). The new experimental structures now also enable inclusion of active structures (PDB codes: 3CAP, 3P0G, 3QAK) as templates for modeling (e.g., ligand-free native opsin structure (3CAP) was recently used (21, 80)).

5. Ligand-specific binding in the same receptor (a residue that is important for one ligand may not be important for another)

Table 2
Summary of binding-site residues

Position	Receptor	Inferred role	Reference	Residue number in human β2ADR
1.35	Adenosine A$_{2A}$	Water-mediated agonist contact (X-ray)	(5)	M36
2.60	CXCR4	Antagonist contact (X-ray)	(71)	F89
2.61	β1ADR	Ligand-specific partial agonist contact (X-ray)	(72)	G90
	Adenosine A$_{2A}$	Water-mediated agonist contact (X-ray)	(5)	
	D3R	Antagonist contact (X-ray)	(73)	
2.63	CXCR4	Antagonist contact (X-ray)	(71)	A92
2.64	Adrenergic	Antagonist specificity	(61)	H93
	β1ADR	Ligand-specific partial agonist contact (X-ray)	(72)	
2.65	β1ADR	Ligand-specific partial agonist contact (X-ray)	(72)	I94
	Taste T2R	Agonist specificity	(12)	
3.25	Muscarinic M1	Interaction with ligand	(68)	C106
3.28	Bioamine receptors	Agonist binding	(69)	F108
	β1ADR	Ligand contact (X-ray)	(72)	
	β2ADR	Ligand contact (X-ray)	(37)	
	D3R	Antagonist contact (X-ray)	(73)	
	CXCR4	Antagonist contact (X-ray)	(71)	
3.29	Muscarinic M1	Interaction with ligand	(68)	W109
	β1ADR	Ligand contact (X-ray)	(72)	
3.32	Dopamine, serotonin, histamine	Interaction with ligand	(61)	D113
	Acetylcholine, adrenergic	Interaction with ligand	(61)	
	CCR5 chemokine	Antagonist binding	(65)	
	Adenosine A$_{2A}$	Interaction with ligand (X-ray)	(5, 19)	
	Taste T2R	Agonist-induced activation	(12)	
	Taste T2R	Agonist-induced activation	(63)	
	β1ADR	Ligand contact (X-ray)	(72)	
	β2ADR	Ligand contact (X-ray)	(37, 66)	
	D3R	Antagonist contact (X-ray)	(73)	
	CXCR4	Antagonist contact (X-ray)	(71)	
3.33	CCR2 chemokine	Antagonist binding	(65)	V114
	β1ADR	Ligand contact (X-ray)	(72)	
	β2ADR	Inverse agonist-specific contact (X-ray)	(37)	
	Adenosine A$_{2A}$	Interaction with ligand (X-ray)	(5, 67)	
	D3R	Antagonist contact (X-ray)	(73)	
	Muscarinic M1	Interaction with ligand	(68)	

(continued)

Table 2
(continued)

Position	Receptor	Inferred role	Reference	Residue number in human β2ADR
3.36	Some aminergic receptors	Interaction with some ligands	(61)	V117
	Muscarinic M1	Agonist binding	(68)	
	Bioamine receptors	Agonist binding	(69)	
	β1ADR	Ligand contact (X-ray)	(72)	
	β2ADR	Ligand contact (X-ray)	(37)	
	D3R	Antagonist contact (X-ray)	(73)	
	Adenosine A$_{2A}$	Agonist contact (X-ray)	(5, 19)	
3.37	Muscarinic M1	Interaction with ligand	(68)	T118
	Adenosine A$_{2A}$	Interaction with ligand	(19)	
	β2ADR	Inverse agonist-specific contact (X-ray)	(37)	
	Taste T2R	Agonist-induced activation	(21)	
3.40	Histamine	Antagonist specificity	(61)	I121
4.53	Muscarinic M1	Interaction with ligand	(68)	S161
4.57	Muscarinic M1	Interaction with ligand	(68)	T164
4.61	Muscarinic M1	Interaction with ligand	(68)	P168
	Taste T2R	Agonist binding and activation	(64)	
5.29	Adenosine A$_{2A}$	Antagonist contact (X-ray)	(67)	C190
5.30	Adenosine A$_{2A}$	Antagonist contact (X-ray)	(67)	C191
5.32	β2ADR	Ligand contact (X-ray)	(37)	F193
5.38	β2ADR	Inverse agonist-specific contact (X-ray)	(37)	Y199
	Adenosine A$_{2A}$	Ligand contact (X-ray)	(5, 67)	
5.39	β1ADR	Ligand contact (X-ray)	(72)	A200
	β2ADR	Ligand contact (X-ray)	(37)	
	D3R	Antagonist contact (X-ray)	(73)	
	Histamine	Interaction with ligand	(61)	
	Muscarinic M1	Agonist binding	(68)	
5.42	Adrenergic	Interaction with ligand	(61)	S203
	β1ADR	Ligand contact (X-ray)	(72)	
	β2ADR	Ligand contact (X-ray)	(37, 66)	
	D3R	Antagonist contact (X-ray)	(73)	
	Muscarinic M1	Agonist binding	(68)	
	Bioamine receptors	Agonist binding	(69)	
	Adenosine A$_{2A}$	Antagonist contact (X-ray)	(19, 67)	

(continued)

Table 2
(continued)

Position	Receptor	Inferred role	Reference	Residue number in human β2ADR
5.43	Adrenergic	Interaction with ligand	(61)	S204
	Adenosine A$_{2A}$	Interaction with ligand	(19)	
	β1ADR	Ligand-specific partial agonist contact (X-ray)	(72)	
	β2ADR	Ligand contact (X-ray)	(37)	
	D3R	Antagonist contact (X-ray)	(73)	
	Taste T2R	Agonist binding and activation	(64)	
5.46	Adrenergic	Interaction with ligand	(61)	S207
	β1ADR	Agonist-specific hydrogen bond (X-ray)	(72)	
	β2ADR	Ligand contact (X-ray)	(37, 66)	
	D3R	Antagonist contact (X-ray)	(73)	
	Muscarinic M1	Agonist binding	(68)	
5.47	Muscarinic M1	Antagonist binding	(68)	F208
6.48	Many aminergic receptors	Ligand binding or activation	(61)	W286
	Muscarinic M1	Interaction with ligand	(68)	
	β1ADR	Antagonist-specific contact (X-ray)	(72)	
	β2ADR	Inverse agonist-specific contact (X-ray)	(37)	
	D3R	Antagonist contact (X-ray)	(73)	
	Adenosine A$_{2A}$	Interaction with ligand (X-ray)	(5, 67)	
6.51	Many aminergic receptors	Ligand binding or activation	(61)	F289
	Taste T2R	Agonist binding and activation	(64)	
	β1ADR	Ligand contact (X-ray)	(72)	
	β2ADR	Ligand contact (X-ray)	(37)	
	D3R	Antagonist contact (X-ray)	(73)	
	Adenosine A$_{2A}$	Interaction with ligand (X-ray)	(5, 67)	
	Muscarinic M1	Interaction with ligand	(68)	
	Taste T2R	Agonist-induced activation	(12)	P288
6.52	Many aminergic receptors	Ligand binding or activation	(61)	F290
	Muscarinic M1	Interaction with ligand	(68)	
	β1ADR	Ligand contact (X-ray)	(72)	
	β2ADR	Ligand contact (X-ray)	(37)	
	D3R	Antagonist contact (X-ray)	(73)	
	Adenosine A$_{2A}$	Agonist contact (X-ray), water-mediated antagonist contact (X-ray)	(5, 67)	

(continued)

Table 2
(continued)

Position	Receptor	Inferred role	Reference	Residue number in human β2ADR
6.55	Adrenergic	Ligand specificity	(61)	N293
	Muscarinic M1	Interaction with ligand	(68)	
	β1ADR	Ligand contact (X-ray)	(72)	
	β2ADR	Ligand contact (X-ray)	(37, 66)	
	D3R	Antagonist contact (X-ray)	(73)	
	Adenosine A_{2A}	Interaction with ligand (X-ray)	(5, 67)	
6.56	D3R	Antagonist contact (X-ray)	(73)	I294
6.58	GnRH	Agonist binding	(62)	H296
6.59	Adenosine A_{2A}	Interaction with ligand	(19)	V297
7.32	Adenosine A_{2A}	Antagonist contact (X-ray)	(67)	K305
7.35	Adrenergic	Agonist selectivity	(61)	Y308
	β1ADR	Ligand-specific full agonist contact (X-ray)	(72)	
	β2ADR	Ligand contact (X-ray)	(37, 66)	
	D3R	Antagonist contact (X-ray)	(73)	
	Adenosine A_{2A}	Interaction with ligand (X-ray)	(5, 67)	
7.36	β1ADR	Full and partial agonist contact (X-ray)	(72)	I309
	β2ADR	Agonist-specific contact (X-ray)	(37)	
	Adenosine A_{2A}	Interaction with ligand (X-ray)	(5, 67)	
7.39	Adrenergic	Ligand specificity	(61)	N312
	Chemokine	Ligand-specific binding	(65)	
	Taste T2R	Antagonist binding	(20)	
	Taste T2R	Agonist specificity (with 7.42)	(12)	
	β1ADR	Ligand contact (X-ray)	(72)	
	β2ADR	Ligand contact (X-ray)	(37, 66)	
	D3R	Antagonist contact (X-ray)	(73)	
	CXCR4	Antagonist contact (X-ray)	(71)	
	Muscarinic M1	Interaction with ligand	(68)	
	Adenosine A_{2A}	Interaction with ligand (X-ray)	(5, 67)	
7.40	β1ADR	Ligand-specific partial agonist contact (X-ray)	(72)	W313
7.42	Taste T2R	Antagonist binding	(20)	G315
	Taste T2R	Agonist specificity (with 7.39)	(12)	
	Adenosine A_{2A}	Agonist specificity	(5, 19)	
	Muscarinic M1	Interaction with ligand	(68)	

(continued)

Table 2
(continued)

Position	Receptor	Inferred role	Reference	Residue number in human β2ADR
7.43	β1ADR	Ligand contact (X-ray)	(72)	Y316
	β2ADR	Ligand contact (X-ray)	(37, 66)	
	D3R	Antagonist contact (X-ray)	(73)	
	Muscarinic M1	Interaction with ligand	(68)	
	Adenosine A$_{2A}$	Interaction with ligand (X-ray)	(5)	
7.46	Adenosine A$_{2A}$	Interaction with ligand	(19)	S319
7.49	β2ADR	Antagonist binding	(9)	N322

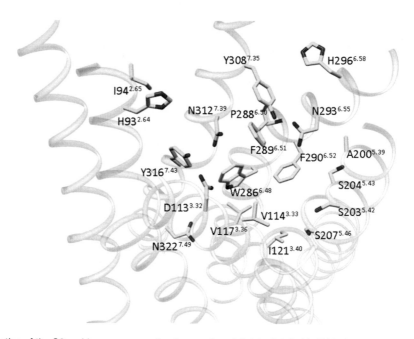

Fig. 8. Visualization of the β2 residues corresponding to experimental data detailed in Table 2.

has been shown (64, 81), as well as in a recent X-ray study of β2ADRs (66). Thus, involvement of a residue in the binding of one ligand does not necessarily imply that it is involved in the binding of another. This is true not only for agonist vs. antagonist, but also for ligands with the same pharmacological effect.

6. Lack of activation in a mutant receptor can be due to a variety of reasons, such as steric hindrance of agonist binding/receptor activation or loss of a crucial agonist interaction site, but it can also be due to receptor misfolding and other complications

in the biosynthetic pathway of the mutated receptor. Immunocytochemical experiments can give some indication of whether such nonspecific event, leading to loss of function, has taken place. If specific antibodies are available or the receptor has been tagged with an epitope, cells can be transiently transfected with receptor constructs and subjected to immunocytochemical analysis. Since GPCRs must reach the plasma membrane to be activated, additional staining with an appropriate cell-surface marker, such as concanavalin A, is strongly recommended. By comparing the staining patterns of the mutant receptor and the corresponding functional wild-type receptor, gross changes in plasma membrane association, and apparent differences in expression levels and intracellular accumulation can be monitored. Nevertheless, conclusive evidence for the proper folding of a nonfunctional receptor is extremely difficult to obtain. An elegant way of addressing this problem is to assemble all those amino-acid residues shown to interact with the agonist in a receptor with a different set of agonists to generate a receptor mutant exhibiting gain of function for the original agonist. However, as this requires the availability of a reasonably closely related, yet pharmacologically distinct, receptor, this approach may not be feasible in all cases.

7. Site-directed mutagenesis effects are observed despite a lack of actual ligand–receptor contact

In some cases, mutation of a residue affected binding affinity or activation by the ligand, but when the structure was solved, it showed no direct contact. For example (taken from (9)), $V^{3.32}$, $F^{5.43}$, and $H^{7.43}$, which are indicated by site-directed mutagenesis data to be involved in antagonist binding (82, 83), are in fact not in close contact with ZM241385 in the A_{2A} adenosine receptor crystal structure (PDB: 3EML). $D^{2.50}$ in the aminergic receptor, whose mutation reduces affinity for agonists but not for antagonists, is relatively far from the binding site and is likely to influence activation and modulatory effects of sodium on activation (61). Naturally occurring (35, 36) and engineered (21) variations in hTAS2R38 affect its activation even when it is not predicted to be in the binding site. A thermally stable variant of β1ADR carries mutations in residues which are not part of the agonist-binding pocket, but stabilize the inactive state, thus shifting the equilibrium away from the active state (84). This is in line with the ensemble view of allostery in which a certain constituent of the conformational assembly is biologically active (e.g., in signal transduction) and the overall activity is related to the relative occupancy of this conformational state. The occupancy of any conformational state is determined by the differential stability of that state with respect to the overall conformational distribution and is therefore amenable to modification by mutations that change the conformational distribution (85).

A recent study used advanced sequence analysis and residue swapping to unravel specificity determinants in serotonin and dopamine receptors. Interestingly, the four residue swaps giving the largest enhancement of serotonin responsiveness, $I48T^{1.46}$, $M117F^{3.35}$, $N124H^{3.42}$, and $T205M^{5.54}$, do not reside in ligand-contacting positions, and, except for $M117F^{3.35}$, are at positions that are at least 10 Å away from the ligands in the β2ADR structures. These are proposed to trigger conformational changes leading to distinct G-protein activation (69).

8. Validation vs. experimental 3D structures

There are a few cases in which homology model-assisted binding-site prediction and site-directed mutagenesis could be evaluated in hindsight using experimental structures. For example, the structure of carazolol in complex with β2ADR is in line with earlier site-directed mutagenesis results supporting the involvement of the three polar residues $D^{3.32}$, $S^{5.42}$, and $N^{7.39}$ in binding of agonists and antagonists, and the spatial distribution of those three critical residues in a b-Rho-based homology model is quite close to the solved crystal structure (9). Recently, an important initiative involved a community-wide modeling and docking experiment *prior* to the release of the structure of A_{2A} adenosine receptor (86). The most accurate model in terms of ligand RMSD and correct contacts was selected and ranked on the basis of docking scores and agreement with available mutational data, interpreted on the basis of previous modeling studies by S. Costanzi (86).

A new GPCRdock experiment (http://gpcr.scripps.edu/GPCRDock2010/index.html) provided assessment of the current status of GPCR ligand-binding site modeling (87). GPCR-ligand complex details could be accurately predicted using closely related templates and incorporation of experimental data (87). These results are very encouraging in view of the constantly increasing numbers of available GPCR templates, as more and more Xray structures are being solved (88).

Acknowledgments

We thank the Niedersachsen-Israeli research foundation (M.Y.N.) and the German Research Foundation, Deutsche Forschungsgemeinschaft (ME1024/2-3) (W.M.) and (ME 2014/8-1) (W.M., M.B. and M.Y.N), for funding and Dr. Talia Yarnitzky and Dr. Merav Fichman for helpful comments.

References

1. Heilker R et al (2009) G-protein-coupled receptor-focused drug discovery using a target class platform approach. Drug Discov Today 14:231–240

2. Rosenbaum DM, Rasmussen SGF, Kobilka BK (2009) The structure and function of G-protein-coupled receptors. Nature 459: 356–363

3. Topiol S, Sabio M (2009) X-ray structure breakthroughs in the GPCR transmembrane region. Biochem Pharmacol 78:11–20

4. Sprang SR (2011) Cell signalling: binding the receptor at both ends. Nature 469:172–173

5. Xu F et al (2011) Structure of an agonist-bound human A2A adenosine receptor. Science 332(6027):322–327

6. Mobarec JC, Sanchez R, Filizola M (2009) Modern homology modeling of G-protein coupled receptors: which structural template to use? J Med Chem 52:5207–5216

7. Yarnitzky T, Levit A, Niv MY (2010) Homology modeling of G-protein-coupled receptors with X-ray structures on the rise. Curr Opin Drug Discov Devel 13:317–325

8. Senderowitz H, Marantz Y (2009) G Protein-coupled receptors: target-based in silico screening. Curr Pharm Des 15:4049–4068

9. de Graaf C, Rognan D (2009) Customizing G Protein-coupled receptor models for structure-based virtual screening. Curr Pharm Des 15: 4026–4048

10. Reisert J, Restrepo D (2009) Molecular tuning of odorant receptors and its implication for odor signal processing. Chem Senses 34: 535–545

11. Cui M et al (2006) The heterodimeric sweet taste receptor has multiple potential ligand binding sites. Curr Pharm Des 12:4591–4600

12. Brockhoff A et al (2010) Structural requirements of bitter taste receptor activation. Proc Natl Acad Sci U S A 107:11110–11115

13. Lagerstrom MC, Schioth HB (2008) Structural diversity of G protein-coupled receptors and significance for drug discovery. Nat Rev Drug Discov 7:339–357

14. Harmar AJ et al (2009) IUPHAR-DB: the IUPHAR database of G protein-coupled receptors and ion channels. Nucleic Acids Res 37:D680–D685

15. Petrel C et al (2004) Positive and negative allosteric modulators of the Ca^{2+}-sensing receptor interact within overlapping but not identical binding sites in the transmembrane domain. J Biol Chem 279:18990–18997

16. Bhattacharya S et al (2010) Allosteric antagonist binding sites in class B GPCRs: corticotropin receptor 1. J Comput Aided Mol Des 24:659–674

17. Niv MY et al (2006) Modeling activated states of GPCRs: the rhodopsin template. J Comput Aided Mol Des 20:437–448

18. Niv MY, Filizola M (2008) Influence of oligomerization on the dynamics of G-protein-coupled receptors as assessed by normal mode analysis. Proteins 71:575–586

19. Ivanov AA, Barak D, Jacobson KA (2009) Evaluation of homology modeling of G-protein-coupled receptors in light of the A(2A) adenosine receptor crystallographic structure. J Med Chem 52:3284–3292

20. Slack JP et al (2010) Modulation of bitter taste perception by a small molecule hTAS2R antagonist. Curr Biol 20:1104–1109

21. Biarnes X et al (2010) Insights into the binding of phenyltiocarbamide (PTC) agonist to its target human TAS2R38 bitter receptor. PLoS One 5:e12394

22. Vaidehi N, Pease JE, Horuk R (2009) Modeling small molecule-compound binding to G-protein-coupled receptors. Methods Enzymol 460:263–288

23. Simms J et al (2009) Homology modeling of GPCRs. Methods Mol Biol 552:97–113

24. Zhang Y (2009) I-TASSER: fully automated protein structure prediction in CASP8. Proteins 77(Suppl 9):100–113

25. Ballesteros JA, Weinstein H (1995) Integrated methods for the construction of three dimensional models and computational probing of structure function relations in G protein-coupled receptors. Methods Neurosci 25: 366–428

26. Okuno Y et al (2008) GLIDA: GPCR ligand database for chemical genomics drug discovery database and tools update. Nucleic Acids Res 36:D907–D912

27. Ivetac A, McCammon JA (2010) Mapping the druggable allosteric space of G-protein-coupled receptors: a fragment-based molecular dynamics approach. Chem Biol Drug Des 76:201–217

28. Beukers MW et al (1999) TinyGRAP database: a bioinformatics tool to mine G-protein-coupled receptor mutant data. Trends Pharmacol Sci 20:475–477

29. Horn F, Lau AL, Cohen FE (2004) Automated extraction of mutation data from the literature: application of MuteXt to G protein-coupled receptors and nuclear hormone receptors. Bioinformatics 20:557–568

30. Costanzi S (2008) On the applicability of GPCR homology models to computer-aided drug discovery: a comparison between in silico

204   A. Levit et al.

and crystal structures of the beta2-adrenergic receptor. J Med Chem 51:2907–2914

31. Reynolds KA, Katritch V, Abagyan R (2009) Identifying conformational changes of the beta(2) adrenoceptor that enable accurate prediction of ligand/receptor interactions and screening for GPCR modulators. J Comput Aided Mol Des 23:273–288

32. Laurie AT, Jackson RM (2005) Q-SiteFinder: an energy-based method for the prediction of protein-ligand binding sites. Bioinformatics 21:1908–1916

33. Cherezov V et al (2007) High-resolution crystal structure of an engineered human beta2-adrenergic G protein-coupled receptor. Science 318:1258–1265

34. Hanson MA et al (2008) A specific cholesterol binding site is established by the 2.8 A structure of the human beta2-adrenergic receptor. Structure 16:897–905

35. Floriano WB et al (2006) Modeling the human PTC bitter-taste receptor interactions with bitter tastants. J Mol Model 12:931–941

36. Miguet L, Zhang Z, Grigorov MG (2006) Computational studies of ligand-receptor interactions in bitter taste receptors. J Recept Signal Transduct Res 26:611–630

37. Rasmussen SG et al (2011) Structure of a nanobody-stabilized active state of the beta(2) adrenoceptor. Nature 469:175–180

38. Horn F et al (2003) GPCRDB information system for G protein-coupled receptors. Nucleic Acids Res 31:294–297

39. Jones G et al (1997) Development and validation of a genetic algorithm for flexible docking. J Mol Biol 267:727–748

40. Friesner RA et al (2004) Glide: a new approach for rapid, accurate docking and scoring. 1. Method and assessment of docking accuracy. J Med Chem 47:1739–1749

41. Halgren TA et al (2004) Glide: a new approach for rapid, accurate docking and scoring. 2. Enrichment factors in database screening. J Med Chem 47:1750–1759

42. Venkatachalam CM et al (2003) LigandFit: a novel method for the shape-directed rapid docking of ligands to protein active sites. J Mol Graph Model 21:289–307

43. Wu G et al (2003) Detailed analysis of grid-based molecular docking: a case study of CDOCKER-A CHARMm-based MD docking algorithm. J Comput Chem 24:1549–1562

44. Wallace AC, Laskowski RA, Thornton JM (1995) LIGPLOT: a program to generate schematic diagrams of protein-ligand interactions. Protein Eng 8:127–134

45. Fang G et al (1999) PCR-mediated recombination: a general method applied to construct chimeric infectious molecular clones of plasma-derived HIV-1 RNA. Nat Med 5:239–242

46. Reichling C, Meyerhof W, Behrens M (2008) Functions of human bitter taste receptors depend on N-glycosylation. J Neurochem 106:1138–1148

47. Behrens M et al (2004) Molecular cloning and characterisation of DESC4, a new transmembrane serine protease. Cell Mol Life Sci 61:2866–2877

48. Cunningham BC, Wells JA (1989) High-resolution epitope mapping of hGH-receptor interactions by alanine-scanning mutagenesis. Science 244:1081–1085

49. Ueda T et al (2003) Functional interaction between T2R taste receptors and G-protein alpha subunits expressed in taste receptor cells. J Neurosci 23:7376–7380

50. Di Virgilio F, Steinberg TH, Silverstein SC (1990) Inhibition of Fura-2 sequestration and secretion with organic anion transport blockers. Cell Calcium 11:57–62

51. Bufe B et al (2002) The human TAS2R16 receptor mediates bitter taste in response to beta-glucopyranosides. Nat Genet 32:397–401

52. Behrens M et al (2004) The human taste receptor hTAS2R14 responds to a variety of different bitter compounds. Biochem Biophys Res Commun 319:479–485

53. Kuhn C et al (2004) Bitter taste receptors for saccharin and acesulfame K. J Neurosci 24:10260–10265

54. Sela I et al (2010) G protein coupled receptors-in silico drug discovery and design. Curr Top Med Chem 10:638–656

55. Brockhoff A et al (2007) Broad tuning of the human bitter taste receptor hTAS2R46 to various sesquiterpene lactones, clerodane and labdane diterpenoids, strychnine, and denatonium. J Agric Food Chem 55:6236–6243

56. de Graaf C et al (2008) Molecular modeling of the second extracellular loop of G-protein coupled receptors and its implication on structure-based virtual screening. Proteins 71:599–620

57. Shi L, Javitch JA (2004) The second extracellular loop of the dopamine D2 receptor lines the binding-site crevice. Proc Natl Acad Sci U S A 101:440–445

58. Mustafi D, Palczewski K (2009) Topology of class A G protein-coupled receptors: insights gained from crystal structures of rhodopsins, adrenergic and adenosine receptors. Mol Pharmacol 75:1–12

59. Bokoch MP et al (2010) Ligand-specific regulation of the extracellular surface of a G-protein-coupled receptor. Nature 463:108–112

60. Ahuja S et al (2009) Helix movement is coupled to displacement of the second extracellular

loop in rhodopsin activation. Nat Struct Mol Biol 16:168–175

61. Shi L, Javitch JA (2002) The binding site of aminergic G protein-coupled receptors: the transmembrane segments and second extracellular loop. Annu Rev Pharmacol Toxicol 42:437–467

62. Coetsee M et al (2008) Identification of Tyr(290(6.58)) of the human gonadotropin-releasing hormone (GnRH) receptor as a contact residue for both GnRH I and GnRH II: importance for high-affinity binding and receptor activation. Biochemistry 47:10305–10313

63. Pronin AN et al (2004) Identification of ligands for two human bitter T2R receptors. Chem Senses 29:583–593

64. Sakurai T et al (2010) Characterization of the beta-D-glucopyranoside binding site of the human bitter taste receptor hTAS2R16. J Biol Chem 285(36):28373–28378

65. Hall SE et al (2009) Elucidation of binding sites of dual antagonists in the human chemokine receptors CCR2 and CCR5. Mol Pharmacol 75:1325–1336

66. Wacker D et al (2010) Conserved binding mode of human beta2 adrenergic receptor inverse agonists and antagonist revealed by X-ray crystallography. J Am Chem Soc 132:11443–11445

67. Jaakola V-P et al (2008) The 2.6 Angstrom crystal structure of a human A2A adenosine receptor bound to an antagonist. Science 322:1211–1217

68. Goodwin JA et al (2007) Roof and floor of the muscarinic binding pocket: variations in the binding modes of orthosteric ligands. Mol Pharmacol 72:1484–1496

69. Rodriguez GJ et al (2010) Evolution-guided discovery and recoding of allosteric pathway specificity determinants in psychoactive bioamine receptors. Proc Natl Acad Sci U S A 107:7787–7792

70. Gloriam DE et al (2009) Definition of the G protein-coupled receptor transmembrane bundle binding pocket and calculation of receptor similarities for drug design. J Med Chem 52:4429–4442

71. Wu B et al (2010) Structures of the CXCR4 chemokine GPCR with small-molecule and cyclic peptide antagonists. Science 330:1066–1071

72. Warne T et al (2011) The structural basis for agonist and partial agonist action on a beta(1)-adrenergic receptor. Nature 469:241–244

73. Chien EY et al (2010) Structure of the human dopamine D3 receptor in complex with a D2/D3 selective antagonist. Science 330:1091–1095

74. Jaakola VP et al (2010) Ligand binding and subtype selectivity of the human A(2A) adenosine receptor: identification and characterization of essential amino acid residues. J Biol Chem 285:13032–13044

75. Hofmann KP et al (2009) A G protein-coupled receptor at work: the rhodopsin model. Trends Biochem Sci 34:540–552

76. de Graaf C, Rognan D (2008) Selective structure-based virtual screening for full and partial agonists of the beta2 adrenergic receptor. J Med Chem 51:4978–4985

77. Vaidehi N (2010) Dynamics and flexibility of G-protein-coupled receptor conformations and their relevance to drug design. Drug Discov Today 15(21–22):951–957

78. Bhattacharya S, Vaidehi N (2010) Computational mapping of the conformational transitions in agonist selective pathways of a G-protein coupled receptor. J Am Chem Soc 132:5205–5214

79. Provasi D, Filizola M (2010) Putative active states of a prototypic g-protein-coupled receptor from biased molecular dynamics. Biophys J 98:2347–2355

80. Ishikawa M et al (2010) Investigation of the histamine H3 receptor binding site. Design and synthesis of hybrid agonists with a lipophilic side chain. J Med Chem 53:6445–6456

81. Shapiro DA et al (2002) Evidence for a model of agonist-induced activation of 5-hydroxytryptamine 2A serotonin receptors that involves the disruption of a strong ionic interaction between helices 3 and 6. J Biol Chem 277:11441–11449

82. Jiang Q et al (1997) Mutagenesis reveals structure-activity parallels between human A2A adenosine receptors and biogenic amine G protein-coupled receptors. J Med Chem 40:2588–2595

83. Kim J et al (1995) Site-directed mutagenesis identifies residues involved in ligand recognition in the human A2a adenosine receptor. J Biol Chem 270:13987–13997

84. Balaraman GS, Bhattacharya S, Vaidehi N (2010) Structural insights into conformational stability of wild-type and mutant beta1-adrenergic receptor. Biophys J 99:568–577

85. Hilser VJ (2010) An ensemble view of allostery. Science 327:653–654

86. Michino M et al (2009) Community-wide assessment of GPCR structure modelling and ligand docking: GPCR Dock 2008. Nat Rev Drug Discov 8:455–463

87. Kufareva I et al (2011) Status of GPCR modeling and docking as reflected by community-wide GPCR dock 2010 Assessment, Structure 19:1108–1126

88. Jacobson KA and Costanzi S (2012) New insights for drug design from the X-ray crystallographic structures of GPCRs. Mol Pharmacol doi:10.1124/mol.112.079335

Chapter 12

Comparative Modeling of Lipid Receptors

Abby L. Parrill

Abstract

Comparative modeling is a powerful technique to generate models of proteins from families already represented by members with experimentally characterized three-dimensional structures. The method is particularly important for modeling membrane-bound receptors in the G Protein-Coupled Receptor (GPCR) family, such as many of the lipid receptors (such as the cannabinoid, prostanoid, lysophosphatidic acid, sphingosine 1-phosphate, and eicosanoid receptor family members), as these represent particularly challenging targets for experimental structural characterization methods. Although challenging modeling targets, these receptors have been linked to therapeutic indications that vary from nociception to cancer, and thus are of interest as therapeutic targets. Accurate models of lipid receptors are therefore valuable tools in the drug discovery and optimization phases of therapeutic development. This chapter describes the construction and evaluation of comparative structural models of lipid receptors beginning with the selection of template structures.

Key words: Comparative modeling, Homology modeling, GPCR, Sequence alignment, Lipid receptors

1. Introduction

Comparative modeling is a technique to construct a tertiary structure model of a target protein from its primary structure on the basis of anticipated structural similarity to a template protein tertiary structure that shares substantial primary structural and functional identity with the target (1). Comparative modeling (also called homology modeling) depends on the theoretical relationship expected between the three-dimensional structures of proteins that have evolved from a common ancestor protein while retaining substantial similarity in both amino acid sequence and overall function. It is expected that any changes to the amino acid sequence that alter three-dimensional structure in a dramatic fashion would concomitantly alter protein function. Comparative modeling became widely applicable to members of the G Protein-Coupled Receptor (GPCR)

Nagarajan Vaidehi and Judith Klein-Seetharaman (eds.), *Membrane Protein Structure and Dynamics: Methods and Protocols*, Methods in Molecular Biology, vol. 914, DOI 10.1007/978-1-62703-023-6_12, © Springer Science+Business Media, LLC 2012

family, which includes numerous receptors specific for lipids, when the crystallographic structure of bovine rhodopsin was published in 2000 (2). Since that time, the comparative modeling strategy has been used to construct predictive models of numerous lipid receptors that have provided insights into lipid recognition (3–13) and guided the discovery of antagonists that effectively block lipid-stimulated receptor functions (14–17).

The GPCR superfamily, particularly the class A or rhodopsin-like subfamily, includes numerous members that signal in response to lipids. The cannabinoid receptors (CB1 and CB2) signal in response to natural endocannabinoids such as anandamide and have long been of interest as therapeutic targets for analgesics (18), and have recently gained interest as targets in the treatment of neurodegenerative and neuroinflammatory diseases such as Alzheimer's disease (19). A second widely studied subfamily of lipid receptors are those responsive to two phospholipids, lysophosphatidic acid (LPA) and sphingosine 1-phosphate (S1P) (20, 21). Receptors responsive to LPA or S1P have generated substantial interest as therapeutic targets for cancer (22, 23), neuropathic pain (24–27), and multiple sclerosis (28–30). In particular, fingolimod (FTY-720), which is phosphorylated in vivo by sphingosine kinase to its bioactive form, has recently been approved for the treatment of relapsing multiple sclerosis through its action at the $S1P_1$ receptor (31). Additional bioactive lipids with signaling pathways involving GPCR, such as resolvin E1 and the eoxins are being reported on a regular basis (21), suggesting that modeling studies of lipid receptors will be of value for some time to drive the discovery of both therapeutics and pharmacological probes that can aid in elucidating the functional roles of lipid receptors.

Comparative modeling methods typically construct the tertiary structure of a target protein using the exact positions of identical amino acids in the template structure, and backbone atom positions of differing amino acids in the template structure. Amino acids that align against gaps in the target (template insertions) or in the template (template deletions) are modeled based on protein fragments of appropriate length that overlap on the backbone atom positions surrounding the insertion or deletion (32). Side-chain conformations for amino acids differing between the template and target are then added from a library of side-chain rotamers to optimize packing. It is often necessary to produce multiple models differing in the inserted loop structures and side-chain rotamers in order to select good models for subsequent studies. In this chapter, the method to construct and analyze a tertiary structural model for lipid receptors is presented, with particular focus on the rationale to be applied for template and model selection. Although some of the notes are specific to lipid receptors, the majority of the method described is applicable not only to lipid-responsive GPCR but to other classes of proteins as well.

2. Materials

2.1. Protein Sequences and Structures

1. Obtain the primary structure of your target protein from GenBank (http://www.ncbi.nlm.nih.gov/genbank) in FASTA format (see Note 1).

2. Obtain the tertiary structure of your template protein from the Protein Data Bank (33) (http://www.rcsb.org) in PDB format (see Note 2).

2.2. Software

Numerous software packages and online servers can be used to accomplish the alignment and comparative modeling steps described in the method; however, the description is based on options implemented in the MOE software package from the Chemical Computing Group (version 2009.10, Montreal, Canada). Suitable alternatives are described in Note 3.

Comparative models of GPCR are often (but not always) relaxed using geometry optimization and molecular dynamics using explicit solvated bilayer environments. Careful environmental modeling is essential if dynamic events such as lipid entry into the binding pocket from the surrounding bilayer as examined for the cannabinoid receptor (11) are of interest. The impact of the solvated membrane on the overall structure of lipid receptors implicitly carries through the comparative modeling process from the template structures, which were crystallized from a variety of lipid phases. Therefore, the value of modeling the membrane explicitly when using the comparative models as docking targets is not as high as for studies of dynamic events. The construction and simulation of such fully solvated and membrane-embedded systems is not discussed here.

2.3. Hardware

Hardware requirements for alignment and comparative modeling depend on the software selected. The MOE software runs on a variety of hardware from desktop PCs and laptops to high-performance linux computing clusters. The sequence alignment and comparative modeling functions do not require specialized computers or operating systems, and run quickly on typical desktop computing configurations. Likewise, the free tools available through numerous servers can also be utilized from any desktop computer. The overall flow of activities involved in comparative modeling of lipid receptors is shown in Fig. 1.

3. Methods

3.1. Alignment

1. Open three-dimensional structures of template protein structures appropriate for modeling either inactive receptor conformations (currently β2 adrenergic receptor (β2, PDB (33))

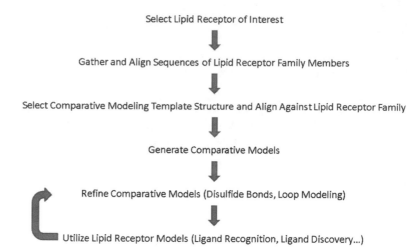

Fig 1. Flowchart of lipid receptor comparative modeling process.

entry 2RH1 (34), see Note 4), rhodopsin (PDB (33) entry 1U19 (35), chain A), β1 adrenergic receptor (β1, PDB (33) entry 2VT4 (36), chain B), and adenosine A2A receptor (A2A, PDB (33) entry 3EML (37), see Note 4)) or active receptor conformations (currently β2 adrenergic receptor (β2, PDB (33) entries 3P0G and 3PDS (38, 39)), opsin (PDB (33) entry 3DQB (40), chain A), and a theoretical rhodopsin model (PDB (33) entry 1BOJ (41)) (see Note 2)).

2. Open sequences of the target protein and closely related family members (see Note 1).

3. Perform a group-to-group alignment defining the template sequences in one group and the target sequence (and its close relatives) in the other (see Note 1 for details on performing group-to-group alignments in both the MOE software and using the free tools in the Biology Workbench server).

4. Manually close alignment gaps in helical segments 1–7, ensuring that the most conserved positions in each helical segment (42) are aligned (see Note 5). Alignments in the MOE software can be manually altered by right-clicking over a residue in the sequence window and pulling it to the desired position in the alignment.

5. One hallmark of the lipid receptor sequences that differentiates them from the currently available template structure sequences is the lack of the conserved proline residue in the fifth transmembrane domain (TM5 residue P5.50), as indicated in Note 5. The lack of proline in TM5 can lead to errors in alignment that have profound impact on the quality of comparative models of lipid receptors. As an example, comparative models of the LPA and S1P receptors were used for several years that were able to

identify residues playing key roles in agonist recognition and selectivity (3–8, 14, 43) that later had to be modified due to misalignment of the fifth transmembrane domain against the template during the model-building process (9). The alignment was adjusted based on the observation that the initial model was consistent with experimental mutagenesis results for all mutation sites except those in the fifth TM. Consistency between a modified model and the mutations to residues in TM5 became very good when the sequence alignment was manually adjusted by one residue, although the overall alignment score based on standard amino acid substitution matrices decreased slightly. Several different alignments of TM5 could be used to generate sets of candidate models for evaluation when modeling new classes of lipid receptors for which limited experimental data on the roles of residues in TM5 are available.

3.2. Model Generation

1. Build homology models (see Note 6) of the target sequence based on the most closely related template sequence or the template sequence that best represents the activation state of interest for the target receptor. For example, if the structural model is being generated to find new agonists it is imperative to use the crystal structure of an active state of the GPCR as template rather than its inactive state. It is fairly typical for cases with high sequence identity between target and template structure (>50 %) to build relatively few models (on the order of 10) due to the high level of confidence that the target structure closely resembles the template structure, and for cases with lower sequence identity to build a greater number of models. These models will differ in side-chain rotamers at unconserved positions, as well as backbone structure and side-chain rotamers of inserted loops.

2. Critically evaluate the resulting models. Tools that can be used include Ramachandran plots (ideally few φ/ψ combinations will fall in disallowed regions of the Ramachandran plot) and contact reports (hydrogen bond networks involving residues at polar conserved positions (44) should be present in a good model). Information from the literature such as mutagenesis or spectroscopic studies, when available, can also be used to select better models from among the complete set produced by homology modeling. For example, in the S1P receptors, mutagenesis studies have been performed that demonstrate importance of residues in the third, fifth, and seventh transmembrane (TM) segments in S1P-induced receptor activation (5, 6, 9, 45). If the essential residues in these domains do not cluster to form a candidate binding pocket in some of the generated models, those models might be eliminated from consideration. Spectroscopic studies are also emerging that indicate a shift between hydrophobic and polar environments as a function of

activation state for particular residues (46, 47). Placement of an amino acid side chain expected to reside in a polar environment in the activation state of interest outward toward the surrounding lipid would be sufficient reason to eliminate a candidate model based on comparison to spectroscopic results. The number of candidate models can usually be limited to 5–10 using a combination of experimental results from the literature and computational measures of model quality.

3. Carefully evaluate candidate disulfide bonds. Many of the lipid receptors lack the cysteine residue at the top of TM3 that forms a disulfide bond stabilizing the second extracellular loop geometry in many of the currently available template structures. In general, if the cysteine residues involved in forming disulfide bonds in a carefully selected template structure are conserved in the target sequence, these disulfide bonds should be formed in the models of the target protein. If the cysteine residues involved in disulfide bonds in the template structure are not conserved in the template, disulfide bonds should only be formed on the basis of experimental information. In prior studies of both LPA and S1P receptors (3, 5, 6, 8–10, 14–16, 45), our group has chosen not to include any disulfide bonds due to poor conservation of cysteine residues observed to form disulfide bonds in the crystallized GPCR examples combined with a lack of experimental evidence of which alternate disulfide bonds might occur in these lipid receptors.

4. Determine the importance of alternative loop modeling strategies. In the event that binding of agonists/antagonists is expected to involve interactions with residues in the extracellular loops, the comparative model(s) can serve as a starting point for loop optimization methods. Numerous options can be used to generate alternative loop models for comparison. One example is to use experimentally characterized segment models from lipid receptors characterized by NMR spectroscopy. Segment structures are available in the PDB for the third intracellular loop (IL3) of the CB1 receptor (48, 49) and the first extracellular loop (EL1) of the $S1P_4$ receptor (50). These segments could be used as comparative modeling templates for the corresponding segments of another target lipid receptor using the methods described in steps 1–3. Alternatively, loop modeling techniques such as those available in MODELLER through the more user-friendly SWIFT MODELLER interface described at http://bitmesra.ac.in/swift-modeller/swift.htm can be utilized. Such methods can be used to develop alternative models of particular loops, such as the second extracellular loop, which is often quite different in both sequence and length between different members of the class A GPCR, by entering the amino acid sequence range and number of

models to be generated. Step 3 would then be repeated to analyze the resulting loop models to select models to be used in subsequent modeling activities.

5. Use the best model (or subset of models) as a starting point for other investigations. Comparative models can be used to investigate complexes formed between the lipid receptor and agonists/antagonists (a good tutorial on docking has been published by Muegge and Rarey (51)), as targets for *in silico* screening to identify novel agonists/antagonists, and as starting points to investigate the membrane-embedded lipid receptor structure and function. Many of these subsequent studies can be used to provide important validation and feedback into the modeling process as shown in Fig. 1 and should be used to guide model optimization studies.

4. Notes

1. Target lipid receptors that are part of a closely related subfamily of GPCR (such as the cannabinoid receptor subfamily with two human homologs, CB1 and CB2) may align better against the template primary sequences using a group-to-group alignment strategy. In the MOE software package, this can be done by choosing the Align option from the Homology menu in the sequence window, selecting the chains of the subfamily sequences, activating the Partition option at the top of the dialog box, and then clicking the OK button. A similar group-to-group alignment strategy can be performed using the free tools in the Biology Workbench server (http://workbench.sdsc.edu) by generating separate alignments for the families, and then choosing the CLUSTALWPROF option (Align Two Existing Alignments) option from the Alignment Tools menu. Therefore, download primary sequences for all closely related members of the target receptor subfamily.

2. As of April 2011, there are over 20 atomic-resolution (2, 34–40, 52–62) crystallographic structures of six different class A GPCR (rhodopsin, β1 adrenergic, β2-adrenergic, adenosine A2a receptors, dopamine D3, and CXCR4 chemokine) from which to choose a template for modeling a lipid receptor. The Stephen White laboratory at UC Irvine maintains a Web page (http://blanco.biomol.uci.edu/Membrane_Proteins_xtal. html) that tabulates membrane proteins of known structure by categories including G Protein-Coupled Receptors and is a good place to check for additional available structural templates. The GPCR network also maintains a Web site that tracks their progress toward crystallization of numerous GPCR targets, with PDB references provided for their successfully

characterized targets (http://cmpd.scripps.edu/tracking_status.htm), often available prior to publication. The group-to-group alignment mentioned in Note 1 may benefit from using at least one crystal structure of each unique GPCR available.

3. Numerous servers and downloadable software packages have been developed that provide the functionality described in the process of building comparative models of lipid receptors. The alternatives identified here are those that can be used free of charge by academic researchers. A variety of other excellent software options are available with the purchase of licenses, but these options are not identified in these notes. Sequence alignments can be performed using the Biology Workbench at the San Diego Supercomputer Center (http://workbench.sdsc.edu/). Common alignment tools available through this server include ClustalW, which is also available through other servers. Comparative models can be developed using Swiss-Model (63, 64) (http://swissmodel.expasy.org/), MODELLER (65) (http://www.salilab.org/modeller/), and the BioInfoBank Metaserver (http://meta.bioinfo.pl/submit_wizard.pl) coupled with 3D Jury (66) consensus analysis. While each is capable of generating quality models, a few details on each are provided here for comparative purposes. The SwissModel server provides options for both automatic modeling from a sequence as well as an alignment mode in which the user provides a sequence alignment of the target to the template. The automatic modeling option may be of particular interest for investigators new to comparative modeling as a useful benchmark for models generated using more user-guided methods. MODELLER provides additional specialized features including de novo modeling of loops connecting the helical segments of GPCR. The added loop modeling feature would be of particular interest for lipid GPCR due to the high variability observed so far in loop structures among the six class A GPCR crystallized and the low sequence identity between the loop regions of the lipid GPCR and the currently available template structures. More computational expertise is required, however, as the software must be downloaded and installed in contrast to SwissModel which operates from a Web page interface. The BioInfoBank MetaServer can be used to combine the ease of use provided by a Web server with the advanced functions of many different comparative modeling and profile-based alignment tools. The advantage of the MetaServer is its use of multiple servers to construct comparative models, followed by extensive comparative analysis of the resulting models. One of the servers used by the MetaServer, ESyPred3D, actually uses the MODELLER package to construct its models.

4. The 2RH1 (β2), 3EML (A2A), 3OCU (CXCR4, also entries 3OE8, 3OE0, 3)E6, and 3OE0, and 3PBL (dopamine D3)

structures include residues ASN1002-ALA1160, ASN1002-TYR1161, ASN1002-SER1201, or ASN1002-TYR1161, respectively, from T4 lysozyme in place of the third intracellular loop residues (37, 67). These residues should be deleted from all structures as they do not reflect the correct structure or sequence necessary for the alignment and model building steps.

5. Class A GPCR exhibit very high conservation of residues in each helical segment, a feature that has been used to describe an index-based numbering system useful for comparing different GPCR family members.(42) These residues are conserved for both structural and functional reasons, and their alignment between the template and target should be carefully verified prior to model construction. Positions of these helical index residues in the 1U19 structure are N55 (N1.50), D83 (D2.50), R135 (R3.50), W161 (W4.50), P214 (P5.50), P267 (P6.50), and P303 (P7.50). It is common for lipid receptors not to show conservation of P5.50 and alignment within helix 5 will have to be guided based on other aspects of the sequence alignment.

6. It is important in this step to ensure that the potential energy function used in scoring and minimizing intermediate homology/comparative models does not impose an aqueous solvation environment (either explicitly or implicitly). This can be ensured by setting exterior dielectric values to 3 or by neglecting solvation entirely. This will ensure that the greatest errors in the contributions to the relative energies will be in the extracellular and intracellular loops, where the weakest homology and greatest errors are inherently localized. It is also important in the absence of bound ligands as induced fit environments not to minimize the energy of the intermediate models to very low root mean square gradient values (no lower than 0.1 kcal/mol Å is recommended) as any open pockets within the helical bundle may collapse to promote greater interactions across the pocket.

Acknowledgment

This work was supported by NIH grant HL 084007.

References

1. Esposito EX, Tobi D, Madura JD (2006) Comparative protein modeling. In: Lipkowitz KB, Cundari TR, Gillet VJ (eds) Reviews in computational chemistry. Wiley, Hoboken, N. J, pp 57–167

2. Palczewski K, Kumasaka T, Hori T, Behnke CA, Motoshima H, Fox BA, Le Trong I, Teller DC, Okada T, Stenkamp R, Yamamoto M, Miyano M (2000) Crystal structure of rhodopsin: a G protein-coupled receptor. Science 289:739–745

3. Parrill AL, Baker DL, Wang D, Fischer DJ, Bautista DL, van Brocklyn J, Spiegel S, Tigyi G (2000) Structural features of EDG1 receptor-ligand complexes revealed by computational modeling and mutagenesis. In: Goetzl EJ,

Lynch KR (eds) Lysophospholipids and eicosanoids in biology and pathophysiology. New York Academy of Sciences, New York, pp 330–339

4. Parrill AL, Wang D-A, Bautista DL, Van Brocklyn JR, Lorincz Z, Fischer DJ, Baker DL, Liliom K, Spiegel S, Tigyi G (2000) Identification of Edg1 receptor residues that recognize sphingosine 1-phosphate. J Biol Chem 275:39379–39384

5. Wang D, Lorincz Z, Bautista DL, Liliom K, Tigyi G, Parrill AL (2001) A single amino acid determines ligand specificity of the $S1P_1$ (EDG1) and LPA_1 (EDG2) phospholipid growth factor receptors. J Biol Chem 276: 49213–49220

6. Holdsworth G, Osborne DA, Pham TT, Fells JI, Hutchinson G, Milligan G, Parrill AL (2004) A single amino acid determines preference between phospholipids and reveals length restriction for activation of the S1P4 receptor. BMC Biochem 5:12

7. Fujiwara Y, Sardar V, Tokumura A, Baker D, Murakami-Murofushi K, Parrill A, Tigyi G (2005) Identification of residues responsible for ligand recognition and regioisomeric selectivity of lysophosphatidic acid receptors expressed in mammalian cells. J Biol Chem 280:35038–35050

8. Inagaki Y, Pham TT, Fujiwara Y, Kohno T, Osborne DA, Igarashi Y, Tigyi G, Parrill AL (2005) Sphingosine-1-phosphate analog recognition and selectivity at $S1P_4$ within the endothelial differentiation gene family of receptors. Biochem J 389:187–195

9. Fujiwara Y, Osborne DA, Walker MD, Wang DA, Bautista DA, Liliom K, Van Brocklyn JR, Parrill AL, Tigyi G (2007) Identification of the hydrophobic ligand binding pocket of the S1P1 receptor. J Biol Chem 282:2374–2385

10. Valentine WJ, Fells JI, Perygin DH, Mujahid S, Yokoyama K, Fujiwara Y, Tsukahara R, Van Brocklyn JR, Parrill AL, Tigyi G (2008) Subtype-specific residues involved in ligand activation of the endothelial differentiation gene family lysophosphatidic acid receptors. J Biol Chem 283:12175–12187

11. Hurst DP, Grossfield A, Lynch DL, Feller S, Romo TD, Gawrisch K, Pitman MC, Reggio PH (2010) A lipid pathway for ligand binding is necessary for a cannabinoid G protein-coupled receptor. J Biol Chem 285:17954–17964

12. Ruan KH, Wijaya C, Cervantes V, Wu J (2008) Characterization of the prostaglandin H2 mimic: binding to the purified human thromboxane A2 receptor in solution. Arch Biochem Biophys 477:396–403

13. Stitham J, Stojanovic A, Merenick BL, O'Hara KA, Hwa J (2003) The unique ligand-binding pocket for the human prostacyclin receptor. Site-directed mutagenesis and molecular modeling. J Biol Chem 278:4250–4257

14. Jo E, Sanna MG, Gonzalez-Cabrera PJ, Thangada S, Tigyi G, Osborne DA, Hla T, Parrill AL, Rosen H (2005) $S1P_1$-Selective in vivo-active agonists from high throughput screening: off-the-shelf chemical probes of receptor interactions, signaling and fate. Chem Biol 12:703–715

15. Fells JI, Tsukahara R, Fujiwara Y, Liu J, Perygin DH, Osborne DA, Tigyi G, Parrill AL (2008) Identification of non-lipid LPA3 antagonists by virtual screening. Bioorg Med Chem 16:6207–6217

16. Fells JI, Tsukahara R, Liu J, Tigyi G, Parrill AL (2009) Structure-based drug design identifies novel LPA3 antagonists. Bioorg Med Chem 17:7457–7464

17. Fells JI, Tsukahara R, Liu J, Tigyi G, Parrill AL (2010) 2D binary QSAR modeling of LPA3 receptor antagonism. J Mol Graph Model 28:828–833

18. Guindon J, Hohmann AG (2009) The endocannabinoid system and pain. CNS Neurol Disord Drug Targets 8:403–421

19. Martin-Moreno AM, Reigada D, Ramirez BG, Mechoulam R, Innamorato N, Cuadrado A, de Ceballos ML (2011) Cannabidiol and other cannabinoids reduce microglial activation in vitro and in vivo: relevance to Alzheimers' disease. Mol Pharmacol 79(6):964–973

20. Parrill AL (2008) Lysophospholipid interactions with protein targets. Biochim Biophys Acta 1781:540–546

21. Im DS (2009) New intercellular lipid mediators and their GPCRs: an update. Prostaglandins Other Lipid Mediat 89:53–56

22. Murph M, Mills GB (2007) Targeting the lipids LPA and S1P and their signalling pathways to inhibit tumour progression. Expert Rev Mol Med 9:1–18

23. Pua TL, Wang FQ, Fishman DA (2009) Roles of LPA in ovarian cancer development and progression. Future Oncol 5:1659–1673

24. Inoue M, Rashid MH, Fujita R, Contos JJA, Chun J, Ueda H (2004) Initiation of neuropathic pain requires lysophosphatidic acid receptor signaling. Nat Med 10:712–718

25. Inoue M, Ma L, Aoki J, Chun J, Ueda H (2008) Autotaxin, a synthetic enzyme of lysophosphatidic acid (LPA), mediates the induction of nerve-injured neuropathic pain. Mol Pain 4:6

26. Inoue M, Xie W, Matsushita Y, Chun J, Aoki J, Ueda H (2008) Lysophosphatidylcholine induces neuropathic pain through an action of autotaxin to generate lysophosphatidic acid. Neuroscience 152:296–298

27. Okudaira S, Yukiura H, Aoki J (2010) Biological roles of lysophosphatidic acid signaling through its production by autotaxin. Biochimie 92:698–706

28. Brinkmann V, Davis MD, Heise CE, Albert R, Cottens S, Hof R, Bruns C, Prieschl E, Baumruker T, Hiestand P, Foster CA, Zollinger M, Lynch KR (2002) The immune modulator, FTY720, targets sphingosine 1-phosphate receptors. J Biol Chem 277:21453–21457

29. Hammack BN, Fung KY, Hunsucker SW, Duncan MW, Burgoon MP, Owens GP, Gilden DH (2004) Proteomic analysis of multiple sclerosis cerebrospinal fluid. Mult Scler 10:245–260

30. Herr DR, Chun J (2007) Effects of LPA and S1P on the nervous system and implications for their involvement in disease. Curr Drug Targets 8:155–167

31. Brinkmann V, Billich A, Baumruker T, Heining P, Schmouder R, Francis G, Aradhye S, Burtin P (2010) Fingolimod (FTY720): discovery and development of an oral drug to treat multiple sclerosis. Nat Rev Drug Discov 9:883–897

32. Fechteler T, Dengler U, Schomburg D (1995) Prediction of protein three-dimensional structures in insertion and deletion regions: a procedure for searching data bases of representative protein fragments using geometric scoring criteria. J Mol Biol 253:114–131

33. Berman HM, Westbrook J, Feng Z, Gilliland G, Bhat TN, Weissig H, Shindyalov IN, Bourne PE (2000) The protein data bank. Nucleic Acids Res 28:235–242

34. Cherezov V, Rosenbaum DM, Hanson MA, Rasmussen SG, Thian FS, Kobilka TS, Choi HJ, Kuhn P, Weis WI, Kobilka BK, Stevens RC (2007) High-resolution crystal structure of an engineered human beta2-adrenergic G protein-coupled receptor. Science 318:1258–1265

35. Okada T, Sugihara M, Bondar AN, Elstner M, Entel P, Buss V (2004) The retinal conformation and its environment in rhodopsin in light of a new 2.2 A crystal structure. J Mol Biol 342:571–583

36. Warne T, Serrano-Vega MJ, Baker JG, Moukhametzianov R, Edwards PC, Henderson R, Leslie AG, Tate CG, Schertler GF (2008) Structure of a beta1-adrenergic G-protein-coupled receptor. Nature 454:486–491

37. Jaakola VP, Griffith MT, Hanson MA, Cherezov V, Chien EY, Lane JR, Ijzerman AP, Stevens RC (2008) The 2.6 angstrom crystal structure of a human A2A adenosine receptor bound to an antagonist. Science 322:1211–1217

38. Rasmussen SG, Choi HJ, Fung JJ, Pardon E, Casarosa P, Chae PS, Devree BT, Rosenbaum DM, Thian FS, Kobilka TS, Schnapp A, Konetzki I, Sunahara RK, Gellman SH, Pautsch A, Steyaert J, Weis WI, Kobilka BK (2011) Structure of a nanobody-stabilized active state of the beta(2) adrenoceptor. Nature 469:175–180

39. Rosenbaum DM, Zhang C, Lyons JA, Holl R, Aragao D, Arlow DH, Rasmussen SG, Choi HJ, Devree BT, Sunahara RK, Chae PS, Gellman SH, Dror RO, Shaw DE, Weis WI, Caffrey M, Gmeiner P, Kobilka BK (2011) Structure and function of an irreversible agonist-beta(2) adrenoceptor complex. Nature 469:236–240

40. Scheerer P, Park JH, Hildebrand PW, Kim YJ, Krauss N, Choe HW, Hofmann KP, Ernst OP (2008) Crystal structure of opsin in its G-protein-interacting conformation. Nature 455:497–502

41. Pogozheva ID, Lomize AL, Mosberg HI (1997) The transmembrane 7-a-bundle of rhodopsin: distance geometry calculations with hydrogen bonding constraints. Biophys J 70:1963–1985

42. Ballesteros JA, Weinstein H (1995) Chapter 19. Integrated methods for the construction of three-dimensional models and computational probing of structure function relations in 6 protein-coupled receptors. In: Conn PM, Sealfon SC (eds) Methods in neurosciences. Academic, San Diego, pp 366–428

43. Sardar VM, Bautista DL, Fischer DJ, Yokoyama K, Nusser N, Virag T, Wang D, Baker DL, Tigyi G, Parrill AL (2002) Molecular basis for lysophosphatidic acid receptor antagonist selectivity. Biochim Biophys Acta 1582:309–317

44. Zhang D, Weinstein H (1994) Polarity conserved positions in transmembrane domains of G-protein coupled receptors and bacteriorhodopsin. FEBS Lett 337:207–212

45. Naor MM, Walker MD, Van Brocklyn JR, Tigyi G, Parrill AL (2007) Sphingosine 1-phosphate pKa and binding constants: intramolecular and intermolecular influences. J Mol Graph Model 26:519–528

46. Ye S, Zaitseva E, Caltabiano G, Schertler GF, Sakmar TP, Deupi X, Vogel R (2010) Tracking G-protein-coupled receptor activation using

genetically encoded infrared probes. Nature 464:1386–1389

47. Ye S, Huber T, Vogel R, Sakmar TP (2009) FTIR analysis of GPCR activation using azido probes. Nat Chem Biol 5:397–399

48. Ulfers AL, McMurry JL, Kendall DA, Mierke DF (2002) Structure of the third intracellular loop of the human cannabinoid 1 receptor. Biochemistry 41(38):11344–11350

49. Ulfers AL, McMurry JL, Miller A, Wang L, Kendall DA, Mierke DF (2002) Cannabinoid receptor-G protein interactions: G(alphai1)-bound structures of IC3 and a mutant with altered G protein specificity. Protein Sci 11:2526–2531

50. Pham TT, Kriwacki RW, Parrill AL (2007) Peptide design and structural characterization of a GPCR loop mimetic. Biopolymers 86:298–310

51. Muegge I, Rarey M (2001) Small molecule docking and scoring. In: Lipkowitz KB, Boyd DB (eds) Reviews in computational chemistry. Wiley, New York

52. Okada T, Fujiyoshi Y, Silow M, Navarro J, Landau EM, Shichida Y (2002) Functional role of internal water molecules in rhodopsin revealed by X-ray crystallography. Proc Natl Acad Sci U S A 99:5982–5987

53. Li J, Edwards PC, Burghammer M, Villa C, Schertler GF (2004) Structure of bovine rhodopsin in a trigonal crystal form. J Mol Biol 343:1409–1438

54. Standfuss J, Xie G, Edwards PC, Burghammer M, Oprian DD, Schertler GF (2007) Crystal structure of a thermally stable rhodopsin mutant. J Mol Biol 372:1179–1188

55. Salom D, Lodowski DT, Stenkamp RE, Le Trong I, Golczak M, Jastrzebska B, Harris T, Ballesteros JA, Palczewski K (2006) Crystal structure of a photoactivated deprotonated intermediate of rhodopsin. Proc Natl Acad Sci U S A 103:16123–16128

56. Park JH, Scheerer P, Hofmann KP, Choe HW, Ernst OP (2008) Crystal structure of the ligand-free G-protein-coupled receptor opsin. Nature 454:183–187

57. Murakami M, Kouyama T (2008) Crystal structure of squid rhodopsin. Nature 453:363–367

58. Shimamura T, Hiraki K, Takahashi N, Hori T, Ago H, Masuda K, Takio K, Ishiguro M, Miyano M (2008) Crystal structure of squid rhodopsin with intracellularly extended cytoplasmic region. J Biol Chem 283:17753–17756

59. Rasmussen SG, Choi HJ, Rosenbaum DM, Kobilka TS, Thian FS, Edwards PC, Burghammer M, Ratnala VR, Sanishvili R, Fischetti RF, Schertler GF, Weis WI, Kobilka BK (2007) Crystal structure of the human beta2 adrenergic G-protein-coupled receptor. Nature 450:383–387

60. Hanson MA, Cherezov V, Griffith MT, Roth CB, Jaakola VP, Chien EY, Velasquez J, Kuhn P, Stevens RC (2008) A specific cholesterol binding site is established by the 2.8 A structure of the human beta2-adrenergic receptor. Structure 16:897–905

61. Wu B, Chien EY, Mol CD, Fenalti G, Liu W, Katritch V, Abagyan R, Brooun A, Wells P, Bi FC, Hamel DJ, Kuhn P, Handel TM, Cherezov V, Stevens RC (2010) Structures of the CXCR4 chemokine GPCR with small-molecule and cyclic peptide antagonists. Science 330: 1066–1071

62. Chien EY, Liu W, Zhao Q, Katritch V, Han GW, Hanson MA, Shi L, Newman AH, Javitch JA, Cherezov V, Stevens RC (2010) Structure of the human dopamine D3 receptor in complex with a D2/D3 selective antagonist. Science 330:1091–1095

63. Kiefer F, Arnold K, Kunzli M, Bordoli L, Schwede T (2009) The SWISS-MODEL Repository and associated resources. Nucleic Acids Res 37:D387–D392

64. Arnold K, Bordoli L, Kopp J, Schwede T (2006) The SWISS-MODEL workspace: a web-based environment for protein structure homology modelling. Bioinformatics 22: 195–201

65. Eswar N, Webb B, Marti-Renom MA, Madhusudhan MS, Eramian D, Shen MY, Pieper U, Sali A (2006) Comparative protein structure modeling using Modeller. Curr Protoc Bioinformatics Chapter 5, Unit 5.6

66. Ginalski K, Elofsson A, Fischer D, Rychlewski L (2003) 3D-Jury: a simple approach to improve protein structure predictions. Bioinformatics 19:1015–1018

67. Rosenbaum DM, Cherezov V, Hanson MA, Rasmussen SG, Thian FS, Kobilka TS, Choi HJ, Yao XJ, Weis WI, Stevens RC, Kobilka BK (2007) GPCR engineering yields high-resolution structural insights into beta2-adrenergic receptor function. Science 318:1266–1273

Chapter 13

Quantification of Structural Distortions in the Transmembrane Helices of GPCRs

Xavier Deupi

Abstract

A substantial part of the structural and much of the functional information about G protein-coupled receptors (GPCRs) comes from studies on rhodopsin. Thus, analysis tools for detailed structure comparison are key to see to what extent this information can be extended to other GPCRs. Among the methods to evaluate protein structures and, in particular, helix distortions, HELANAL has the advantage that it provides data (local bend and twist angles) that can be easily translated to structural effects, as a local opening/tightening of the helix.

In this work I show how HELANAL can be used to extract detailed structural information of the transmembrane bundle of GPCRs, and I provide some examples on how these data can be interpreted to study basic principles of protein structure, to compare homologous proteins and to study mechanisms of receptor activation. Also, I show how in combination with the sequence analysis tools provided by the program GMoS, distortions in individual receptors can be put in the context of the whole Class A GPCR family. Specifically, quantification of the strong proline-induced distortions in the transmembrane bundle of rhodopsin shows that they are not standard proline kinks. Moreover, the helix distortions in transmembrane helix (TMH) 5 and TMH 6 of rhodopsin are also present in the rest of GPCR crystal structures obtained so far, and thus, rhodopsin-based homology models have modeled correctly these strongly distorted helices. While in some cases the inherent "rhodopsin bias" of many of the GPCR models to date has not been a disadvantage, the availability of more templates will clearly result in better homology models.

This type of analysis can be, of course, applied to any protein, and it may be particularly useful for the structural analysis of other membrane proteins. A detailed knowledge of the local structural changes related to ligand binding and how they are translated into larger-scale movements of transmembrane domains is key to understand receptor activation.

Key words: GPCRs, Rhodopsin, β_2 adrenergic receptor, Transmembrane helices, Structural bioinformatics

---

1. Introduction

Class A G protein-coupled receptors (GPCRs) are one of the most prevalent protein families in the human genome. These proteins transduce extracellular signals across the cell membrane by responding

Nagarajan Vaidehi and Judith Klein-Seetharaman (eds.), *Membrane Protein Structure and Dynamics: Methods and Protocols*, Methods in Molecular Biology, vol. 914, DOI 10.1007/978-1-62703-023-6_13, © Springer Science+Business Media, LLC 2012

to a remarkable diversity of ligands, such as hormones, peptides, and neurotransmitters, as well as mediate the sense of vision, smell, and taste (see ref. 1 for a review). Rhodopsin, the photoreceptor of the retinal rod cells, and the β_2 adrenergic receptor, which responds to adrenaline and noradrenaline, are commonly used as model systems to study Class A (or rhodopsin-like) GPCRs.

There is a vast amount of information about many aspects of GPCRs deposited in specialized databases, as IUPHAR-DB (http://www.iuphar-db.org), with pharmacological, functional and pathophysiological information, GLIDA (http://pharminfo.pharm.kyoto-u.ac.jp/services/glida/), with biological information of GPCRs as well as chemical information of their known ligands, the GPCR NaVa database (http://nava.liacs.nl/) which describes natural sequence variants, as rare mutations or polymorphisms, including SNPs, or the GPCRdb (http://www.gpcr.org/7tm/) that stores a large amount of heterogeneous data, such as sequence alignments, ligand binding constants, and mutations. However, structural data of GPCRs is still scarce. Presently, the structures of only seven Class A GPCRs (bovine and squid rhodopsin, human β_2-adrenergic, turkey β_1-adrenergic, human A_{2A} adenosine (*see* ref. 2 for a review), human CXCR4 chemokine receptor (3) and dopamine D_3 receptor (4)) are known. These structures show a common fold of seven transmembrane helices (TMHs) connected by intervening loops and arranged in a compact bundle that is likely to be conserved throughout the family.

Much of the information about the mechanism of activation of GPCRs is based on spectroscopic studies on rhodopsin (see ref. 5 for a review). Thus, detailed structural comparisons with rhodopsin are important to understand the similarities and differences of other class A GPCRs to this receptor. This is a common problem in structural biology: when we refer to "similar" structures, what do we exactly mean? How similar is "similar"? It is very common to use root mean square deviation (RMSD) values to compare related protein structures. While this parameter can provide an overall idea of similarity/difference, particularly for very high (or very low) RMSD values, this parameter compresses too much the structural information encoded in a 3D structure. As a result, the interpretation of this parameter may be sometimes overstated. For instance, an RMSD of 0.6 Å is used as an argument to substantiate the similarity of the four crystal structures of the chemokine receptor CXCR4 bound to the small antagonist IT1t, while, in the same work, an RMSD of 0.9 Å is considered sufficient to report conformational differences induced by the binding of the CVX15 cyclic antagonist peptide (3). In this case, it is difficult to visualize how a mere difference of 0.3 Å separates structural similarities from differences. Even when the RMSD values are computed per-residue, this parameter cannot be directly interpreted in structural terms; for example, a difference in RMSD of 0.6 Å does not contain any information about the details of the structures being compared.

There are several ways to translate structural changes to parameters that can be more easily interpreted in terms of protein structure. For instance, calculation of the backbone φ and ψ dihedral angles (6), of the intrahelical hydrogen bond distance (7), or of rise per residue and spoke angles (8) are methods to measure explicitly helix distortions that are ideal for helical membrane proteins. However, some of these parameters are still hard to interpret. As a classic example, it can be very difficult to grasp the impact of a change in φ and ψ on the overall protein structure.

An easy and straightforward interpretation of structural distortions in alpha helices is as changes in the overall bend angle. For instance, the program HBEND (9), developed by DJ Barlow and JM Thornton, calculates a global helix axis as the path followed by a model probe that slides along the real helix. For each residue, the program fits the backbone atoms of the probe and real helices, and projects the C_α atom of the real helix on to the probe helix axis. The set of points thus generated describes the course of the real helix axis, and its global bend is measured as the radius of curvature. Thus, HBEND basically provides a global measure of the helix distortion.

Helix distortions can also be described as the combination of a bend and a twist component of the N- and C-terminal helix fragments relative to each other (10). The impact of the twist component on the overall helix distortion is sometimes overlooked, although it can be of key structural and functional importance. For instance, this twist results in a local helix opening or tightening that can modify the orientation of nearby amino acid side chains (11). Such changes in the helix twist can arise from the presence of prolines, of small polar residues (as Ser or Thr) or from combinations thereof (12). For instance, TMH 5 of the β_2 adrenergic receptor presents a local bend and opening of the helix induced or stabilized by Ser204(5.43) (13). This distortion results in a tightening of the binding pocket that is required for productive binding of small-molecule agonists.

More sophisticated algorithms take into account helix twist or related parameters. For instance , the program ProKink (14), developed by I. Visiers and H. Weinstein, quantifies the structure of proline-containing helices in terms of bend angle, wobble angle, and face shift, between user-defined pre-proline and a post-proline segments. The bend angle is the angle between the pre- and post-proline segments. The wobble angle defines the orientation of the post-proline segment with respect to the proline alpha carbon. The face shift measures the distortion that causes a twisting of the helix face, resulting in an overwounded or an underwounded helix. Thus, while the bend and wobble angles describe the overall helix distortion, the face shift describes the internal rearrangement of the atoms involved directly in the distortion. While these three parameters provide a throughout characterization of the helix structure, they are global parameters, i.e., each helix is defined by a single bend, wobble, and face shift.

The program HELANAL (15), developed by M. Bansal and S. Kumar, overcomes this restriction by calculating local bend and twist angles for a sliding window of four C_α atoms along the helix (see refs. 15, 16 for a detailed explanation of the algorithm). These parameters are a particularly useful measure of helix structure at a local level, as they are easy to translate into observable distortions that can be discussed in terms of protein structure, i.e., a local increase/decrease in the helix bend and/or an opening/tightening of a single helix turn. While this program was originally used to classify helices in globular proteins as linear, curved, or kinked (16), its use is not restricted to globular proteins (see refs. 13, 17–19 for examples of application to GPCRs). HELANAL does not require any hint from the user regarding the position of the structural distortion (similarly to HBEND) and includes the calculation of the helix twist in the analysis (similarly to ProKink). However, the main difference with these two programs is that in this case the structural parameters used to quantify the geometry of the helix are calculated for each residue, which allows a finer detail of description.

In the following sections, I explain how to use HELANAL to perform a detailed structural analysis of the transmembrane bundle of GPCRs, and I also provide some examples on how these data can be interpreted. The combination of structure and sequence analysis (for instance, using the tools provided by the program GMoS, available at lmc.uab.cat/gmos) allows to hypothesize about the presence of specific structural features in GPCRs for which high-resolution structural information is not yet available. For instance, the local unwinding of TMH 5 induced by Pro5.50 (see Example 2), present in all the GPCR crystal structures obtained so far, and stabilized by a bulky hydrophobic residue at position 3.40, is likely to be a conserved feature of most GPCRs, due to the high degree of conservation of these residues (17).

2. Materials

1. The source code of HELANAL (written in Fortran 77) can be downloaded from http://mdanalysis.googlecode.com/svn/trunk/doc/html/documentation_pages/analysis/helanal.html and compiled for local use. HELANAL can also be executed through a web server at http://nucleix.mbu.iisc.ernet.in/helanal/helanal.shtml.

2. The program needs two input files: a PDB file containing the structure to be analyzed (see Note 1) and a free-format file containing the position and length of the helices (see Note 2).

3. Methods

1. The program was designed by its developers to be executed interactively in the command line through a series of questions and answers where the user provides the names and format of the input files; for instance,

   (the arrows highlight the answers that must be provided by the user):

   ```
   > ./helanal
   > QUESTIONS AND THEIR ANSWERS
   > Do You wish to use HELIX records in PDB files?
   > n                    < ——— (see Note 1)
   > Key in name of the INPUT file containing
   helix start/end information
   > e.g. ***.inp :
   > rhodopsin.inp   < ——— (see Note 2)
   > Is the information in the same format as
   HELIX records in PDB files?
   > n                    < ——— (see Note 2)
   > If not, key in the input data format:
   > (Specify fields for coordinate file name,
   helix name, name, chain and number
   > for first and last residues. Use three letter
   code for residue names.)
   > (a8,1x,a3,1x,a3,1x,a1,3x,i3,2x,a3,1x,a1,3x
   ,i3) <— (see Note 2)
   > Is 1GZM.pdb    a PDB file?
   > y                        < ———
   > WRITING THE RESULTS TO : rhodopsin.prm
   > AND IN TABULAR FORM TO : rhodopsin.tab
   ```

2. The program generates several files with structural and geometrical information for each helix segment defined in the additional input file (see ref. 15 for a complete description of the information contained in these files). The most relevant output files are *angle.out*, which contains the angles between successive local helix axes, i.e., the local bend angles, and *nh. out*, which contains, among other information, the local twist angles. From these files it is easy to extract the individual values to a spreadsheet or graphics program.

3. Interpretation of the local bend angles (*angle.out*). For each helix segment specified in the input file, HELANAL calculates the bend angle between successive local helix axes computed for sets of four C_α atoms (i.e., for helix turns). Throughout this paper, individual residues from GPCRs will be referred to using their sequence number (when appropriate) and the Ballesteros–Weinstein number (10) in parenthesis, which allows an easy

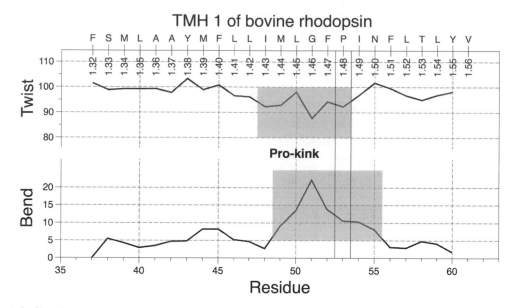

Fig. 1. Profiles of local bend (*bottom panel*) and twist (*top panel*) angles for TMH 1 of bovine rhodopsin (PDBid: 1GZM), built using the values shown in Notes 3 and 4. Pro53(1.48), delimited by solid vertical lines, stabilizes a local bend of ~20° (*bottom panel*) and a modest decrease of the twist (i.e., a local opening, see Note 5 and Fig. 2) (*top panel*) in the preceding residues. These distortions span almost two turns of the helix (*grey area*).

comparison among residues in the TM segments of different receptors. In this general numbering scheme, the position of each residue is described by two numbers: the first (1 through 7) corresponds to the helix in which the residue is located; the second indicates its position relative to the most conserved residue in that helix, arbitrarily assigned to 50. For instance, in TMH 1 of rhodopsin (PDBid: 1GZM (20)), HELANAL reports a peak in the local bend of 22° centered in residue Gly51(1.46) (see Note 3 and Fig. 1, bottom panel) that corresponds to the angle between the axis calculated between the helix turn Ile48(1.43)-Gly51(1.46) and the turn Gly51(1.46)-Ile54(1.49). This bend corresponds to the proline kink stabilized by Pro53(1.48), but the detailed HELANAL output shows that the maximum distortion, in terms of helix bend, is not centered in the Pro, and also shows that the proline-induced distortion extends from Met49(1.44) to Asn55(1.50), i.e., roughly two helix turns (see Note 3 and Fig. 1, bottom panel). The peak value of the local bend corresponds remarkably well to the mean value of a Pro-kink calculated using other methods in structure databases (21) and molecular dynamics simulations (22).

4. Interpretation of the local twist angles (*nh.out*). For each helix segment specified in the input file, HELANAL calculates the twist angle for each set of four successive Cα atoms, i.e., for each helical turn (see Note 4). For instance, in TMH 1 of

rhodopsin (PDBid: 1GZM) HELANAL reports a minimum in the local twist of 88° in the turn Gly51(1.46)-Ile54(1.49) (see Note 4), i.e., in the proline kink stabilized by Pro53(1.48). Again, the detailed HELANAL output shows that the maximum distortion, which corresponds to a minimum value of the helix twist, is not centered in the Pro and that the proline-induced distortion extends along two helix turns, from Ile48(1.43) to Phe56(1.51) (Fig. 1). The decrease in the helix turns preceding Pro53(1.48) are interpreted as an opening in the helix (see Note 5).

In summary, an analysis using the values of local bend and twist allows a detailed description of helix structural distortions. For instance, the example discussed above shows that TMH 1 of rhodopsin represents a "canonical" example of proline-kinked α helix. While many times the distortion induced by a proline is measured simply as an increase of the bend angle of ~20° at the position of the proline, the more detailed per-residue analysis of HELANAL (Fig. 1) reveals that the distortion spans almost two helical turns preceding the proline residue and that, in addition to the increase in bend, there is a small local opening in the helix.

4. Examples of Use

This section provides three short examples of how HELANAL can be used to quantify structural distortions in the transmembrane segments of GPCRs. In combination with the sequence analysis tools provided by GMoS (lmc.uab.cat/gmos), these distortions can be put in context of the whole Class A GPCR family.

1. Proline residues in the TMHs of rhodopsin stabilize different types of distortions

    The crystal structures of rhodopsin (20, 23) revealed that its TMHs contain some strong structural irregularities, many of them related to the presence of proline residues (24). A detailed analysis in terms of bend and twist (Fig. 3) allows to quantify and compare these distortions. As shown in the Methods section, Pro53(1.48) in TMH 1 (Fig. 3, black lines) stabilizes a standard Pro-kink that can be identified by an increase in the bend (up to 22°) and a mild opening (i.e., an decrease in the local twist of 2–8°) in the preceding turns of the helix. In TMH 5, Pro215(5.50) (Fig. 3, red lines) stabilizes a similar bend angle (peak of 23°) than a standard Pro-kink, but a much more pronounced decrease in the local twist (down to 73°), resulting in a very noticeable local opening of the helix (see Fig. 2, left panel) that is stabilized by Leu125(3.40) (11, 17). In TMH 6, Pro267(6.50) (Fig. 3, green lines)

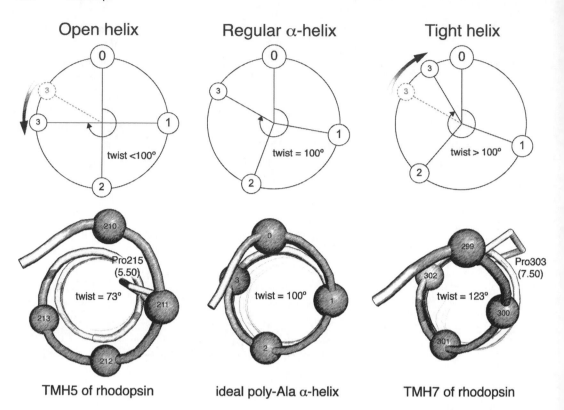

Open helix   **Regular α-helix**   **Tight helix**

TMH5 of rhodopsin   ideal poly-Ala α-helix   TMH7 of rhodopsin

Fig. 2. Depiction of the twist parameter in HELANAL. The top panel shows a schematic helix turn of a regular α helix (*middle*) with ~3.6 residues per turn and a twist angle of ~100° (360°/3.6), of a tightened helix (*right*) with less residues per turn and an increased twist angle, and of an open helix (*left*) with more residues per turn and a decreased twist angle. The bottom panel shows actual helix turns of an ideal poly-Ala α helix (*middle*) with a twist of 100°, TMH 7 of rhodopsin (*right*), with a tightening in the preceding turn from Pro303(7.50) that can be detected as an increased twist value of 123° in the Ala299(7.46)-Asn302(7.49) turn, and TMH 5 of rhodopsin (*left*), with an opening in the preceding turn from Pro215(5.50) that can be detected as a decreased twist value of 73° in the Val210(5.45)-Ile213(5.48) turn.

stabilizes a much larger bend angle (peak of 44°) than a standard Pro-kink, while the pronounced helix opening (decrease in local twist down to 78°) is highly localized in the turn 6.49–6.52. This distortion is stabilized by a water molecule that links TMH 6 and TMH 7. Finally, in TMH 7, Pro303(7.50) (Fig. 3, blue lines) also stabilizes a much larger bend angle (peak of 43°) than a standard Pro-kink, but, in this case, the presence of the proline is translated into a strong helix tightening (increase in the local twist up to 123°) (see Fig. 2, right panel) that results in the short 3_{10} segment observed in the crystal structures. This strong distortion in TMH 7 is also stabilized by a number of water molecules (20).

In summary, the local bend and twist analysis performed with HELANAL clearly shows the differences between the structural distortions stabilized by Pro215(5.50), Pro267(6.50), and Pro303(7.50), and how they do not correspond to standard Pro-kinks (e.g., Pro53(1.48)). This analysis also allows to

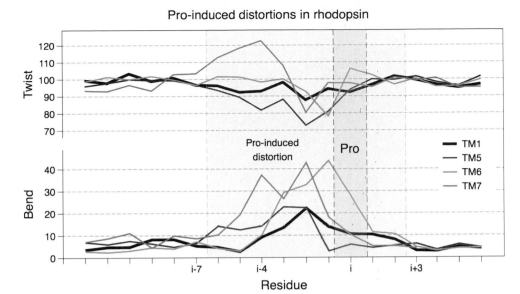

Fig. 3. Profiles of bend (*bottom panel*) and twist (*top panel*) angles for TMH 1, TMH 5, TMH 6 and TMH 7 of bovine rhodopsin. The profiles have been superimposed so Pro53(1.48), Pro215(5.50), Pro267(6.50), and Pro303(7.50) are in the same position i (*dark grey area*). This graphic shows how all proline-induced distortions span roughly two helix turns, from positions i-7 to i + 3 (light grey area). However, the proline-induced distortions in TMH 5 (*red*), TMH 6 (*green*), and TMH 7 (*blue*) are very different from the standard Pro-kink of TMH 1 (*black*).

visualize and quantify the fundamental difference between the distortion stabilized by Pro303(7.50) and the rest of proline-induced distortions. Finally, Fig. 3 also shows how all these distortions extend over two helix turns, mostly preceding the Pro residue.

2. Comparison between related protein structures: rhodopsin vs. β_2 adrenergic receptor

The analysis tools provided by HELANAL can be easily used to perform a detailed per-residue comparison between related proteins that can be readily interpreted in structural terms. For instance, comparison between the structures of rhodopsin (20) and the β_2 adrenergic receptor (25) allows to visualize and quantify the similarities and differences in the structures of TMH 1, TMH 5, TMH 6, and TMH 7.

Pro(1.48) is only present in 12 % of Class A GPCRs, and is conserved only in some subfamilies, as leukotriene B4 or proteinase-activated like receptors and some vertebrate opsin subtypes (http://lmc.uab.cat/gmos/index.php?subset=1&family=001&position=1.48&type=P&show=&button=Search). Thus, the Pro-kink in TMH 1 of rhodopsin is likely to be a family specific feature. Accordingly, the crystal structures of β_1 (26) and β_2 (25) adrenergic receptors, which lack a Pro residue at this position, feature almost a straight TMH 1. However,

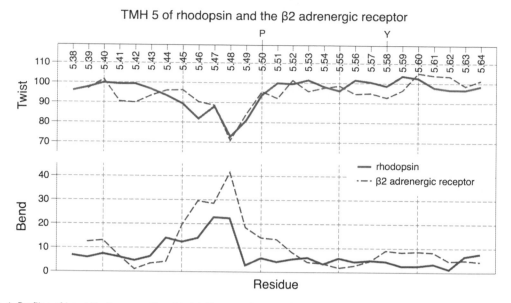

Fig. 4. Profiles of bend (*bottom panel*) and twist (*top panel*) angles for TMH 5 of bovine rhodopsin (*solid line*) and the β_2 adrenergic receptor (*broken line*). The different position of Tyr(5.58) in the β_2 adrenergic receptor compared to rhodopsin does not arise from a change in the helix twist around Pro(5.50), but from a change in the bend angle and small cumulative structural changes in both bend and twist in the cytoplasmic side of the helix.

the structure of the chemokine receptor CXCR4 (3) presents a kinked TMH1 without featuring a Pro at position 1.48. It has been shown that helix packing can stabilize kinks without prolines (7). In this structure, Thr51(1.45) in the g- conformation is hydrogen-bonding the backbone carbonyl at position i-4, which stabilizes the helix bend (13).

In TMH 5, the distortions stabilized by the highly conserved Pro(5.50) (http://lmc.uab.cat/gmos/index.php?position=5.50&type=&subset=1&family=&button=Search) in rhodopsin and the β_2 adrenergic receptor are very similar (Fig. 4). In the two preceding helix turns from Pro(5.50), both receptors present a strong and almost identical decrease in the helix twist, i.e., a opening in the helix (see Fig. 2, left panel), and a strong bend, which is larger in the β_2 adrenergic receptor. Usually, the detection of this type of distortions is an excellent starting point for further analyses. For instance, a detailed inspection of the 3D crystal structures reveals that, in both receptors, this distortion of TMH 5 is stabilized by a strong packing with a highly conserved bulky hydrophobic residue at position 3.40 (http://lmc.uab.cat/gmos/index.php?position=3.40&type=&subset=1&family=&button=Search) (11). This distortion is key for ligand-induced structural changes that lead to receptor activation in the histamine H1 receptor, and possibly other Class A GPCRs (17).

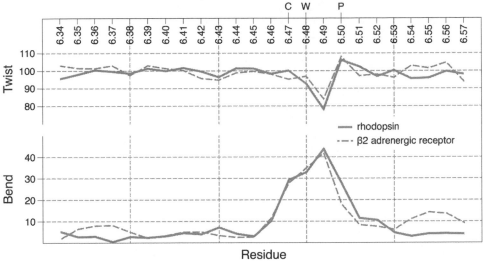

Fig. 5. Profiles of bend (*bottom panel*) and twist (*top panel*) angles for TMH 6 of bovine rhodopsin (*solid line*) and the β_2 adrenergic receptor (*broken line*). This graphic shows how these two receptors feature an almost identical structure of TMH 6, particularly around the highly conserved CWxP motif (http://lmc.uab.cat/gmos/index.php?position=6.47&type=cwxp&subset=1&family=&button=Search).

On the other hand, the twist and bend profiles of TMH 6 in rhodopsin and the β_2 adrenergic receptor are virtually identical (Fig. 5). A water molecule present in both structures stabilizes this distortion of TMH 6. The conservation of the residues around the highly conserved Pro(6.50) (http://lmc.uab.cat/gmos/index.php?position=6.50&type=&subset=1&family=&button=Search) suggests that this water molecule is also conserved in Class A GPCRs (18).

Finally, the bend and twist profiles of TMH 7 in rhodopsin and the β_2 adrenergic receptor (Fig. 6) reveal some interesting structural differences. In rhodopsin (solid thick line), there is a noticeable increase in the local twist (helix tightening) centered in the 7.46–7.49 turn (see Fig. 2, right panel), that is translated into a short 3_{10} helix segment (20), accompanied by a large bend (up to 40°). Although Pro(7.50) is highly conserved (http://lmc.uab.cat/gmos/index.php?position=7.50&type=&subset=1&family=&button=Search), this proline-induced distortion in the β_2 adrenergic receptor (broken line) is milder and less extensive, both in bend and twist. Strikingly, the bend and twist patterns for the β_2 adrenergic receptor, bound to an inverse agonist and thus stabilized in an inactive state, are similar to those of activated opsin bound to the Ct-Gα peptide (27) (solid thin line). Thus, while the β_2 adrenergic receptor is in an inactive conformation, its TMH 7 seems to be in a "preactivated" conformation that allows Tyr326(7.53)

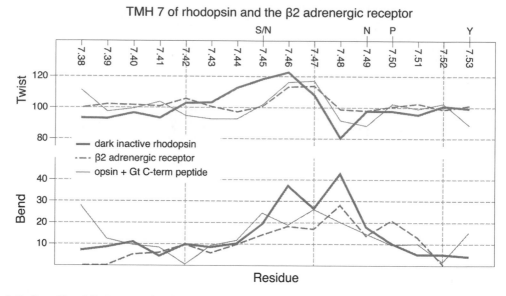

Fig. 6. Profiles of bend (*bottom panel*) and twist (*top panel*) angles for TMH 7 of bovine rhodopsin (*solid thick line*), opsin (*solid thin line*), and the β₂ adrenergic receptor (*broken line*). Strikingly, the β₂ adrenergic receptor features a TMH 7 more similar to opsin than to rhodopsin.

to orient towards the transmembrane bundle. The turkey β1 adrenergic receptor (26), crystallized using a completely different approach, presents a similar profile of bend and twist in TMH 7 (data not shown), which strongly suggests that the structure of TMH 7 in adrenergic receptors "is a feature, not a bug." In this case, the structure of TMH 7 in rhodopsin might be an exception, being hold in a particularly inactive conformation by a different interhelix packing, involving, for instance, the aromatic–aromatic interaction between Y306(7.53) and Phe313 in helix 8.

In summary, the combination of the detailed structural analysis performed with HELANAL with the family specific sequence analysis of GMoS allows to put the structural data in the context of Class A GPCRs. Thus, while the nonconserved Pro(1.48) is likely related to family specific structural distortions, the highly conserved Pro(5.50), Pro(6.50) and Pro(7.50) probably stabilize similar structures in Class A GPCRs, although TMH 7 in rhodopsin may be atypical.

3. Structural changes of TMH 6 during rhodopsin activation correspond to a rigid body movement.

The detailed structural analysis provided by HELANAL can also be used to interpret conformational changes during protein activation. For instance, Fig. 7 shows how TMH 6 of dark inactive rhodopsin (solid line) presents a very localized helix opening and bend, stabilized by a water molecule (see above). When the same analysis is applied to the structure of

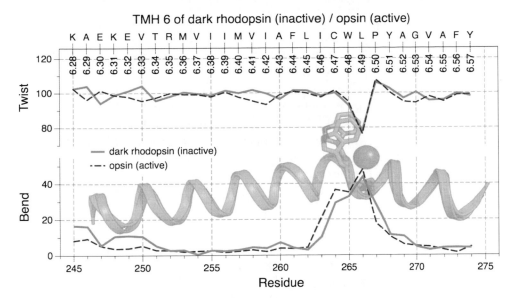

Fig. 7. Profiles of bend (*bottom panel*) and twist (*top panel*) angles for TMH 6 of bovine rhodopsin (*solid line, light green*) and opsin (*broken line, dark green*). The background shows these two helices (in the same color scheme) superimposed, with Trp(6.48) as sticks and the water molecule that stabilizes this distortion in rhodopsin as a sphere. This graphic shows how the conformational change of TMH 6 during rhodopsin activation corresponds to a rigid body movement, including the side chain of Trp265(6.48), which does not change conformation. A water molecule in rhodopsin stabilizes the strong and localized distortion in the preceding turn of Pro(5.50). Although this water is not resolved in the structure of opsin, the remarkable preservation of this distortion strongly suggests that this water is present throughout the activation process.

opsin bound to the Ct-Gα peptide (27), a model for the active state of rhodopsin, TMH 6 displays virtually the same patterns of helix bend and twist (Fig. 7, broken line). These data demonstrate that the shape of the helix is maintained during activation of rhodopsin, without neither a modulation of the bend nor a local opening/tightening of the helix.

5. Notes

1. In principle, a complete PDB file contains all the information required for HELANAL to run, as the program is able to recognize HELIX records in the PDB file. However, there are a number of reasons that recommend including an additional input file with the position and length of the helices:

   (a) Relatively distorted helices may have been automatically split in several segments in the HELIX records and, in this case, HELANAL would not be able to quantify their distortion;

   (b) Structures coming from homology modeling or molecular dynamics simulations will not have HELIX records; and

(c) Any change in the HELIX record format in the PDB compared to the format hardcoded in the HELANAL source code will result in HELANAL not being able to recognize the helix segments.

2. Information about the position and length of the helices can be provided in an additional input file. Importantly, the user has to provide the format in which this information is written, by providing the specific format statement (in Fortran 77 style) during the interactive execution of the program. In the following example, the format is (a8,1x,a3,1x,a3,1x,a1,3x,i3,2x,a3,1x,a1,3x,i3):

```
1GZM.pdb    H1  TRP A     35  HIS A     65
1GZM.pdb    H2  ARG A     70  HIS A    100
1GZM.pdb    H3  GLY A    106  VAL A    139
1GZM.pdb    H4  GLU A    150  VAL A    173
1GZM.pdb    H5  ASN A    200  LYS A    231
1GZM.pdb    H6  GLN A    244  GLN A    279
1GZM.pdb    H7  ILE A    286  MET A    309
1GZM.pdb    H8  LYS A    311  LEU A    321
```

Each helical segment has to be at least nine residues long.

3. How to read the results in the *angle.out* file. In this example (TMH1 1 of rhodopsin (PDBid: 1GZM)) the first value of 4.994° corresponds to the bend angle between the axes computed for residues 1 to 4 (i.e., the helix turn from Pro34(1.29) to Phe37(1.32)) and 4 to 7 (i.e., the helix turn from Phe37(1.32) to Leu40(1.35)). Thus, this value of 4.994° measures the local bend angle at residue 4 in the user-defined helix, which corresponds to Phe37(1.32) in the sequence. In each of the following rows, each helix turn is slid one residue.

```
1GZM.pdb    HELIX    H1    PRO A 34      GLN A 64
    1          4     4   7              4.994
    2          5     5   8              5.535
    3          6     6   9              4.266
    4          7     7  10              2.947
    5          8     8  11              3.530
    6          9     9  12              4.729
    7         10    10  13              4.785
    8         11    11  14              8.190
    9         12    12  15              8.181
   10         13    13  16              5.218
   11         14    14  17              4.670
   12         15    15  18              2.689
   13         16    16  19              9.128 -> Met49(1.44)
   14         17    17  20             13.482
   15         18    18  21             22.181 -> Gly51(1.46)
   16         19    19  22             13.817
   17         20    20  23             10.573 -> Pro53(1.48)
   18         21    21  24             10.333
   19         22    22  25              8.108 -> Asn55(1.50)
```

```
        20          23       23   26          3.074
        21          24       24   27          2.863
        22          25       25   28          4.839
        23          26       26   29          4.092
        24          27       27   30          1.688
        25          28       28   31          4.130
```

4. How to read the results in the *nh.out* file. In this example (TMH 1 of rhodopsin (1GZM)) the first value of 104.825° corresponds to the local twist of residues 1 to 4 (i.e., Pro34(1.29) to Phe37(1.32)). Assignation of the twist value to a single residue is less straightforward than in the case of the bend as, by definition, the twist measures distortion in a helix turn. I have decided to assign the value of twist to the first residue in the helix turn, so, in this example, the value of 104.825° is assigned to residue 1 in the user-defined helix, which corresponds to Pro34(1.29) in the sequence. When these values are interpreted in terms of protein structure, it is important to remember that we are mapping the distortion of each helix turn to the first residue of the turn.

```
1GZM.pdb   HELIX       H1      PRO A 34    GLN A 64
1           4       104.825     3.434       1.578
2           5       102.694     3.506       1.454
3           6       103.238     3.487       1.511
4           7       101.536     3.546       1.570
5           8        98.817     3.643       1.421
6           9        99.222     3.628       1.459
7          10        99.252     3.627       1.403
8          11        99.326     3.624       1.549
9          12        97.753     3.683       1.429
10         13       103.297     3.485       1.654
11         14        98.881     3.641       1.427
12         15       100.783     3.572       1.550
13         16        96.574     3.728       1.414
14         17        96.165     3.744       1.456
15         18        92.295     3.901       1.238
16         19        92.918     3.874       1.591
17         20        98.144     3.668       1.810
18         21        87.719     4.104       1.270 -> Gly51(1.46)-Ile54(1.49) turn
19         22        94.256     3.819       1.663
20         23        92.280     3.901       1.378 -> Pro53(1.48)-Phe56(1.51) turn
21         24        96.758     3.721       1.455
22         25       101.657     3.541       1.490
23         26        99.530     3.617       1.445
24         27        96.727     3.722       1.372
25         28        95.009     3.789       1.507
26         29        96.939     3.714       1.524
27         30        98.165     3.667       1.586
28         31        95.637     3.764       1.451
```

5. Changes in local twist are interpreted as follows: a regular α helix, with approximately 3.6 residues per turn, has a twist angle of approximately 100° (360°/3.6) (Fig. 2, middle panels), a tightened helix segment, with <3.6 residues per turn, has a twist >100° (Fig. 2, right panels), and an open helix segment, with >3.6 residues per turn, has a twist <100° (Fig. 2, left panels).

Acknowledgments

I am grateful for financial support from the Swiss National Science Foundation (SNSF) [grant number 31003A_132815] and the ETH Zürich within the framework of the National Center for Competence in Research in Structural Biology Program.

References

1. Rosenbaum DM, Rasmussen SGF et al (2009) The structure and function of G-protein-coupled receptors. Nature 459:356–363

2. Tate CG, Schertler GF (2009) Engineering G protein-coupled receptors to facilitate their structure determination. Curr Opin Struct Biol 19:386–395

3. Wu B, Chien EY et al (2010) Structures of the CXCR4 Chemokine GPCR with Small-Molecule and Cyclic Peptide Antagonists. Science. doi:10.1126/science.1194396

4. Chien EY, Liu W et al (2010) Structure of the human dopamine D3 receptor in complex with a D2/D3 selective antagonist. Science 330:1091–1095

5. Hofmann KP, Scheerer P et al (2009) A G protein-coupled receptor at work: the rhodopsin model. Trends Biochem Sci 34:540–552

6. Devillé J, Rey J et al (2009) An indel in transmembrane helix 2 helps to trace the molecular evolution of class A G-protein-coupled receptors. J Mol Evol 68:475–489

7. Yohannan S, Faham S et al (2004) The evolution of transmembrane helix kinks and the structural diversity of G protein-coupled receptors. Proc Natl Acad Sci USA 101:959–963

8. Riek RP, Rigoutsos I et al (2001) Non-alpha-helical elements modulate polytopic membrane protein architecture. J Mol Biol 306:349–362

9. Barlow DJ, Thornton JM (1988) Helix geometry in proteins. J Mol Biol 201:601–619

10. Ballesteros J, Weinstein H (1995) Integrated methods for the construction of three dimensional models and computational probing of structure-function relations in G protein-coupled receptors. Methods Neurosci 25:366–428

11. Deupi X, Dolker N et al (2007) Structural models of class a G protein-coupled receptors as a tool for drug design: insights on transmembrane bundle plasticity. Curr Top Med Chem 7:991–998

12. Deupi X, Govaerts C et al (2005) Conformational plasticity of GPCR binding sites. In: Devi L.A. (ed) The G protein-coupled receptors handbook. Humana Press, New Jersey

13. Deupi X, Olivella M et al (2010) Influence of the g- conformation of Ser and Thr on the structure of transmembrane helices. J Struct Biol 169:116–123

14. Visiers I, Braunheim BB et al (2000) Prokink: a protocol for numerical evaluation of helix distortions by proline. Protein Eng 13:603–606

15. Bansal M, Kumar S et al (2000) HELANAL: a program to characterize helix geometry in proteins. J Biomol Struct Dyn 17:811–819

16. Kumar S, Bansal M (1998) Geometrical and sequence characteristics of alpha-helices in globular proteins. Biophys J 75:1935–1944

17. Sansuk K, Deupi X et al (2011) A structural insight into the reorientation of transmembrane domains 3 and 5 during family A G protein-coupled receptor activation. Mol Pharmacol 79:262–269

18. Pardo L, Deupi X et al (2007) The role of internal water molecules in the structure and

function of the rhodopsin family of G protein-coupled receptors. Chembiochem 8:19–24

19. Springael JY, de Poorter C et al (2007) The activation mechanism of chemokine receptor CCR5 involves common structural changes but a different network of interhelical interactions relative to rhodopsin. Cell Signal 19: 1446–1456

20. Li J, Edwards PC et al (2004) Structure of bovine rhodopsin in a trigonal crystal form. J Mol Biol 343:1409–1438

21. Cordes FS, Bright JN et al (2002) Proline-induced distortions of transmembrane helices. J Mol Biol 323:951–960

22. Deupi X, Olivella M et al (2004) Ser and Thr residues modulate the conformation of pro-kinked transmembrane alpha-helices. Biophys J 86:105–115

23. Palczewski K, Kumasaka T et al (2000) Crystal structure of rhodopsin: A G protein-coupled receptor. Science 289:739–745

24. Ballesteros JA, Shi L et al (2001) Structural mimicry in G protein-couple receptors: implications of the high-resolution structure of rhodopsin for structure-function analysis of rhodopsin-like receptors. Mol Pharmacol 60:1–19

25. Rosenbaum DM, Cherezov V et al (2007) GPCR engineering yields high-resolution structural insights into beta2-adrenergic receptor function. Science 318:1266–1273

26. Warne T, Serrano-Vega MJ et al (2008) Structure of a beta1-adrenergic G-protein-coupled receptor. Nature 454:486–491

27. Scheerer P, Park JH et al (2008) Crystal structure of opsin in its G-protein-interacting conformation. Nature 455:497–502

Chapter 14

Structure Prediction of G Protein-Coupled Receptors and Their Ensemble of Functionally Important Conformations

Ravinder Abrol, Adam R. Griffith, Jenelle K. Bray, and William A. Goddard III

Abstract

G protein-coupled receptors (GPCRs) are integral membrane proteins whose "pleiotropic" nature enables transmembrane (TM) signal transduction, amplification, and diversification via G protein-coupled and β arrestin-coupled pathways. GPCRs appear to enable this by being structurally flexible and by existing in different conformational states with potentially different signaling and functional consequences. We describe a method for the prediction of the three-dimensional structures of these different conformations of GPCRs starting from their amino acid sequence. It combines a unique protocol of computational methods that first predict the TM regions of these receptors and TM helix shapes based on those regions, which is followed by a locally complete sampling of TM helix packings and their scoring that results in a few (~10–20) lowest energy conformations likely to play a role in binding to different ligands and signaling events. Prediction of the structures for multiple conformations of a GPCR is starting to enable the testing of multiple hypotheses related to GPCR activation and binding to ligands with different signaling profiles.

Key words: GPCR, Protein structure prediction, Transmembrane helix, Hydrophobicity, Helix kinks, BiHelix, GPCR conformations, Activation, Transmembrane signaling

1. Introduction

G protein-coupled receptors (GPCRs) are intrinsic membrane proteins with seven transmembrane (TM) helices. As their name suggests, these receptors were thought to be G protein-coupled, but enough evidence has accumulated in favor of their exclusive coupling to β arrestins (1, 2) that these are being referred to as 7TM proteins. They are the largest superfamily in human genome with ~800 GPCRs identified, including ~340 nonolfactory receptors organized into six families (GRAFTS) (3): Glutamate, Rhodopsin, Adhesion, Frizzled, Taste2 (Bitter), and

Nagarajan Vaidehi and Judith Klein-Seetharaman (eds.), *Membrane Protein Structure and Dynamics: Methods and Protocols*, Methods in Molecular Biology, vol. 914, DOI 10.1007/978-1-62703-023-6_14, © Springer Science+Business Media, LLC 2012

Secretin. A variety of bioactive molecules (including biogenic amines, peptides, lipids, nucleotides, and proteins) modulate GPCR activity to affect regulation of essential physiological processes (e.g., neurotransmission, cellular metabolism, secretion, cell growth, immune defense, and differentiation). Thus, many important cell recognition and communication processes involve GPCRs. Due to their mediation of numerous critical physiological functions, GPCRs are involved in all major disease areas including cardiovascular, metabolic, neurodegenerative, psychiatric, cancer, and infectious diseases.

In the last few years there has been a rapid increase in the solution of crystal structures for many GPCRs (4–15) due mainly to a technological revolution in membrane protein structure determination methods (16). The topological comparison of these crystal structures and their implications for GPCR activation has been reviewed extensively (17–21). All GPCR conformations observed so far have corresponded to the inactive state of the respective receptor, except for bovine opsin (9) and nanobody stabilized human β2 adrenergic receptor (β2AR) bound to an agonist (12). Structure determination efforts are moving towards the stabilization of GPCRs in different functional conformations (e.g., bound to agonists, G proteins or β arrestins), and structural computational biology can help by mapping the energy landscape sampled by GPCRs during their life-cycle and characterizing the critical conformations along the way to link with those that are observed in experimental structures.

It has been well established that during the process of GPCR activation a sequence of conformational changes takes place that take the receptor from an inactive state to an active state. This enables the cell to convert extracellular signals into intracellular signaling cascades through G protein-coupled and/or β arrestin-coupled pathways resulting in specific physiological responses (1, 2). Strong experimental evidence for this flexibility includes fluorescence lifetime measurements that have identified distinct agonist-induced conformational states for the β_2 Adrenergic Receptor (22). It has also been recently shown that conformational coupling between the extracellular surface (ECS) and orthosteric binding site in the TM region can stabilize different conformations of a GPCR (23). It is expected that functionally different ligands (agonists, antagonists, inverse agonists, and allosteric modulators) bind to different receptor conformations among the continuum of conformational states that lie along the activation pathway, so if there was a method to identify all low-energy conformations accessible to GPCRs, it can provide a starting point to test hypotheses related to the binding of different ligands and associating specific functions with different active state conformations some of which may only couple to G protein pathways and some to β arrestin pathways.

Protein structure prediction and modeling is playing an increasingly important role in providing detailed structural information that is relevant to GPCR activation and holds the most promise to describe the continuum of conformational states involved in GPCR function. Membrane proteins and their environment have been the focus of structure prediction and dynamics simulations for some years now (24). The interaction of these proteins with their lipid environment is considered critical to their in vivo folding and many recent studies have attempted to quantify this interaction on an absolute thermodynamic basis (25) by providing, for example, thermodynamic costs for the insertion of amino acids (that make up the TM helices) into the lipid bilayer (26). The three-dimensional structure of these α-helical membrane proteins, to which GPCRs belong, is strongly affected by interhelical interactions (mainly H-bonds and salt-bridges) (27). An accurate GPCR structure prediction methodology needs to be able to sample and describe these interhelical interactions very thoroughly. The conformational changes that accompany GPCR activation are known to occur on the millisecond or higher time scales, suggesting that explicit all-atom molecular dynamics will not be able to describe these large conformational changes for many years and even then cannot ensure complete sampling of available conformational space.

Many methods have been used to obtain model structures for membrane proteins due to their pharmacological importance. These methods have been reviewed elsewhere (24, 28). For GPCRs the main approach has been homology modeling (using the X-ray structure of Bovine Rhodopsin as a reference till 2007, and others more recently). Because of their low homology to other GPCRs of pharmacological interest, most studies have used constraints based on mutation and binding experiments coupled to the homologous Rhodopsin or later structures to guide additional mutation experiments. These structures have not generally been sufficiently accurate for predicting binding sites of ligands. Methods are also available for predicting structures of membrane proteins in general (see e.g., (29)). Some recent computational studies have started from inactive structures (from crystals) and provided a more detailed view of ligand-stabilized GPCR conformations (30, 31) which is consistent with experimental observations. Our group has been developing de novo computational approaches (not based on pure homology) such as MembStruk and HierDock (32), for predicting the 3D structure of a GPCR, and its ligand binding sites that have been used in the past (33–38).

This review focuses on the new generation of methods developed in our group aimed at generating multiple conformations of a GPCR. It is referred to as the GEnSeMBLE (GPCR Ensemble of Structures in Membrane BiLayer Environment) method, and includes a highly efficient procedure to completely sample the available conformational space. It is described in the next section followed by practical caveats in the Notes section.

2. Methods

In order to provide the 3D structures for these various conformations needed to understand the function of GPCRs and to help design new ligands, we developed the GEnSeMBLE (*GPCR Ensemble of Structures in Membrane BiLayer Environment*) method to predict the 3D structure (without using homology to known 3D structures). GEnSeMBLE predicts the ensemble of low-energy conformational states for a GPCR. The methodology consists of the following key steps:

1. *PredicTM*: A new method to predict the TM regions for a helical membrane protein that doesn't use any fitted parameters and also accounts for TM region extensions beyond the hydrophobic region.

2. *OptHelix*: A new and unbiased approach using molecular dynamics to generate relaxed helices with their natural kinks.

3. *BiHelix/SuperBiHelix*: A novel and highly efficient sampling algorithm to sample multiple conformations of a GPCR, which divides the N-helix interaction problem into a limited number of 2-helix interactions and uses the SCREAM method for side-chain optimization (39).

4. *CombiHelix/SuperCombiHelix*: An algorithm that takes the results from BiHelix step, generates the multiple rotational combinations, optimizes the side-chains using SCREAM, evaluates an implicit membrane solvation contribution, and evaluates the total energy of the optimized multihelix bundle. This step provides an ensemble of multiple low-energy structures and docking of agonists, antagonists, and inverse agonists to all these conformations will provide a unique route to understand GPCR activation.

GEnSeMBLE provides a unique protocol to predict not only the lowest energy structure but also other low-energy conformations that will be relevant to understand the function of GPCRs in relation to their interaction with different ligands as well as to their relevance for GPCR activation. The individual steps in the GEnSeMBLE method are now described in detail.

2.1. PredicTM

Accurate ab initio prediction of helical membrane proteins like GPCRs begins with the identification of the location of the TM regions of the protein within the amino acid sequence. There are many methods already available for predicting the location of TM regions. These methods encompass simple hydrophobicity-based procedures, hidden Markov models, and neural networks. PredicTM combines hydrophobic analysis with helicity predictions to generate TM regions that encompass hydrophobic regions and can also potentially extend beyond them. The difficulty with this approach is sufficiently reducing noise in the hydrophobicity analysis without losing data.

The steps of the PredicTM procedure are broken into six parts:

1. Retrieval of similar protein sequences from a database.
2. Multiple sequence alignment of similar sequences.
3. Hydrophobic profile generation and noise removal.
4. Initial transmembrane region predictions.
5. Helicity predictions and extension of hydrophobic regions.
6. Identification of hydrophobic centers.

Step 1: *Similarity search*. The first step of the procedure, retrieving sequences related to the target sequence, is carried out through a modified BLAST search. BLAST, the Basic Local Alignment Search Tool, is a bioinformatics program used to search databases for related sequences and is widely used (2). One of the standard implementations of BLAST, and the implementation used by PredicTM, is at the Expert Protein Analysis System (ExPASy) server at the Swiss Institute of Bioinformatics (SIB). Typically, one gives a sequence or sequence identifier as an input to a Web-based form, along with some parameters, and the BLAST program will return either all sequences satisfying a statistical cutoff threshold or up to a user-specified number of sequences, whichever is reached first. The goal of this step for PredicTM is to always retrieve all sequences satisfying the statistical cutoff. For this reason, PredicTM incrementally increases the requested number of sequences until no new sequences are returned by BLAST. Once this completeness has been achieved, the sequences are returned in FASTA format for the multiple sequence alignment step (see Note 1).

Step 2: *Multiple sequence alignment*. The second step of the PredicTM procedure is multiple sequence alignment of the sequences returned from BLAST. This is executed using the MAFFT method. It has several different algorithms available, but the one used by PredicTM is the "E-INS-i" method, which is best suited for sequences with multiple aligning segments separated by nonaligning segments, which perfectly describes the situation for GPCRs (see Note 2).

Step 3: *Hydrophobicity Profile*. The third step of PredicTM is the generation of the hydrophobicity profile from the multiple sequence alignment. Because the sequences in the alignment typically have at least some nonaligning portions, there will almost always be gaps in the alignment of the target sequence. PredicTM is only concerned with the portions of the alignment that correspond directly to the target sequence, thus the first step of the profile generation procedure is to remove all portions of the alignment that align to gaps in the target sequence. The result is called a condensed alignment.

242   R. Abrol et al.

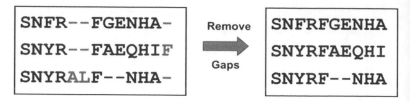

Fig. 1. Example showing the procedure to generate a condensed alignment.

Note that while there are no longer any gaps in the target sequence, there may still be gaps in the aligned sequences. This step is illustrated in Fig. 1.

The next step is to replace the amino acids in the condensed alignment with their corresponding hydrophobicity values. PredicTM uses the Wimley–White whole-residue octanol scale (25), which is a thermodynamic scale derived from transfer of residues from water into *n*-octanol. Unresolved amino acids in the alignment (B, Z, J, X) are replaced with gaps rather than potentially using an incorrect hydrophobicity value. With hydrophobicity values assigned to each residue in the alignment, a raw average hydrophobicity profile is generated. This is done by taking the average of the hydrophobicity values for all nongap residues for each position in the condensed alignment (Fig. 2).

It should be noted that by removing gaps in the alignment to create the condensed alignment and by ignoring gaps when averaging hydrophobicities, PredicTM is a gap-tolerant and gap-unbiased method. The raw average hydrophobicity is too noisy to serve for final prediction of helices. Left panel in Fig. 3 shows a sample raw average hydrophobicity profile (with a flipped sign) taken from a default application of PredicTM to the dopamine D1 receptor. Moving window averages are applied to reduce the noise in the raw average profile. Hydrophobicity averages are calculated using window sizes 7, 9, 11, ..., 19, 21 and then averaged over these window sizes to obtain a final profile shown in the right panel of Fig. 3 (see Note 3).

Step 4: Raw TM Predictions. The initial prediction of helices based on the final profile is quite simple. Because a thermodynamic scale is being used, hydrophobicity values greater than zero are hydrophobic and thus are part of a TM region, and values less than zero are hydrophilic and are part of the loops or termini. These are referred to as raw TM regions. Typically a rule is applied that eliminates helices less than a certain length, which by default is 10 residues in PredicTM. These eliminations are noted for the user so that they can be visually inspected to ensure that they are not actually part of another helix (see Note 4).

Step 5: TM region capping. The raw helices predicted from the hydrophobicity profile in the previous step represent the core of

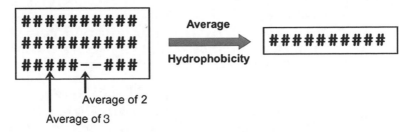

Fig. 2. Scheme to calculate average hydrophobicities along a condensed alignment.

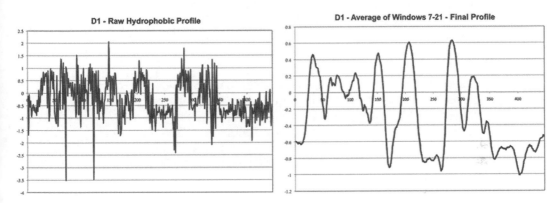

Fig. 3. Raw (*left*) and averaged (*right*) hydrophobic profile for Dopamine D1 receptor.

the TM region and do not account for their extensions beyond the membrane, though many TM helices extend beyond their hydrophobic cores. For an accurate representation of these extensions, we first use secondary structure prediction servers (APSSP2, Porter, PsiPred, Jpred, and SSPro) to obtain helicity predictions for the whole sequence. These predictions are then used to extend the raw TM regions using consensus among the five servers. An example of such predictions is shown in Fig. 4.

Step 6: Hydrophobic Centers. Later steps in the ab initio prediction of GPCR structures require the orientation of helices with respect to a plane representing the center of the membrane. This requires that a predefined point on the helix be placed on that plane. Two different hydrophobic centers are calculated for this reason. First is the raw midpoint center, which corresponds to the geometric center of the raw predicted helices. Second is the "area" or "centroid" center, which corresponds to the position in the raw helix where the area under the hydrophobicity profile curve is equal on both sides. This physically corresponds to the buoyant center of the TM helix in the lipid bilayer.

2.2. OptHelix

OptHelix is a program for the optimization of TM helix shapes, particularly with respect to proline kinks. Helices are individually

TMR1	PredicTMRaw	37	FSMLAAYMFLLIMLGFPINFLTLYV	61
	PredicTM2Hel	34	PWQFSMLAAYMFLLIMLGFPINFLTLYVTVQ	64
	CrystalRhod	33	EPWQFSMLAAYMFLLIMLGFPINFLTLYVTVQH	65
	CrystalOpsin	33	EPWQFSMLAAYMFLLIMLGFPINFLTLYVTVQH	65
TMR2	PredicTMRaw	73	NYILLNLAVADLFMVFGGFTTTLYT	97
	PredicTM2Hel	71	PLNYILLNLAVADLFMVFGGFTTTLYTSLH	100
	CrystalRhod	70	TPLNYILLNLAVADLFMVFGGFTTTLYTSLHG	101
	CrystalOpsin	70	TPLNYILLNLAVADLFMVFGGFTTTLYTSLHG	101
TMR3	PredicTMRaw	111	NLEGFFATLGGEIALWSLVV	130
	PredicTM2Hel	107	PTGCNLEGFFATLGGEIALWSLVVLAIERYVVV	139
	CrystalRhod	105	FGPTGCNLEGFFATLGGEIALWSLVVLAIERYVVVC	140
	CrystalOpsin	105	FGPTGCNLEGFFATLGGEIALWSLVVLAIERYVVVCK	141
TMR4	PredicTMRaw	155	MGVAFTWVMALACAAPPLV	173
	PredicTM2Hel	150	ENHAIMGVAFTWVMALACAAPPLVGW	175
	CrystalRhod	149	GENHAIMGVAFTWVMALACAAPPLV	173
	CrystalOpsin	149	GENHAIMGVAFTWVMALACAAPPLV	173

Fig. 4. Comparison of Raw and Helicity Capped TM region predictions for Rhodopsin against Retinal-bound and Retinal-free crystal forms.

optimized through molecular dynamics (MD) in vacuum using the following steps:

(a) The method begins by identifying the location of prolines in the TM regions that were output in the previous PredicTM step. When located near a terminus, these prolines can have excessively strong bending characteristics in the dynamics simulation. To alleviate this bending and to somewhat mimic the presence of the rest of the protein, alanines are added to the terminus until the proline is 8 residues from the terminus. As an additional option, 8 alanines can be added to the termini of all helices, regardless of proline position, or no alanines can be added.

(b) A canonical helix consisting solely of alanines is produced matching the length of the sequence. Any prolines or glycines are placed in the helix using SCREAM (39). A conjugate gradient minimization, typically down to a 0.5 kcal/mol/Å RMS force threshold, is performed on this structure. At this point, any serines or threonines are placed on the helix using SCREAM. These residues have been shown to interact with the backbone hydrogen bonding network and can influence proline kinks (40). The helix is now ready for molecular dynamics.

(c) Short 10 ps (5,000 steps of 2 fs) dynamics runs are performed at 50 K, 100 K, etc., up to 250 K to warm up the system. Equilibrium dynamics is performed at 300 K following the warm-up and is typically 2 ns in length (1,000,000 steps of 2 fs) and snapshots are taken at 10 ps intervals.

(d) Two structures are derived from the equilibrium dynamics. First, the lowest potential energy snapshot is selected from the last 1.5 ns of the dynamics. Second, the snapshot with an RMSD closest to that of the average structure during the last 1.5 ns of dynamics is selected.

(e) Finally, the original side chains are replaced using SCREAM, the structure is minimized, and any alanines that were added are removed. Application of OptHelix to the TMs of human β2 and bovine rhodopsin yields the RMSD values shown in Fig. 5. These values are taken from only one MD run. However, all RMSD values are less than the resolution of the crystal structures being compared to.

(f) Both MinRMSD and MinEnergy helix conformations are produced as output for input into the helix bundle sampling procedure.

2.3. BiHelix/ SuperBiHelix

1. *Starting Template*: To pack the seven helices from OptHelix into a bundle, requires the definition of six quantities for each

	Beta2		Bovine Rhodopsin	
	MinRMSD	MinEnergy	MinRMSD	MinEnergy
TM1	0.78	0.86	0.82	0.70
TM2	1.07	1.52	2.02	1.81
TM3	1.14	1.34	0.98	1.18
TM4	0.63	0.64	1.24	0.93
TM5	1.61	1.95	1.62	1.45
TM6	1.55	1.37	1.00	0.79
TM7	1.95	2.05	1.77	1.89
Min	0.63	0.64	0.82	0.70
Max	1.95	2.05	2.02	1.89
Avg	1.25	1.39	1.35	1.25

Fig. 5. CαRMSD of the seven TM helices predicted by OptHelix for β2 and rhodopsin relative to their crystals using the two criteria described in the text.

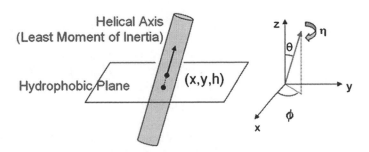

Fig. 6. Coordinates specifying the orientation of a TM helix in a membrane.

helix ($6 \times 7 = 42$ total; see Fig. 6): the x, y, h of some reference point (like hydrophobic center h from PredicTM), the tilt (θ) of each axis from the z axis, the azimuthal orientation (φ) of this tilt; and the rotation (η) of the helix about the helical axis. We choose $z = 0$ for the hydrophobic center (from PredicTM step) for each helix and the choice for the rotation angle η will be described below.

The remaining $4 \times 7 = 28$ coordinates are taken from a template structure. Currently we have multiple choices for this template structure:

(a) The 7.5 electron density map of frog rhodopsin (41).

(b) The bovine rhodopsin structure (4, 14).

(c) The structure for the CCR1 receptor with an antagonist BX 471 bound that was subjected to 10 ns of MD using an infinite membrane and full solvent (36).

(d) The structure for the DP receptor with the CDP2 agonist bound that was subjected to 2 ns of MD using an infinite membrane and full solvent (37).

(e) The structure for the MrgC11 receptor with an agonist FdMRFa bound that was subjected to 7 ns of MD using an infinite membrane and full solvent (42).

(f) Human β2 Adrenergic Receptor (5).

(g) Turkey β1 Adrenergic Receptor (7).

(h) Human A_{2A} Adenosine Receptor (6).

(i) Bovine Opsin (9).

GEnSeMBLE allows for each of these templates to be used in separate predictions, providing an ensemble of bundles among which we can select on the basis of energy or binding energy. As new structures are predicted, validated, and subjected to full MD, they will be added to the ensemble of templates.

To specify an initial rotation (η) of each helix, we use a conserved residue to match the rotation angle of its Cα projection on the $x-y$ plane, to that of the corresponding one in the template structure. The helical axis for rotation is defined as the one corresponding to the least moment of inertia axis obtained using all backbone atoms.

2. *Conformational Sampling of Helices*: Critical to defining the binding sites for ligands is the rotations and tilts of each helix about its axis (θ, ϕ, η), which are determined by the interhelical interactions between the various residues in different helices. The procedure for a complete sampling of all helical rotational combinations is called BiHelix, and even with a 30° increment for each of the seven helices, requires the sampling of $12^7 \sim 35$ million conformations, where the side chains need to be optimized for each helical rotation combination. The procedure for

a complete sampling of all helical tilt and rotation combinations is called SuperBiHelix and requires the sampling of trillions of conformations. This is computationally intractable and actually not necessary as in any given template some helices do not interact with each other. This led to the idea of the BiHelix sampling method described below.

The Sampling Method: In this procedure, first interacting pairs of helices are identified based on the template being considered. The rhodopsin template for Class A GPCRs allows for 12 interacting pairs of helices (H1–H2, H1–H7, H2–H3, H2–H4, H2–H7, H3–H4, H3–H5, H3–H6, H3–H7, H4–H5, H5–H6, H6–H7) as shown with two-way arrows in Fig. 7a. Now, in BiHelix, for each interacting pair of helices, we sample all combinations of a full 360° rotation for each helix with 30° increments leading to $12 \times 12 = 144$ combinations (see Note 5). During this sampling, the other 5 helices are not present, for example shown in Fig. 7b for helix 1–2 pair. In SuperBiHelix, helix tilt angles (θ, ϕ) are also sampled in addition to helix rotation angles η (see Note 6).

For each rotational combination, we optimize the side chains using the rotamer placement method SCREAM (39). SCREAM uses a library of residue conformations ranging from a heavy atom RMS diversity of 0.4A–1.6A in conjunction with a Monte Carlo sampling using full valence, hydrogen bond and electrostatic interactions, but special vdW potentials that reduce somewhat the penalty for contacts that are slightly too short while retaining the normal attractive interactions at full strength. With SCREAM, we find that we can now base the selections on the total energy, without separate considerations of valence, electrostatic, hydrogen bond, and vdw. Because SCREAM does not do energy minimization, it may still lead to vdW interactions slightly too large, hence we minimize each bihelical conformation for ten steps and evaluate the total energy. We take the interhelical and intrahelical components of this energy for the 144 combinations for each helix pair and use them in a pairwise addition equation scheme 1 shown above to obtain energies for

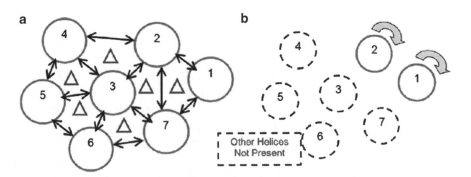

Fig. 7. BiHelix sampling scheme, where sampling is done two helices at a time for all interacting helix pairs. (a) All interacting helix pairs shown with a *double arrow*. (b) Helix1-Helix2 optimization shown in the absence of other helices.

$$e_{total}^{intra}(\eta_1,\eta_2,\eta_3,\eta_4,\eta_5,\eta_6,\eta_7) = \sum_{i=1}^{7}\frac{1}{N_i}\sum_{j=J_1}^{J_{N_i}}\left[e_{ij}^{j,intra}(\eta_i,\eta_j)\right]$$

$$e_{total}^{inter}(\eta_1,\eta_2,\eta_3,\eta_4,\eta_5,\eta_6,\eta_7) = \sum_{i=1}^{6}\sum_{j=J_1}^{J_{N_i}}\left[e_{ij}^{inter}(\eta_i,\eta_j)\right]$$

$$E_{total}^{bundle}(\eta_1,\eta_2,\eta_3,\eta_4,\eta_5,\eta_6,\eta_7) = e_{tot}^{intra} + e_{tot}^{inter}$$

Scheme 1. Pairwise summation of BiHelix energies to obtain seven-helix bundle energies.

all possible $12^7 \sim 35$ million conformational combinations and output the top 2,000 structures with lowest total energy.

2.4. CombiHelix/ SuperCombiHelix

(a) *Starting Bundles*: The top 2,000 structures coming out of BiHelix/SuperBiHelix analysis are built explicitly using the rotations and/or tilts specified for each helix in the combination. The helical axis for rotation is the same as used in BiHelix analysis.

(b) *BiHelix/CombiHelix Bundle Optimization*: For each of the bundles built in the previous step, the side chains are optimized using SCREAM, like in the BiHelix method. The resulting bundle is minimized for ten steps and total energy reported for each bundle.

For each of the six available GPCR crystal structures, the application of this BiHelix/CombiHelix procedure resulted in the crystal packing being ranked first. These results show that the energies used are reliable enough that, upon complete sampling of the conformational space they can still identify the lowest energy conformation assumed to be the crystal conformation. Since the crystal ligand is absent during sampling (except for Bovine Opsin case which is the ligand-free form of Rhodopsin), these results indicate that the GPCR conformations are in their minimum energy state for both apo- and cocrystal forms at least in the helix rotation angle η space.

First we will look at top ten conformations for Bovine Rhodopsin after the application of BiHelix/CombiHelix procedure using 30° sampling of helix rotation angle. Top 2,000 conformations from BiHelix were taken through CombiHelix. Figure 8 shows the top ten conformations after CombiHelix, along with their energy (SCH-Energy) in kcal/mol, BiHelix rank (SBHRnk), and BiHelix energy (SBH-Energy) also in kcal/mol. Since the starting template was the crystal conformation, that conformation corresponds to all seven η angles being "zero" and according to the Fig. 8 (first row), it is ranked first after conformational sampling using BiHelix/ CombiHelix. This is again a validation of the energies being used to rank/score the conformations as well as additional validation of the side-chain placement program SCREAM (39).

SCHRnk	SCH-Energy	SBHRnk	SBH-Energy	Eta1	Eta2	Eta3	Eta4	Eta5	Eta6	Eta7	CRmsd
1	250	1	211	0	0	0	0	0	0	0	0.0
2	252	3	246	0	0	0	0	0	330	0	0.7
3	268	12	267	0	0	0	0	150	0	0	1.6
4	284	4	250	0	0	0	0	60	0	0	0.9
5	285	20	275	0	0	0	0	180	0	0	1.7
6	287	11	267	0	0	0	0	300	0	0	0.9
7	292	335	338	0	0	330	0	0	0	0	0.6
8	293	2	241	0	0	0	0	90	0	0	1.2
9	302	5	253	0	0	0	0	330	0	0	0.5
10	305	7	255	0	0	0	0	0	90	0	1.7

Fig. 8. Top 10 Bovine Rhodopsin conformations after BiHelix/CombiHelix.

Even though only helix rotation angle was sampled during this procedure, the top ten conformations provide signatures for additional conformations that may be important as part of the function of Rhodopsin. First thing that is evident is the high conformational flexibility of TM5 in the helix rotational space, which is not inconsistent with the flexibility associated with TM5 for class A GPCRs. In addition, we see counterclockwise 30° rotations in TM3 and TM6 as shown in Fig. 8 by red cells, which is consistent with the helix rotations observed in the Bovine Opsin (the retinal-free form of Bovine Rhodopsin) crystal structure, –28° rotation for TM3 and –31° rotation for TM6, relative to Bovine Rhodopsin. We do not see a single conformation in the top 2,000 that shows both TM3 and TM6 with this rotation, because these rotations are also associated with changes in helix tilts which were not sampled in this case. The results for Bovine Rhodopsin show that: a. BiHelix/CombiHelix procedure can identify the crystal conformations; b. This procedure has the potential to identify multiple GPCR conformations that may have a physiological role in their function.

Figure 9 shows the ensemble of conformations that result from the application of SuperBiHelix/SuperCombiHelix procedure to adenosine A_{2A} receptor, which sampled helix tilt and rotation angles simultaneously. The sampling grid was: θ (–10, 0, +10), ϕ (–30, –15, 0, +15, +30), and η (–30, –15, 0, +15, +30) (see Note 7). Figure 9a shows the top ten conformations after SuperBiHelix and the conformation observed in the crystal (shaded cells) is ranked second out of ~10 trillion conformations. Figure 9b shows the top ten conformations after SuperCombiHelix on the top 2,000 conformations out of SuperBiHelix. The conformation observed in the crystal (shaded cells) is ranked sixth and all the top conformations are near-native. These results highlight the accuracy of the

a

							SuperBiHelix																	Energy	RMSD
θ	H1	H2	H3	H4	H5	H6	H7	φ	H1	H2	H3	H4	H5	H6	H7	η	H1	H2	H3	H4	H5	H6	H7	(kcal/mol)	(Å)
	0	0	0	-10	10	0	0		0	0	15	-30	0	0	0		0	0	0	-30	30	0	0	392.6	1.3
	0	0	0	0	0	0	0		0	0	0	0	0	0	0		0	0	0	0	0	0	0	396.9	0
	0	0	0	-10	0	0	0		0	0	15	-30	0	-15	0		0	0	0	-30	-15	0	0	398.3	1
	0	0	0	-10	10	0	0		0	0	15	-30	0	0	0		0	0	15	-30	30	0	0	400.6	1.3
	0	0	0	-10	0	0	0		0	0	15	-30	0	0	0		0	0	0	-30	0	0	0	400.7	0.9
	0	0	0	-10	0	0	0		0	0	15	-30	0	0	0		0	0	0	-30	-15	0	0	401.2	0.9
	0	0	0	-10	10	0	0		0	0	15	-30	-15	0	0		0	0	0	-30	15	0	0	401.3	1.3
	0	0	0	0	0	0	0		0	0	0	15	0	0	0		0	0	0	0	0	0	0	402.4	0.4
	0	0	0	-10	10	0	0		0	-15	15	-30	0	0	0		0	0	0	-30	30	0	0	402.4	1.3
	0	0	0	-10	0	0	0		0	0	15	-30	-15	0	0		0	0	0	-30	0	0	0	402.8	1

b

							SuperComBiHelix																	Energy	RMSD
θ	H1	H2	H3	H4	H5	H6	H7	φ	H1	H2	H3	H4	H5	H6	H7	η	H1	H2	H3	H4	H5	H6	H7	(kcal/mol)	(Å)
	0	0	0	0	0	0	0		0	0	0	0	-15	0	0		0	0	0	0	0	0	0	59.1	0.4
	0	0	0	0	0	0	0		0	0	-15	0	0	0	0		0	0	0	0	0	0	0	71.2	0.4
	0	0	0	0	0	0	0		0	0	-15	0	0	15	0		0	0	0	0	0	0	0	74.7	0.6
	0	0	0	0	0	0	0		0	0	-15	0	0	-15	0		0	0	0	0	0	0	0	76.3	0.7
	0	0	0	0	0	0	0		0	0	0	-15	0	-15	0		0	0	0	0	0	0	0	77	0.6
	0	0	0	0	0	0	0		0	0	0	0	0	0	0		0	0	0	0	0	0	0	78	0
	0	0	0	0	10	0	0		0	0	0	0	0	-15	0		0	0	0	0	15	0	0	79.1	1
	0	0	0	0	0	0	0		0	0	0	0	0	-15	0		0	0	0	0	0	0	0	85.8	0.5
	0	0	0	0	0	0	0		0	0	0	0	0	-15	-15		0	0	0	0	0	0	0	86.2	0.6
	0	0	0	0	0	0	0		0	0	-15	0	0	15	15		0	0	0	0	0	0	0	87.8	0.6

Fig. 9. Ensembles of top ten conformations for adenosine A_{2A} receptor after SuperBiHelix (**a**) and SuperCombihelix (**b**). Crystal conformation is highlighted by shaded cells.

energies used that allow the SuperBiHelix/SuperCombiHelix procedure to identify near-native conformations even out of ~10 trillion conformations.

The GEnSeMBLE procedure brings together methods for TM helix region prediction, helix shape optimization, and exhaustive but efficient TM helix bundle conformational sampling. This exhaustive sampling predicts an ensemble of low-energy GPCR conformations some of which are expected to play different functional roles in GPCR mediated signaling pathways.

3. Materials

1. The GEnSeMBLE suite of programs consists of the following modules: PredicTM, OptHelix, BiHelix, CombiHelix, SuperBiHelix, and SuperCombiHelix. Some of these modules call many helper routines along with SCREAM for optimal side-chain placement.

2. These programs run on a standard linux cluster of PCs running CentOS (a Red Hat Linux Clone). OptHelix through SuperCombiHelix modules run on multiple processors and

require number of processors ranging from 7 (OptHelix) to 24 (SuperCombiHelix).

3. No programming knowledge of scripting languages like Perl and Python is required to run the programs.

4. Notes

1. *PredicTM*: This step also allows the user to select some of the standard BLAST options. The ExPASy implementation of BLAST typically searches the Swiss-Prot and TrEMBL databases. However, TrEMBL is not curated and can be excluded from the searches, which is the default option for PredicTM. Additionally, the database can be restricted to a specific taxon, such as eukaryota, vertebrata, or mammalia. These options allow the user to restrict results to the most relevant species for a given protein. The E threshold is the statistical measure used by BLAST and can also be adjusted by the user. Larger E thresholds allow more loosely related proteins to appear in the results while smaller E thresholds restrict the results to more closely related proteins. Finally, BLAST allows the user to filter sequences for regions of low complexity. These regions can often result in unrelated proteins, skewing the results. However, because nothing is known about the protein beforehand, the low-complexity filter can also mask relevant portions of the sequence and must thus be used with caution.

2. *PredicTM*: MAFFT is used as the multiple sequence alignment program in PredicTM module. MAFFT has several different algorithms available, but the one used by PredicTM is the "E-INS-i" method, which is best suited for sequences with multiple aligning segments separated by nonaligning segments, which perfectly describes the situation for GPCRs. All GPCRs have seven conserved TM regions (the aligning portions) separated by nonconserved loops (the nonaligning portions). MAFFT correctly aligns all helices of the five dopamine receptors. Additionally, MAFFT does well for cases where non-GPCR sequences are introduced into the alignment. The dopamine D4 receptor has a long IC3 loop that is proline-dense. When performing a BLAST search with low-complexity filtering turned on, this loop is completely masked. When the low-complexity filter is not used, the results for a BLAST search against D4 have a large number of non-GPCR sequences. When aligned with MAFFT, these unrelated sequences typically do not align in the TM regions, thus allowing an accurate prediction to be made despite having a large number of unrelated sequences in the alignment.

3. *PredicTM*: The final step in generating a hydrophobicity profile is to average the moving windows. The default in PredicTM is to average windows 7 through 21, with 7 corresponding to roughly one helical turn above and below a residue and 21 corresponding to roughly the length of one TM region.

4. *PredicTM*: Prediction of TM3 is expected to be a problem for GPCRs because it is primarily an internal helix and is generally less hydrophobic in nature than the other helices. Additionally, TMs 2 and 3 and TMs 6 and 7 are often connected by very short loops, which can result in the hydrophobic peaks merging into one signal.

5. *BiHelix/CombiHelix*: This sampling mode is useful if the target GPCR is not similar to any of the starting templates. In this case, it is recommended to first perform sampling of helix rotation angles η to obtain a structure with local minimum in the η space. This can then be followed by the sampling of all helix tilts and rotations (SuperBiHelix/SuperCombiHelix).

6. *SuperBiHelix/SuperCombiHelix*: This sampling mode is useful if the target GPCR is highly homologous to one of the starting templates. In this case, sampling of helix tilts and rotations (SuperBiHelix/SuperCombiHelix) can be performed simultaneously as the starting structure can be assumed to be close to the minimum in the η rotation angle space.

7. *SuperBiHelix/SuperCombiHelix*: During the SuperBiHelix step, the following sampling grids have been found useful: θ (−10, 0, +10), ϕ (−30, −15, 0, +15, +30), and η (−30, −15, 0, +15, +30). These are derived from the known topological diversity seen among all available templates.

Acknowledgments

Our development of predictive methods for structures and function of GPCRs started ~1998 with an ARO-MURI grant oriented toward olfaction, a time when no 3D structures were available. The early advances were made by Dr. N. Vaidehi (now Professor of Immunology at City of Hope, Duarte, California) and Wely Floriano (now Associate Professor of Chemistry at Lakehead University, Ontario, Canada). Other important early contributions were made by Deqiang Zhang, Spencer Hall, Peter Freddolino, Eun Jung Choi, Georgios Zamanakos, Peter Kekenes-Huskey, Peter Freddolino, Yashar Kalani, Rene Trabanino, Victor Kam, Art Cho, John Wendel, P. Hummel, Peter Spijker; Joyce Yao-chun Peng, Youyong Li, and Jiyoung Heo. More recently the team has been led by Ravinder Abrol, with major contributions from Jenelle Bray, Soo-Kyung Kim, Bartosz Trzaskowski, Adam Griffith, Caitlin Scott, Caglar Tanrikulu, and Andrea Kirkpatrick. The codes for

PredicTM and OptHelix were written by Adam Griffith. The codes for BiHelix/SuperBiHelix were written by Ravinder Abrol and Jenelle Bray with some modules written by Bartosz Trzaskowski. The funding sources that have allowed us to continue this project came from Sanofi Aventis, Boehringer-Ingelheim, Pfizer, Schering AG, and PharmSelex. Also some funding came from NIH.

Note added in proof

Since this chapter was submitted the number of crystallized GPCR structures has more than doubled, which along with new predicted structures provide additional starting templates (introduced in Section 2.3) for the structure prediction of GPCRs. These additional templates from newly crystallized GPCRs correspond to human Histamine H1, human Dopamine D3, human chemokine CXCR4, human Muscarinic Acetylcholine M2, rat Muscarinic Acetylcholine M3, mouse Mu-Opioid, human Kappa-Opioid, mouse Delta-Opioid, human N/OFQ-Opioid, Sphingosine-1-Phosphate 1, bovine meta II rhodopsin, agonist-bound human Adenosine 2A, and Gs-bound human Beta2 Adrenergic receptors. Some of these have been compared in Abrol et al. (2011) Methods 55(4):405–414. Additional predicted templates correspond to human GLP-1, human Somatostatin type 5, human Chemokine CCR5, human Serotonin 2B, human Olfactory 1G1, human Urotensin 2, human Cannabinoid 1 and 2, human bitter taste member 38, and human Adenosine A3 receptors.

References

1. Barki-Harrington L, Rockman HA (2008) B-arrestins: Multifunctional cellular mediators. Physiology 23(1):17–22

2. Kenakin T, Miller LJ (2010) Seven transmembrane receptors as shapeshifting proteins: the impact of allosteric modulation and functional selectivity on New drug discovery. Pharmacol Rev 62(2):265–304

3. Lagerstrom MC, Schioth HB (2008) Structural diversity of G protein-coupled receptors and significance for drug discovery. Nat Rev Drug Discov 7(4):339–357

4. Palczewski K et al (2000) Crystal structure of rhodopsin: a G protein-coupled receptor. Science 289(5480):739–745

5. Cherezov V et al (2007) High-resolution crystal structure of an engineered human beta2-adrenergic G protein-coupled receptor. Science 318(5854):1258–1265

6. Jaakola VP et al (2008) The 2.6 angstrom crystal structure of a human A2A adenosine receptor bound to an antagonist. Science 322(5905):1211–1217

7. Warne T et al (2008) Structure of a beta1-adrenergic G-protein-coupled receptor. Nature 454(7203):486–491

8. Murakami M, Kouyama T (2008) Crystal structure of squid rhodopsin. Nature 453(7193):363–U33

9. Park JH et al (2008) Crystal structure of the ligand-free G-protein-coupled receptor opsin. Nature 454(7201):183–187

10. Chien EYT et al (2010) Structure of the Human Dopamine D3 Receptor in Complex with a D2/D3 Selective Antagonist. Science 330(6007):1091–1095

11. Wu BL et al (2010) Structures of the CXCR4 Chemokine GPCR with Small-Molecule and Cyclic Peptide Antagonists. Science 330 (6007):1066–1071

12. Rasmussen SG et al (2011) Structure of a nanobody-stabilized active state of the beta(2) adrenoceptor. Nature 469(7329):175–180

13. Rosenbaum DM et al (2011) Structure and function of an irreversible agonist-beta(2) adrenoceptor complex. Nature 469(7329):236–240

14. Okada T et al (2004) The retinal conformation and its environment in rhodopsin in light of a new 2.2 A crystal structure. J Mol Biol 342(2):571–583

15. Scheerer P et al (2008) Crystal structure of opsin in its G-protein-interacting conformation. Nature 455(7212):497–U30

16. Blois TM, Bowie JU (2009) G-protein-coupled receptor structures were not built in a day. Protein Sci 18(7):1335–1342

17. Kobilka B, Schertler GFX (2008) New G-protein-coupled receptor crystal structures: insights and limitations. Trends Pharmacol Sci 29(2):79–83

18. Mustafi D, Palczewski K (2009) Topology of class a G protein-coupled receptors: insights gained from crystal structures of rhodopsins, adrenergic and adenosine receptors. Mol Pharmacol 75(1):1–12

19. Rosenbaum DM, Rasmussen SG, Kobilka BK (2009) The structure and function of G-protein-coupled receptors. Nature 459(7245):356–363

20. Hanson MA, Stevens RC (2009) Discovery of New GPCR biology: one receptor structure at a time. Structure 17(1):8–14

21. Worth CL, Kleinau G, Krause G (2009) Comparative sequence and structural analyses of G-protein-coupled receptor crystal structures and implications for molecular models. PLoS One 4(9):e7011

22. Kobilka BK (2007) G protein coupled receptor structure and activation. Biochimica Et Biophysica Acta-Biomembranes 1768(4):794–807

23. Bokoch MP et al (2010) Ligand-specific regulation of the extracellular surface of a G-protein-coupled receptor. Nature 463(7277):108–U121

24. Fleishman SJ, Unger VM, Ben-Tal N (2006) Transmembrane protein structures without X-rays. Trends Biochem Sci 31(2):106–113

25. Wimley WC, Creamer TP, White SH (1996) Experimentally determined hydrophobicity scales for membrane proteins. Biophys J 70(2):Tuam1–Tuam1

26. Hessa T et al (2007) Molecular code for transmembrane-helix recognition by the Sec61 translocon. Nature 450(7172):1026–U2

27. White SH (2006) How hydrogen bonds shape membrane protein structure. Peptide Solvation H-Bonds 72:157–172

28. Fanelli F, De Benedetti PG (2005) Computational Modeling approaches to structure-function analysis of G protein-coupled receptors. Chem Rev 105(9):3297–3351

29. Yarov-Yarovoy V, Schonbrun J, Baker D (2006) Multipass membrane protein structure prediction using Rosetta. Proteins-Struct Func Bioinform 62(4):1010–1025

30. Bhattacharya S, Hall SE, Vaidehi N (2008) Agonist-induced conformational changes in bovine rhodopsin: Insight into activation of G-protein-coupled receptors. J Mol Biol 382(2):539–555

31. Bhattacharya S et al (2008) Ligand-stabilized conformational states of human beta(2) adrenergic receptor: Insight into G-protein-coupled receptor activation. Biophys J 94(6):2027–2042

32. Vaidehi N et al (2002) Prediction of structure and function of G protein-coupled receptors. Proc Natl Acad Sci USA 99(20):12622–12627

33. Kalani MYS et al (2004) The predicted 3D structure of the human D2 dopamine receptor and the binding site and binding affinities for agonists and antagonists. Proc Natl Acad Sci USA 101(11):3815–3820

34. Freddolino PL et al (2004) Predicted 3D structure for the human beta 2 adrenergic receptor and its binding site for agonists and antagonists. Proc Natl Acad Sci USA 101(9):2736–2741

35. Peng JY et al (2006) The predicted 3D structures of the human M1 muscarinic acetylcholine receptor with agonist or antagonist bound. ChemMedChem 1(8):878–890

36. Vaidehi N et al (2006) Predictions of CCR1 chemokine receptor structure and BX 471 antagonist binding followed by experimental validation. J Biol Chem 281(37):27613–27620

37. Li Y et al (2007) Prediction of the 3D structure and dynamics of human DP G-protein coupled receptor bound to an agonist and an antagonist. J Am Chem Soc 129(35):10720–10731

38. Bray JK, Goddard WA 3rd (2008) The structure of human serotonin 2c G-protein-coupled receptor bound to agonists and antagonists. J Mol Graph Model 27(1):66–81

39. Kam VWT, Goddard WA (2008) Flat-bottom strategy for improved accuracy in protein side-chain placements. J Chem Theor Comput 4(12):2160–2169

40. Deupi X et al (2004) Ser and Thr residues modulate the conformation of pro-kinked transmembrane alpha-helices. Biophys J 86(1 Pt 1):105–115

41. Schertler GF (1998) Structure of rhodopsin. Eye (Lond) 12(Pt 3b):504–510

42. Heo J et al (2007) Prediction of the 3-D structure of rat MrgA G protein-coupled receptor and identification of its binding site. J Mol Graph Model 26(4):800–812

Chapter 15

Target Based Virtual Screening by Docking into Automatically Generated GPCR Models

Christofer S. Tautermann

Abstract

Target based virtual screening (VS) combined with high-throughput measurements is an extremely useful tool to identify small molecule hits for proteins and in particular for G-protein coupled receptors (GPCRs). However, this is a quite difficult process for GPCRs due to the paucity of 3D structural information on these receptors. Therefore, the only possibility for target based VS is to build a structural model of the GPCR to be used for docking. However, GPCR model building is a very time consuming process, if the model should be able to explain all experimental findings and this investment is not always justified, if the model is only used for VS. Thus, a fully automated workflow is presented here, where a large number of GPCR models is built, and the best model is identified to be used for docking. The workflow leads to moderate enrichments with a very low effort. The inputs required are the sequence of the targeted GPCR, a reference ligand with experimental information and a database of small molecules to be used for docking. Manual intervention is recommended at various points, but it is strictly speaking not necessary.

Key words: GPCR modeling, Virtual ligand screening, Application of GPCR homology models, GPCR lead identification

1. Introduction

With about 800 members the class of G-protein coupled receptors (GPCRs) constitutes a large and versatile family of receptors of the human genome (1). Out of this large amount about 350 receptors are thought to be potentially druggable (2) although for more than 100 so-called orphan receptors (3, 4) no endogenous ligand is yet identified. Pharmaceutical industry, in most cases, looks for small molecule ligands, which interfere or substitute the endogenous GPCR ligands to enable, enhance, damp, or interrupt signaling into the cytosol. According to estimates, about 50% of all modern drugs act

Nagarajan Vaidehi and Judith Klein-Seetharaman (eds.), *Membrane Protein Structure and Dynamics: Methods and Protocols*, Methods in Molecular Biology, vol. 914, DOI 10.1007/978-1-62703-023-6_15, © Springer Science+Business Media, LLC 2012

on GPCRs showing that this class of receptors is currently strongly in focus, and it can be successfully targeted (5). In silico drug design for GPCRs is a quite difficult problem due to several reasons: first, the lack of crystal structures for most GPCRs (currently only six different GPCRs—β1 (6, 7), β2 (8–10), $A_{2A}R$ (11), CXCR4 (12), rhodopsin (13), and D3DR (14) are resolved) and the very low sequence homology (<20% sequence identity) between GPCR families make homology modeling approaches for most GPCRs very cumbersome. Second, GPCRs can adopt different geometries according to their signaling state and capturing these structural changes induced by various ligands in silico is currently not feasible. However, as a first step in that direction, approaches exist, which investigate the helical rotations triggered by ligands (15, 16). Very recently first structures of activated GPCRs have been published (10, 17), but the very subtle changes in the binding site upon activation are most likely not possible to be reproduced by current in silico methods (18).

Quite often GPCR models are used to identify new chemotypes by virtual screening (19–21), where a database of small molecules is docked into a receptor model and only the top scoring compounds are actually purchased for biological testing. The advantage of doing target based VS rather than ligand based VS by similarity searches, is that completely new scaffolds can be detected by docking without bias by already known ligands.

There is a wealth of studies in the literature where various receptor modeling, receptor model refinement and docking protocols are introduced and evaluated (19, 22–30). There are very pragmatic workflows, which focus on automated procedures to yield moderate enrichments with low effort (29, 31), and there are also very thorough studies, where ligand information is used to shape the binding site or to even rebuild the receptor model (19). As expected, the additional time and effort, which is put into the receptor structural model refinement, pays off usually, but it is always a trade-off between the huge time and effort, which is needed for proper model validation and the increased enrichment. Site directed mutagenesis studies are normally used to validate the ligand-receptor binding site models and this takes at least a few months. From a pragmatic point of view, it may be faster and cheaper to use a crude nonvalidated receptor model, which leads to modest enrichments. Time and money which is saved because no mutants are generated, may be used for the purchase of additional compounds and their testing—which is obviously required due to a lower expected enrichment. In most published applications, a compromise is applied, where some effort is put in receptor model improvement, but quite rarely mutagenesis experiments are done for model validation.

In this chapter I focus on the most pragmatic and still successful approach of automated receptor model generation, guided selection of the best model, and subsequent docking for VS. Similar approaches have been used several times in literature, e.g., Nowak et al. (29) tried to automate the modeling and docking process as far as possible. By automated docking of known ligands into the 5-HT$_{1A}$ receptor, favorable side chain rotamer positions are identified, which are fixed in a second round of homology modeling. The best model from a set of many decoys is picked by docking of known ligands and ranking the models according to the docking scores. With the best scoring model a VS run is done and the results are very good in terms of enrichment. In a subsequent study by Kneissl et al. (31), the same method was transferred to the neurokinin NK1 receptor, which turned out to be a much harder target, due to the lack of known interactions between reference ligands and the receptor. In the case of biogenic amine receptors such as adrenergic receptors, positively charged ligands are assumed to interact with the conserved D3.32 residue and this is also found in most docking runs. In NK1 no such ionic interaction is present, and thus automated docking of known ligands into a homology model of the receptor is not sufficient to identify good models. In this case information from published mutagenesis experiments had to be used to constrain interactions and to guide model selection. Although the effort in model validation was very small, still an enrichment of factor 3 was reached. This is in line with a study by Bissantz et al. (32), where ligand based and structure based VS methods have been compared for the 5-HT$_{2C}$ receptor, with the result that different structure based VS runs gave an enrichment factor of about 2.

However, it is still quite surprising that automated modeling with a crude model selection method can generally achieve enrichments larger than a random selection. The explanation for this may be found in the fact that VS methods are suitable for reducing the large number of inactive molecules rather than actually enriching the active ones, as it has recently been pointed out by Köppen (33).

2. Materials

2.1. General Requirements

2.1.1. Required Hardware and Computational Skills

The procedure described herein requires massive computational effort if large compound databases are docked into the receptor model. This can only be performed if a multicore high-performance Linux or UNIX cluster is available, where the docking runs can be well parallelized. Good shell scripting skills and skills in SVL (scientific vector language) are advantageous.

2.1.2. Sources for Small Molecule Databases

There are various vendors offering large sets of molecules (e.g., http://www.emolecules.com/), and there are also screening collections available free of charge, e.g., by the NCI/DTP Open Chemical Repository (http://dtp.nci.nih.gov/branches/dscb/

repo_open.html). For retrospective analyses the compound dataset can be extracted from ligand databases such as the Aureus database (http://www.aureus-pharma.com/) or ChEMBLdb (https://www.ebi.ac.uk/chembldb/). These databases contain literature and patent data for small molecules, where the specific targets and the corresponding activities of the molecules are annotated. Usually a large set of inactive molecules is spiked with a small fraction of known active ligands. When compiling the test set, it should be ensured that the sets of active and inactive ligands do not differ in their average molecular weight, as many docking scores are known to be biased by molecular weight.

2.2. Software

In the following the required software for the methods of Subheading 3 is listed.

2.2.1. Software for Creating Sequence Alignments

To create alignments, many different programs can be used, but it should be ensured that one can apply alignment constraints. In the program package MOE of Chemical Computing Group Inc. (34) all required alignment modifications can be performed.

2.2.2. Software for Homology Modeling

Most molecular modeling packages also include a homology modeling module. A standard program, which is currently the state of the art is MODELLER by Sali (35, 36), where homology models are constructed by obeying to spatial restraints derived from the templates. Also additional restraints, like experimental information (e.g., distances derived from NMR experiments), can be included in the modeling process with MODELLER.

To generate the models, an alignment file in the pir-format is required (which can be generated by most programs, which can do alignments) and the template(s) are read as structure files in the pdb-format. If the conserved cysteines that form a disulfide bridge between TM3 and ECL2 are aligned between template(s) and target, this disulfide bridge will also be present in the homology model. Otherwise, the user can define additional disulfide bonds manually or include other known experimental data wherever available.

2.2.3. Software for the Preparation and Docking of Reference Ligands

Generally, many docking programs are available, and a wealth of studies validating and comparing these methods is available (37, 38). Here GOLD (39) employing the Chemscore scoring function (40) is used, because it is known to give quite robust results for VS applications (41) and it is fairly easy to apply to the problem of many proteins and only few ligands. Usually docking programs are designed to dock many ligands into a single protein, and thus, a program where both tasks can be done without too much effort should be used. The program GLIDE (42) has also been used for high-throughput homology model docking (31), but the effort for the setup is much higher, with a comparably good quality of the docking results.

Before docking, both, the protonation state of the residues in the structural models and the protonation state of the ligand have to be assessed. The preparation of the ligand is done in MOE (34) by adding hydrogens to the ligand and a subsequent full optimization with standard thresholds and the MMFF94x forcefield. For the preparation of the homology models, a MOE database is generated, where all homology models are included. Hydrogens are added and each model is optimized by keeping the backbone positions fixed. The protonated, optimized models are exported as pdb-files and are available for docking. The whole step can be fully automated by some small SVL scripts.

For the docking runs with the reference ligand, the setup has to be done only for one homology model and the docking itself is done in batch mode by using the same configuration file for all homology models. This can be achieved by shell-scripts, which loop over all the homology models. The docking of the reference ligand is done at standard precision, which is the default in GOLD.

2.2.4. Software for Preparing the Small Molecule Database and Performing the VS Run

To prepare the docking database, MOE (34) is used to build 3D molecules, to protonate them and to do a subsequent optimization. All the steps can be performed on the whole database, starting from an sd-file in a fully automated way, finally exporting an sd-file used for docking. In analogy to the docking of a reference ligand, many different docking programs can be employed. In this study GOLD (39) is applied, in the "library screening" mode using the Chemscore (40) function for scoring. A diverse subset of top scoring molecules can be generated after importing the resulting sd-file in MOE by the clustering functionality within MOE. Depending on the size of the library of small molecules, the docking should be done in a parallel way on a multiprocessor compute cluster.

3. Methods

A procedure is described, where homology models for a targeted GPCR are constructed in an automated fashion. Out of these models the most appropriate one is chosen by docking of known GPCR ligands. For VS purposes, a database of small molecules is docked into the receptor model and ranked by the docking score. In retrospective studies, the enrichments can be reported and in prospective studies, the well scoring compounds are purchased or synthesized and tested for their biological activity.

Although the procedure is as automated as possible, the users' input in assessing the models is vital at several stages to ensure the quality of the procedure.

The overall procedure is displayed in the flow-chart in Fig. 1.

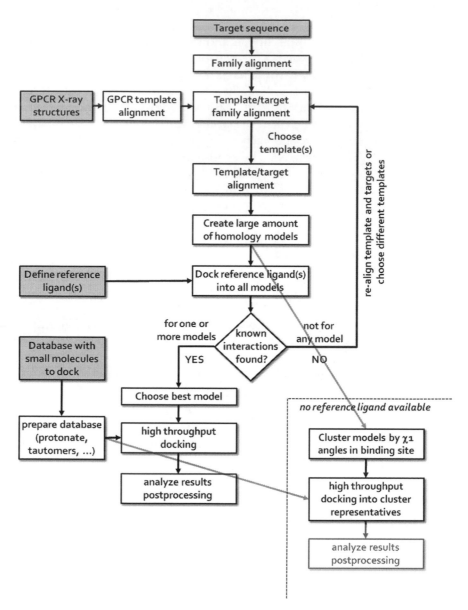

Fig. 1. Flowchart of the overall procedure. The *small insert* describes the situation, when no reference ligand is available. *Grey boxes* denote required input to the workflow.

3.1. Homology Modeling of the Target Receptor

The most common method by far for building GPCR models is homology modeling (36, 43). In this procedure a known structure of a protein related in sequence is used as a template for modeling. The underlying assumption is that the protein structure of a family is more conserved than the sequence, and sequence similarity translates into structural homology. Usually the homology modeling method is used for sequence similarities above 30% to get an unambiguous alignment of target and template, but for GPCRs this homology threshold is shifted to much lower values. The transmembrane

(TM) regions of GPCRs have to be aligned to available GPCR structures with the help of well known conserved motifs (1) in each of the TM helices.

3.1.1. Sequence Alignment

GPCRs can be divided in three classes, where class A, to which all known crystal structures belong, is by far the largest one (2). For class A GPCRs there are conserved motifs in the sequence, identified in all the seven helices and this information is used to align and verify the sequence alignment of GPCRs, even if the sequence similarity between the template and the query sequence is very low (below 30%). The sequence alignment of a target sequence to one or more template sequences is usually generated in few subsequent steps. First, a "family" alignment for the target sequence is done, employing few (3 to 10) very homologous sequences from the same family. The search for homologous sequences can be done at the Uniprot (44) server (http://www.uniprot.org/) employing the Blast-tab. The resulting family alignment is frozen and the sequences of available GPCR templates are aligned to it. Here the user should check if the GPCR specific conserved patterns (1) are aligned. If not, alignment constraints have to be applied in order to enforce the correct alignment. The known patterns are shown in Fig. 2, where the Weinstein–Ballesteros numbering scheme (45) is used for all currently available GPCR templates. For the exceptions, where these patterns are not conserved, for example in class B and class C receptors, the use of the alignment proposed by Surgand et al (46) is recommended.

Once the target-templates alignment is established, the template(s) for homology modeling can be chosen based on the maximal sequence identity found in the TM1–TM7 region. The carboxy and the amino termini may be ignored, as well as excessively long loops as found in the biogenic amine receptors such as D3 dopamine receptor. If a single highly related template can be identified, homology models can be based on a single template model, otherwise multiple template models (with the highest sequence identity) should be chosen.

3.1.2. Homology Modeling Procedure

Using the sequence alignment derived in the previous section, a large number (100–1,000) of homology models is constructed in an automated way. Most homology modeling programs only require the input of the template structure(s) and the alignment of the target sequence to the template(s). The modeling procedure itself is usually done in a fully automated fashion. However, model assessment and the identification of the best model is a non trivial task and will be described in the next section. It has been shown that using more than one template is advantageous (31, 47), especially when trying to use a highly automated workflow. Homology models based on only one template are very similar to the template and the backbones may be nearly identical.

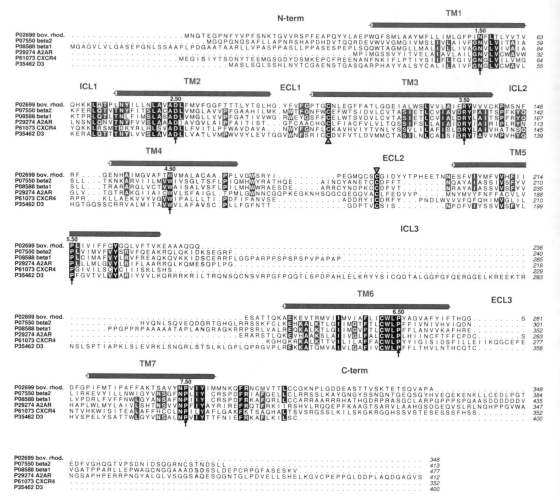

Fig. 2. Sequence alignment of all different (currently available) GPCR templates in the order of their publication: bovine rhodopsin (1F88), β2 adrenergic receptor (2RH1), β1 adrenergic receptor (2VT4), A2AR adenosine receptor (3EML), CXCR4 chemokine receptor (3ODU), and D3 dopamine receptor (3PBL). The sequences shown are full length sequences as accessed in the Uniprot database (Uniprot identifiers given as sequence line headers). The alignment is based upon a combined structural alignment of the crystal structures (pdb accession codes given above in parentheses) and a subsequent sequence alignment of the full length sequences. Conserved residues are marked in black and grey. The x.50 residues as defined by Weinstein and Ballesteros are marked by black arrows and the corresponding enumeration is given. The cysteines, involved in the conserved disulfide bridge between TM3 and the extracellular loop 2 (ECL2) are marked by a triangle. The visualization of the sequence alignment was created employing ALINE (50).

When adding more templates, the accessible conformational space is dramatically enlarged and this provides variability in the backbone structures. This is also the reason why a large number of homology models has to be generated; otherwise, the advantage of more templates cannot be taken.

3.2. Selection of the Most Appropriate Model

The model building procedure yields a large number of homology models in an automated way. The difficult task is now to select the best models out of this large number of decoys. The selection of the "best" model is surely dependent on the goal of the user. In the present case, a workflow for the generation of homology models for VS is presented, thus a "good" model must be able to accommodate known active receptor ligands in a satisfying way. In the ideal case, interactions between the receptor side chains and ligand functional groups are explicitly known, e.g., from single point mutagenesis experiments. For some receptors, for example, the muscarinic receptors, a wealth of mutagenesis information for various ligands is available in the literature (48), and for other cases for, e.g., orphan receptors, there is hardly any such information available. The more information is available, the easier it is to choose models, which are good for docking active ligands.

3.2.1. Docking of Known Reference Ligand(s)

In the case if at least one small molecule ligand to the targeted receptor is known, a ligand is docked into all models. The model quality can be assessed by the docking score, as it is expected, that a "good" model is able to interact well with the ligand, and thus to achieve a better score. If more than one structural class of potent ligands is available, a potent representative of each class can be docked to every homology model and the models can be ranked by a consensus score over all ligands.

If interactions between the ligand and receptor are known, or if some amino acids are known to be involved in ligand binding, this information should be used as a very efficient filter to remove models, where these interactions are not found in the docking poses. This step can be well automated, because the same procedure is done for all homology models in the identical way.

3.2.2. Selection of Models Without Known Reference Ligands

For some receptors, e.g., orphan or recently deorphanized receptors, no small molecule ligands are known. In these cases VS by docking into homology models is one of the few approaches, which can be performed to identify new ligands to these receptors. One may argue that in such cases, other methods like chemogenomics or sequence derived pharmacophores, as recently proposed by Klabunde et al. (27), may be more appropriate. However, ligand docking still provides the opportunity to identify new chemotypes, if the ligand binding site of the homology model includes appropriate amino acid side chains required for a desired receptor-ligand interaction. The problem is to decide, if a homology model is appropriate, when there is no small molecule ligand at hand. Therefore, docking of the large database has to be done into a set of diverse homology models, with the hope to have few appropriate models in the set. This procedure increases the computer time and it will also lower enrichment rates, because many homology models from a diverse set will be "unsuitable" models, due to the lack of correct interaction partners in the binding site.

The generation of a diverse set of homology models may be performed by analyzing the rotamers of bulky amino acids in the proposed binding site. For this a list of $\chi 1$-angles for all homology models has to be generated and based on a clustering (e.g., by principal component analysis) on the amino acid rotamers of the active site, cluster representatives can be chosen to be diverse homology models. There are other possibilities, such as assessing the model diversity by all-atom RMSD of the active site, but these methods require a larger effort, without any guarantee to yield a better set.

Generally ligand docking to homology models without any validation is not recommended and this procedure should be performed with great care. Therefore, a useful, fully automated workflow without intermediate model quality check by a reference compound is still elusive.

3.3. Docking of the Molecule Database: The Virtual Screening Step

Once the best model is defined, a large set of small molecules may be docked in an automated way. The database of small molecules has to be preprocessed and prepared for docking, as the structures are usually retrieved in a 2D format without hydrogens. Thus the first step is to generate the 3D structures for every molecule, and then add the hydrogens. For this, pKa values are calculated, in order to get the protonation state correctly. Also various energetically accessible tautomers should be generated for every compound. Before docking, every structure has to be geometry optimized, because most docking programs are quite sensitive to the quality of the starting structure.

If the database contains several millions of compounds, it may be advisable to reduce the number of structures before docking by applying simple filters, like discarding molecules with unwanted substructures (for e.g., highly reactive species). Another filter may be using the drug-likeness or Lipinski-rules. For every molecule only the top scoring docked pose is stored to keep the complexity as low as possible.

Once the VS run has finished, the molecules are ranked according to their docking score. Theoretically one may purchase or synthesize the desired amount of molecules starting from the top of the list. However, it has been realized, that visual inspection to identify artifacts and unwanted molecules is very useful before proceeding. This is especially important if the number of molecules to be ordered is quite small. To avoid any bias to a particular chemical class of compounds and to identify new chemical scaffolds, it is imperative to include top scoring molecules from different structural classes. Therefore, the first several hundred top scoring compounds can be clustered by their structural similarity and few cluster representatives should be chosen for purchase or synthesis. The molecules are sorted according to the docking score and it is expected, that active compounds are more often found in the high scoring regions. One method to quantify the quality of a VS method is to calculate the improvement of a VS-guided molecule selection over a random compound selection. This is called the enrichment factor, and usually enrichment factors are reported, giving the enrichment of active compounds in the first $x\%$ of a sorted compound list.

In prospective applications, the most important task of a VS-method is to be able to identify new active structural classes rather than having enormous enrichments. Thus, it is very hard to compare different computational methods based on the enrichments, because this may be misleading. If the enrichment is low, but the few discovered structural classes are chemically tractable and interesting for optimization (e.g., good intellectual property situation), it is more valuable than to find many hits from intractable structural classes. Enrichment factors for target based VS-methods on GPCRs are reported to be quite moderate (32), but the strength of these methods is the ability of detecting completely new scaffolds, which are not based on reference ligands.

4. Example for Target Based VS on the Neurokinin NK1 Receptor

The example chosen here describes target based VS based on a NK1 neurokinin receptor homology model.

4.1. Sequence Alignment and Homology Model Construction

To generate the alignment and to choose the appropriate templates, first a family alignment of human NK1, NK2, and NK3 has been done (employing the BLOSUM60 matrix, gap penalty 20) and this has been aligned to the template alignment of the templates from Fig. 2 (employing BLOSUM30 matrix and gap penalty of 20). The N- and C-termini as well as the region of the intracellular loop 3 are removed and the sequence identities between NK1 and the templates are determined. CXCR4 and D3 are shown to be templates with the highest homology to NK1, with about 25% sequence identity. The sequence alignment employed in this example is shown in Fig. 3. Based on this alignment a set of 300 homology models has been constructed with MODELLER (35, 36) in a fully automated way.

4.2. Validation of the Receptor-Ligand Docking Site for NK1

For the NK1-receptor the highly potent ligand CP96345 (49) (Fig. 4) was docked into all homology models employing GOLD (39). For every model the best scoring docking pose is checked for interactions of Gln165 on TM4 and His197 on TM5, which are known to interact with the ligand. If a ligand docking pose does not interact with both the residues, the model is discarded. For Gln165 a hydrogen bond interaction is required and for His197 a hydrophobic/aromatic interaction is postulated. Now, every homology model is checked for the hydrogen bond between one of the nitrogens and Gln165, and if the ligand is located within 5 Å

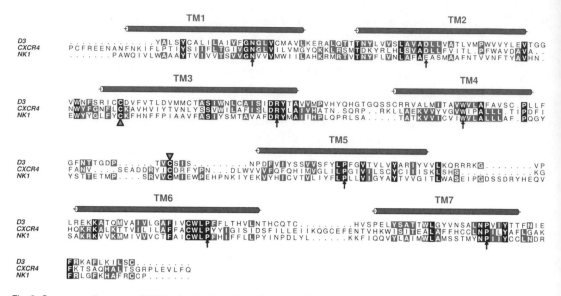

Fig. 3. Sequence alignment of NK1 to the two templates D3 and CXCR4. Weinstein–Ballesteros x.50 residues are marked by *arrows* and the conserved cysteins from the disulfide bridge between TM3 and ECL2 are marked by *green triangles*.

Fig. 4. The NK1 receptor antagonist CP96345.

to His197. Finally, the homology model with the highest docking scoring, which shows both these ligand receptor interactions, is chosen as the model to be used for virtual screening.

4.3. Ligand Database Docking

To test the homology model, a set of 1,509 ligands is extracted from the Aureus GPCR database. Sixty compounds are known to be active on NK1 and the others are various randomly chosen ligands for other GPCR receptors, most probably not active on NK1. This corresponds to a hit rate of 4%. The mean molecular weight of active and inactive molecules is similar and exclusively neutral or negatively charged molecules are compiled, because the reference ligand is positively charged. With these precautions it should be granted, to have an unbiased retrospective VS run. A good opportunity to check whether a ligand set is biased is to perform the retrospective docking experiment with an obviously "bad" or a decoy receptor model. Such "bad" models may be generated by manually rotating the side chains to block the binding site. If an enrichment significantly higher than random is obtained for "bad" models, one may assume that there is a strong bias in the dataset (31).

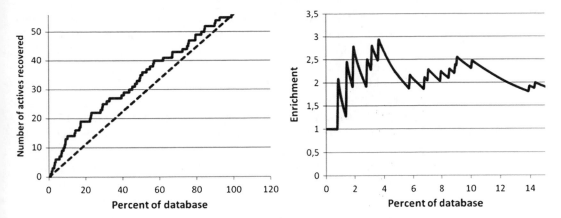

Fig. 5. Enrichment plots for the retrospective NK1 VS-run. *Left*: Number of the active molecules recovered given over the whole range of the database. The *solid line* corresponds to the ranking by docking into the homology model and the *dashed line* corresponds to a random distribution. *Right*: Enrichments calculated for the first top few percent of the database.

The results of the NK1 docking are shown in Fig. 5 as fraction of actives recovered over the whole range of the database and as enrichments for the first few percent of molecules. The latter is much more important to assess the quality of a VS-method, because in a real life situation only a very small fraction of the in silico docked molecules will enter biological testing. Therefore, it is highly desirable to have a high enrichment for the first few percents of molecules of the database. In this example an enrichment of about factor 3 is achieved for the first 2–3% of the database and it drops to an enrichment of factor of 2 when screening up to 5–15% of the database (see Fig. 5, right). This is in the range of what one can expect from docking in GPCR models (32), and it has been achieved with a very low effort, because all steps are fully automated.

5. Notes

A workflow for target based VS for GPCR ligands based on GPCR homology models is presented. All steps can be fully automated, thus providing an approach to achieve moderate enrichments with a very low effort, once the workflow is established. First, the sequence of the target GPCR is aligned to all available GPCR templates and based on the sequence homologies, the templates for homology modeling are chosen. It is advisable to use more than one template, if the maximal sequence identity is below 50%. The two or three X-ray structures with the highest sequence identity to the target are chosen as templates. Based on the alignment a large number of homology models are created in an automated way and the best model is identified by docking of a known reference ligand into all models.

The database used for VS has to be preprocessed properly, in their correct protonated states and with their geometries optimized. In order to dock the preprocessed database into the best homology model, the "high-throughput" adjustment of the docking program should be used to keep CPU time to a moderate level. For prospective applications, it is advisable to purchase or synthesize a diverse subset of the top few percent of the molecules in the hit list.

As already mentioned, the whole workflow can be fully automated. However, some manual intervention is still recommended. Especially a check for plausibility of sequence alignments as well as the choice of the template(s) improves the results significantly. Manual interaction can also be helpful in the visual inspection of the top scoring docked poses of the reference ligand in the homology models in order to check if required interactions with the receptor are formed properly. And last, it is highly advisable to browse manually through hit lists of small molecules, and to discard molecules, which are generally unsuitable for further investigation (due to intellectual property issues, synthetic feasibility, physicochemical parameters, etc.). Although every step in model building, model assessment and docking introduces approximations, several studies in literature show, that docking into GPCR homology models leads to moderate enrichments, and it is a viable approach to identify new chemical classes of ligands for a receptor.

References

1. Lagerstrom MC, Schioth HB (2008) Structural diversity of G protein-coupled receptors and significance for drug discovery. Nat Rev Drug Discov 7:339–357

2. Fredriksson R, Lagerstrom MC, Lundin LG, Schioth HB (2003) The G-protein-coupled receptors in the human genome form five main families. Phylogenetic analysis, paralogon groups, and fingerprints. Mol Pharmacol 63:1256–1272

3. Chung S, Funakoshi T, Civelli O (2008) Orphan GPCR research. Br J Pharmacol 153:S339–S346

4. Lin SH, Civelli O (2004) Orphan G protein-coupled receptors: targets for new therapeutic interventions. Ann Med 36:204–214

5. Lundstrom K (2006) Latest development in drug discovery on G protein-coupled receptors. Curr Protein Pept Sci 7:465–470

6. Warne T, Serrano-Vega MJ, Baker JG, Moukhametzianov R, Edwards PC, Henderson R et al (2008) Structure of a beta1-adrenergic G-protein-coupled receptor. Nature 454:486–491

7. Warne T, Moukhametzianov R, Baker JG, Nehme R, Edwards PC, Leslie AGW et al (2011) The structural basis for agonist and partial agonist action on a beta1-adrenergic receptor. Nature 469:241–244

8. Cherezov V, Rosenbaum DM, Hanson MA, Rasmussen SGF, Thian FS, Kobilka TS et al (2007) High-resolution crystal structure of an engineered human beta2-adrenergic G protein coupled receptor. Science 3181258–1265

9. Rasmussen SGF, Choi HJ, Rosenbaum DM, Kobilka TS, Thian FS, Edwards PC et al (2007) Crystal structure of the human beta2 adrenergic G-protein-coupled receptor. Nature 450:383–387

10. Rasmussen SGF, Choi HJ, Fung JJ, Pardon E, Casarosa P, Chae PS et al (2011) Structure of a nanobody-stabilized active state of the beta2 adrenoceptor. Nature 469:175–180

11. Jaakola VP, Griffith MT, Hanson MA, Cherezov V, Chien EYT, Lane JR et al (2008) The 2.6 Angstrom crystal structure of a human A2A adenosine receptor bound to an antagonist. Science 322:1211–1217

12. Wu B, Chien EY, Mol CD, Fenalti G, Liu W, Katritch V et al (2010) Structures of the CXCR4 chemokine GPCR with small-molecule and cyclic peptide antagonists. Science 330:1066–1071

13. Palczewski K, Kumasaka T, Hori T, Behnke CA, Motoshima H, Fox BA et al (2000) Crystal structure of rhodopsin: a G protein-coupled receptor. Science 289:739–745

14. Chien EYT, Liu W, Zhao Q, Katritch V, Won Han G, Hanson MA et al (2010) Structure of the human dopamine D3 receptor in complex with a D2/D3 selective antagonist. Science 330:1091–1095

15. Bhattacharya S, Hall SE, Vaidehi N (2008) Agonist-induced conformational changes in bovine rhodopsin: insight into activation of G-protein-coupled receptors. J Mol Biol 382: 539–555

16. Bhattacharya S, Vaidehi N (2010) Computational mapping of the conformational transitions in agonist selective pathways of a G-protein coupled receptor. J Am Chem Soc 132:5205–5214

17. Xu F, Wu H, Katritch V, Han GW, Jacobson KA, Gao ZG et al Structure of an agonist-bound human A2A adenosine receptor. Science 332:322–327

18. Tautermann CS, Pautsch A (2011) The implication of the first agonist bound activated G-protein-coupled receptor (GPCR) X-ray structure on GPCR in silico modeling. ACS Med Chem Lett 2:414–418

19. Evers A, Klebe G (2004) Successful virtual screening for a submicromolar antagonist of the neurokinin-1 receptor based on a ligand-supported homology model. J Med Chem 47:5381–5392

20. Kellenberger E, Springael JY, Parmentier M, Hachet-Haas M, Galzi JL, Rognan D (2007) Identification of nonpeptide CCR5 receptor agonists by structure-based virtual screening. J Med Chem 50:1294–1303

21. Sela I, Golan G, Strajbl M, Rivenzon-Segal D, Bar-Haim S, Bloch I et al (2010) G protein coupled receptors—in silico drug discovery and design. Curr Top Med Chem 10: 638–656

22. Becker OM, Marantz Y, Shacham S, Inbal B, Heifetz A, Kalid O et al (2004) G protein-coupled receptors: in silico drug discovery in 3D. Proc Natl Acad Sci U S A 101:11304–11309

23. Bissantz C, Bernard P, Hibert M, Rognan D (2003) Protein-based virtual screening of chemical databases. II. Are homology models of G-protein coupled receptors suitable targets? Proteins 50:5–25

24. Chen JZ, Wang J, Xie XQ (2007) GPCR structure-based virtual screening approach for CB2 antagonist search. J Chem Inf Model 47:1626–1637

25. Evers A, Hessler G, Matter H, Klabunde T (2005) Virtual screening of biogenic amine-binding G-protein coupled receptors: comparative evaluation of protein- and ligand-based virtual screening protocols. J Med Chem 48: 5448–5465

26. Evers A, Klabunde T (2005) Structure-based drug discovery using GPCR homology modeling: successful virtual screening for antagonists of the alpha1A adrenergic receptor. J Med Chem 48:1088–1097

27. Klabunde T, Giegerich C, Evers A (2009) Sequence-derived three-dimensional pharmacophore models for G-protein-coupled receptors and their application in virtual screening. J Med Chem 52:2923–2932

28. McRobb FM, Capuano B, Crosby IT, Chalmers DK, Yuriev E (2010) Homology modeling and docking evaluation of aminergic G protein-coupled receptors. J Chem Inf Model 50:626–637

29. Nowak M, Kolaczkowski M, Pawlowski M, Bojarski AJ (2006) Homology modeling of the serotonin 5-HT1A receptor using automated docking of bioactive compounds with defined geometry. J Med Chem 49:205–214

30. Tikhonova IG, Sum CS, Neumann S, Engel S, Raaka BM, Costanzi S et al (2008) Discovery of novel agonists and antagonists of the free fatty acid receptor 1 (FFAR1) using virtual screening. J Med Chem 51:625–633

31. Kneissl B, Leonhardt B, Hildebrandt A, Tautermann CS (2009) Revisiting automated G-protein coupled receptor modeling: the benefit of additional template structures for a neurokinin-1 receptor model. J Med Chem 52:3166–3173

32. Bissantz C, Schalon C, Guba W, Stahl M (2005) Focused library design in GPCR projects on the example of 5-HT(2c) agonists: comparison of structure-based virtual screening with ligand-based search methods. Proteins 61:938–952

33. Koppen H (2009) Virtual screening—what does it give us? Curr Opin Drug Discov Devel 12:397–407

34. MOE—Molecular Operating Environment (2010) Chemical Computing Group, Montreal. http://www.chemcomp.com/

35. Marti-Renom MA, Stuart AC, Fiser A, Sanchez R, Melo F, Sali A (2000) Comparative protein structure modeling of genes and genomes. Annu Rev Biophys Biomol Struct 29:291–325

36. Sali A, Blundell TL (1993) Comparative protein modelling by satisfaction of spatial restraints. J Mol Biol 234:779–815

37. Cross JB, Thompson DC, Rai BK, Baber JC, Fan KY, Hu Y et al (2009) Comparison of several molecular docking programs: pose prediction and virtual screening accuracy. J Chem Inf Model 49:1455–1474

38. Cummings MD, DesJarlais RL, Gibbs AC, Mohan V, Jaeger EP (2005) Comparison of automated docking programs as virtual screening tools. J Med Chem 48:962–976

39. Verdonk ML, Cole JC, Hartshorn MJ, Murray CW, Taylor RD (2003) Improved protein-ligand docking using GOLD. Proteins 52:609–623

40. Eldridge MD, Murray CW, Auton TR, Paolini GV, Mee RP (1997) Empirical scoring functions: I. The development of a fast empirical scoring function to estimate the binding affinity of ligands in receptor complexes. J Comput Aid Mol Des 11:425–445

41. Bissantz C, Folkers G, Rognan D (2000) Protein-based virtual screening of chemical databases. 1. Evaluation of different docking/scoring combinations. J Med Chem 43:4759–4767

42. Friesner RA, Banks JL, Murphy RB, Halgren TA, Klicic JJ, Mainz DT et al (2004) Glide: a new approach for rapid, accurate docking and scoring. 1. Method and assessment of docking accuracy. J Med Chem 47:1739–1749

43. Baker D, Sali A (2001) Protein structure prediction and structural genomics. Science 294:93–96

44. The UniProt Consortium (2010) The Universal Protein Resource (UniProt) in 2010. Nucleic Acids Res 38:D142–D148

45. Ballesteros JA, Weinstein H (1995) Integrated methods for the construction of three-dimensional models and computational probing of structure–function relations in G protein-coupled receptors. Methods Neurosci 25:366–428

46. Surgand JS, Rodrigo J, Kellenberger E, Rognan D (2006) A chemogenomic analysis of the transmembrane binding cavity of human G-protein-coupled receptors. Proteins 62:509–538

47. Larsson P, Wallner B, Lindahl E, Elofsson A (2009) Using multiple templates to improve quality of homology models in automated homology modeling. Protein Sci 17:990–1002

48. Goodwin JA, Hulme EC, Langmead CJ, Tehan BG (2007) Roof and floor of the muscarinic binding pocket: variations in the binding modes of orthosteric ligands. Mol Pharm 72:1484–1496

49. Lowe JA, Drozda SE, Snider RM, Longo KP, Zorn SH, Morrone J et al (1992) The discovery of (2S,3S)-cis-2-(diphenylmethyl)-N-((2-methoxyphenyl)methyl)-1-azabicyclo(2.2.2)octan-3-amine as a novel, nonpeptide substance P antagonist. J Med Chem 35:2591–2600

50. Bond CS, Schüttelkopf AW (2009) ALINE: a WYSIWYG protein-sequence alignment editor for publication-quality alignments. Acta Cryst D 65:510–512

Chapter 16

Predicting the Biological Activities Through QSAR Analysis and Docking-Based Scoring

Santiago Vilar and Stefano Costanzi

Abstract

Numerous computational methodologies have been developed to facilitate the process of drug discovery. Broadly, they can be classified into ligand-based approaches, which are solely based on the calculation of the molecular properties of compounds, and structure-based approaches, which are based on the study of the interactions between compounds and their target proteins. This chapter deals with two major categories of ligand-based and structure-based methods for the prediction of biological activities of chemical compounds, namely quantitative structure-activity relationship (QSAR) analysis and docking-based scoring. QSAR methods are endowed with robustness and good ranking ability when applied to the prediction of the activity of closely related analogs; however, their great dependence on training sets significantly limits their applicability to the evaluation of diverse compounds. Instead, docking-based scoring, although not very effective in ranking active compounds on the basis of their affinities or potencies, offer the great advantage of not depending on training sets and have proven to be suitable tools for the distinction of active from inactive compounds, thus providing feasible platforms for virtual screening campaigns. Here, we describe the basic principles underlying the prediction of biological activities on the basis of QSAR and docking-based scoring, as well as a method to combine two or more individual predictions into a consensus model. Finally, we describe an example that illustrates the applicability of QSAR and molecular docking to G protein-coupled receptor (GPCR) projects.

Key words: G protein-coupled receptors (GPCRs), Ligand-based drug discovery, Structure-based drug discovery, Molecular Docking, Quantitative structure–activity relationships (QSAR), Comparative molecular field analysis (CoMFA), Comparative molecular similarity index analysis (CoMSIA), Multiple linear regression (MLR), Partial least square (PLS) regression

1. Introduction

The discovery of new drugs involves the expenditure of large amounts of money and manpower. Introducing a compound into clinical trials typically entails the scouting and biological evaluation of a large set of diverse molecules. This lengthy process can now be assisted and accelerated through the integration of a computer-aided

Nagarajan Vaidehi and Judith Klein-Seetharaman (eds.), *Membrane Protein Structure and Dynamics: Methods and Protocols*, Methods in Molecular Biology, vol. 914, DOI 10.1007/978-1-62703-023-6_16, © Springer Science+Business Media, LLC 2012

drug discovery (CADD) strategy that helps the selection of candidate compounds, provides mechanistic hypotheses on their mode of action, and facilitates their development. Notably, CADD is a rapidly growing field and has already experienced a significant advancement since the early days, thanks to the efforts that academic researchers and pharmaceutical companies are putting into the development of new and improved computational methods and to the rapid technological improvement of computers (1, 2).

CADD strategies can be broadly categorized into ligand-based and/or structure-based approaches (2, 3). The former methods rely on the analysis of molecular properties of known ligands without taking into account explicitly the interactions of the ligands with their target protein. Clearly, ligand-based methodologies can only be applied when known ligands exist. Structure-based approaches, instead, are based on the direct calculation of protein–ligand interactions and can be applied only when the structure of the target protein has either been solved experimentally or generated through computational modeling.

In this chapter, after a brief introduction to two specific categories of ligand-based and structure-based CADD approaches to the prediction of biological activities of chemicals, namely quantitative structure–activity relationship (QSAR) analysis and docking-based scoring, we describe the various phases necessary for their implementation (see Fig. 1). We also describe how to generate consensus models that combine the predictions of two or more individual models (see Fig. 2). Moreover, we illustrate the application of QSAR and molecular docking to the prediction of the activity of ligands of G protein-coupled receptors (GPCRs), a superfamily of proteins that, in light of their vast physiological and pathophysiological implications, are among the most pursued targets for pharmacological intervention (4). In particular, we present a case study that deals with the prediction of the activity of ligands for the β_2-adrenergic receptor (5).

1.1. QSAR

QSAR methods encompass a number of ligand-based analyses designed to correlate biological activities with molecular properties calculated using two-dimensional (2D) or three-dimensional (3D) ligand structures (6, 7). QSAR analyses can only be conducted when a set of ligands with known biological activities, known as a training set, is available. Statistical models linking biological activities to molecular properties are built on the basis of such training sets and subsequently applied to the prediction of the activity of novel compounds. In the field of GPCRs, biological activity data have been published for ligands of numerous receptors and can be utilized to generate training sets. For this reason and because of the paucity of information on the 3D structure of GPCRs that, up until recently, has characterized the superfamily, QSAR has been extensively applied to the prediction of the activity of GPCR ligands (3).

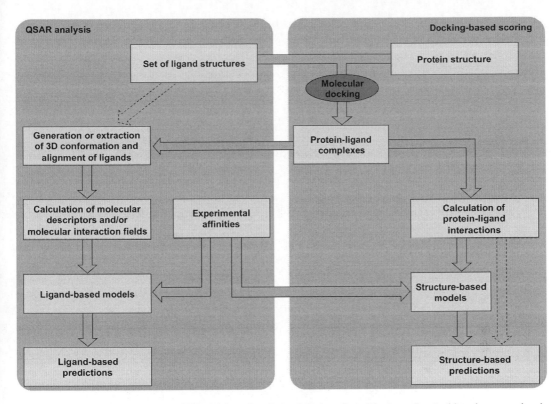

Fig. 1. Flowchart for the construction of ligand-based and structure-based models. According to this scheme, molecular docking plays a key role at the basis of both approaches. Alternatively, as indicated by the *dashed arrow* on the left side of the figure, a pure ligand-based approach can be adopted, in which the conformation and the alignment of the ligands are derived exclusively from their molecular features. Additionally, the scheme also illustrate that, when a training set of ligands with known activity is available, this can be used to train structure-based scoring functions. Alternatively, as indicated by the dashed arrows on the right side of the figure, a pure structure-based approach can be adopted in which prepackaged scoring functions are applied without the need for the use of a training set.

However, for orphan or less-studied receptors, the absence or the paucity of known ligands may prevent or seriously hinder the application of ligand-based modeling.

QSAR analyses require the calculation of molecular descriptors that reflect the topology or the physicochemical properties of molecules. Once such descriptors have been calculated for the whole dataset, the correlation between descriptors and experimental activities is studied through statistical analyses, such as linear regression, multiple linear regression (MLR), or partial least square (PLS) regression.

In 2D-QSAR, molecules are described through properties calculated on the basis of their 2D topology. Instead, 3D-QSAR analyses are based on molecular properties that depend on the 3D structure of the molecules. For the calculation of some of these properties, models of the bioactive 3D conformation of the ligands are sufficient. For others, instead, a 3D alignment of the bioactive conformation of all the ligands is also necessary. In a pure ligand-based

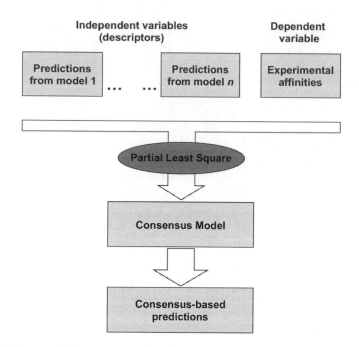

Fig. 2. Flowchart for the construction of a consensus model based on the combination, through PLS regression, of the predictions derived from *n* individual models.

modeling approach, 3D alignments are generated simply by superimposing the ligands on the basis of their common features. However, more effectively, 3D alignments can be obtained through molecular docking, with a strategy that combines structure-based and ligand-based modeling (see Fig. 1).

Within 3D QSAR methodologies, it is worth mentioning two techniques that have been among the most widely applied to the prediction of the activity of GPCR ligands (2), namely Comparative Molecular Field Analysis (CoMFA) (8) and Comparative Molecular Similarity Index Analysis (CoMSIA) (9). CoMFA and CoMSIA are based on the representation of ligands through molecular fields measured in the space that surrounds them. In particular, molecular fields are sampled at each point of a 3D lattice in which the aligned ligands are immersed and used as descriptors in a subsequent QSAR analysis. Due to the high number of descriptors that CoMFA and CoMSIA entail, a fundamental factor that contributed to their development has been the introduction of the PLS regression technique. This statistical method combines characteristics from principal component analysis (PCA) and MLR and reduces the dimensionality of the independent variables into fewer orthogonal components, thus allowing the conduction of regression analyses even when the number of independent variables is very high (10, 11).

Recently, alignment independent 3D-QSAR analyses have also been developed and applied to study of GPCR ligands, for instance the autocorrelation of molecular electrostatic potential approach devised by Moro and coworkers (12, 13) and the grid-independent descriptors (GRIND) approach devised by Clementi, Cruciani, and coworkers (14, 15).

QSAR models are very dependent on the nature of the training set. They are endowed with high predictive power when applied to compounds structurally related to those included in the training set, but perform poorly when applied to structurally diverse molecules. Thus, to ensure a robust predictive power, their use should be circumscribed to the analysis of compounds with molecular characteristics well represented within the training set.

1.2. Docking-Based Scoring

Predictions of biological activities based on docking scores do not require a training set of ligands with known activities. However, an absolute condition for the molecular docking experiments that lie at their foundation and the subsequent calculation of the protein–ligand interactions is that the structure of the target protein be known, possibly experimentally. For a long time, their applicability to the discovery and the development of GPCR ligands has been hampered, although not precluded, by the paucity of structural knowledge that has characterized the superfamily. In particular, for years, rhodopsin has been the only receptor with experimentally derived 3D information and has served as a prototype for the study of the whole GPCR superfamily (16). In recent years, however, breakthroughs in GPCR crystallography led to the solution of additional crystal structures, while many more are expected to be solved in the near future (17). Notably, besides their obvious direct application to structure-based CADD, these crystal structures will also provide an increasingly solid platform for the construction of homology models for the members of the superfamily that have yet to be experimentally elucidated (18–20).

The most accurate structure-based methodologies to rank the binding affinity of a set of given ligands for a target protein are those that rely on first-principle methods for the calculation of their free binding energy, for instance through free energy perturbations (FEP) or thermodynamic integration (TI) (21, 22). However, these techniques are time-consuming and require significant effort for the preparation and optimization of the system to allow high-throughput application. Moreover, they are best suited for the analysis of closely related compounds. For these reasons, in the context of molecular docking, compounds are usually ranked through simpler and faster scoring functions, broadly classifiable into force field-based, empirical or knowledge-based methods (3). Scoring functions are notorious for yielding scores that lack fine correlation with experimental affinities, even when the calculations are based on geometrically accurate complexes (23). Nevertheless, they have been shown capable of effectively distinguishing between active and inactive compounds (1).

Accordingly, many studies have demonstrated the applicability of molecular docking methods to virtual screening campaigns, including those targeting GPCRs, especially when the structure of the receptor is known crystallographically (5, 24–28). Particularly noteworthy are recent studies that reported the discovery, in high yields, of novel ligands of the β_2-adrenergic and the adenosine A_{2A} receptors through docking-based virtual screening targeting the crystal structures of the two receptors (29–32). Moreover, sufficiently accurate in silico models of GPCRs, although outperformed by crystal structures, have been shown to be effectively applicable to virtual screening not only in controlled studies but also through real-life quests for new ligands (25, 26, 28, 33–37).

2. Materials

2.1. Software

Molecular descriptors can be calculated using a plethora of dedicated software, including, but certainly not limited to, DRAGON, CODESSA, MOE, the Schrödinger package, and ICM (38–42). Alternatively, they can be measured experimentally. CoMFA and CoMSIA calculations are implemented in SYBYL (43). Molecular docking experiments can be carried out by means of a variety of modeling software, including, but certainly not limited to, MOE, the Schrödinger package, SYBYL, GOLD, and ICM (40–44). Some modeling packages, including, but certainly not limited to, the Schrödinger package, MOE, SYBYL, and ICM, allow directly performing statistical analyses. Specialized software packages for statistical analyses are also available, including, but certainly not limited to, R (45) and STATISTICA (46).

2.2. Skills

Molecular modeling software can run with a variety of operating systems, including various implementations of Unix and Linux. Some software can be operated through a graphic user interface (GUI). However, most software can also (or exclusively) be operated through command line or through scripts written by the user. For these reasons, some knowledge of the Unix/Linux operating systems as well as the ability of writing scripts and interacting with software through command line is advisable.

2.3. Data

2.3.1. Structure and Biological Activity of Ligands

Sets of known ligands, to be used as training and/or test sets, can be compiled using in-house data or data retrieved from the literature (see Note 1). Instead sets of novel potential ligands, to be evaluated in silico prior to their experimental testing, can be designed through computer-aided or classic medicinal chemistry approaches.

The 3D structures of the retrieved ligands can either be sketched within the chosen molecular modeling package or, if available, downloaded from databases such as PubChem (pubchem. ncbi.nlm.nih.gov). Particular attention to the chirality of the compounds is necessary. Once drawn, the structures should be saved in a format readable by the software chosen for the docking or the QSAR analysis. Some programs require one file per ligand; others allow the use of one file enlisting all the ligands.

For training and test sets, but not for novel putative ligands, biological activities—preferably binding affinities—need to be collected, either from the literature or from in-house data, and properly codified. If the experimental measurements were conducted on different species, it is advisable to verify the presence of sufficient similarity in the amino acid sequence of the target protein across species, especially within the binding cavity. If there are discrepancies between the activities reported for a compound by different articles, the data can either be excluded or an average value can be considered. Of note, information on GPCR ligands, with references to the relevant literature, is available through the GLIDA database (pharminfo.pharm.kyoto-u.ac.jp/services/glida) (47).

For the statistical analyses, a spreadsheet must be compiled, in the format required by the chosen software, listing the biological activity of the ligands along with the values of the molecular descriptors associated to them and/or their docking scores. Several software packages allow the calculation of the descriptors and also the subsequent statistical analyses. Usually, these packages require saving structures and biological activities in a single spreadsheet file.

2.3.2. Protein Structure

A file with the 3D coordinates of the target protein is required in order to perform docking experiments. For proteins that have been solved experimentally, such files can be downloaded from the Protein Data Bank (www.rcsb.org). Alternatively, homology models of the receptor can be constructed (18–20).

3. Methods

3.1. Preparation of the Ligands

Known or candidate ligands, collected or designed as indicated in subheading 2.3.1, need to be subjected to a careful preparation procedure in order to: add hydrogen atoms, if these are not present; generate all ionization and tautomeric states available to the molecules within a certain pH range; generate all possible stereoisomers by varying the configuration of all the chiral centers in a combinatorial manner (unless the dataset contains molecules with known, specified chiralities); and minimize the energy of the ligands through a molecular mechanics engine.

For pure ligand-based QSAR analyses, the most probable ionization and tautomeric state of each compound can be chosen on the basis of energetic considerations. Instead, for docking-based calculations, the ionization and tautomeric state favored by the receptor can be identified for each compound on the basis of the docking scores. All the other states can then be discarded, leading to a dataset containing only a single instance for each molecule.

3.2. Preparation of the Protein

Crystal structures downloaded from the Protein Data Bank and homology models must be carefully inspected and processed in order to: add hydrogen atoms, if they are not present; optimize the geometry and interaction network of the hydrogen atoms; ensure that bond orders are properly assigned; ensure that disulfide bridges are properly connected; delete water molecules, if so desired; and add capping groups to truncated termini (see Note 2). Most docking packages offer automated procedures that help carry out these operations in an automated manner.

3.3. Generation of Protein–Ligand Complexes Through Molecular Docking

As illustrated in Fig. 1, besides being a fundamental step in the calculation of structure-based scores that reflect protein–ligand interactions, molecular docking is also our method of choice for the generation of the structural alignments of ligands necessary for most 3D-QSAR analyses, such as CoMFA and CoMSIA (5, 48).

During the docking procedure, typically, a variety of ligand conformations and orientations are sampled within the target protein by means of specific algorithms. Most docking software requires the user to specify a region of the protein within which to confine the docking of the ligands. Knowledge of the biology of the target is fundamental for a correct identification of the binding site. When data are not available, all the cavities present within a protein and on its surface should be explored. Target proteins are usually treated as rigid molecules, mostly to reduce calculation times. Most software packages, however, allow granting some flexibility to the receptor, if so desired, although usually with a considerable increase of the computational demand. Most docking programs allow usage of constraints on protein–ligand interactions derived from experimental data. For example, required hydrogen bonds or hydrophobic interactions may be specified, or the occupation of a particular region of the binding pocket may be enforced.

3.4. Prediction of Biological Activities Through QSAR Analysis

After the collection of the compounds and, where necessary, their alignment, QSAR analyses can be conducted through the steps described in the following paragraphs.

3.4.1. Calculation of the Independent Variables: QSAR Based on Molecular Descriptors

Different numerical values associated with each molecule, known as descriptors, can be either calculated through a variety of existing software packages or measured experimentally (see Note 3)—for a noncomprehensive list of programs that can calculate molecular descriptors see the Subheading 2. While for the calculation of 2D descriptors only the topology of the ligands is required, the calculation of 3D descriptors requires also the bioactive conformation and, in some instances, the 3D alignment of the ligands. As mentioned, our method of choice for the derivation of bioactive conformations and structural alignments is molecular docking. Alternatively, ligand-based conformational analyses and 3D superimpositions can be applied. Most of the software allows including the descriptors in the same spreadsheet that contains the structures and biological activities of the compounds.

3.4.2. Calculation of the Independent Variables: CoMFA and CoMSIA

CoMFA (8) and CoMSIA (9) are 3D-QSAR analyses implemented in SYBYL (43). Once the molecules have been aligned, our protocol for the development of CoMFA and CoMSIA models proceeds as follows: Gasteiger-Hückel charges are calculated for all the compounds; a 3D cubic lattice is defined, extending 4 Å over the aligned molecules in all directions, with a spacing of 1 Å and a probe atom consisting of an sp3 hybridized carbon (c.3) with a charge of +1.0; two molecular interaction fields (steric and electrostatic) are calculated for CoMFA studies, using a distance dependent dielectric and a cut-off of 30.0 kcal/mol for the calculation of the Coulombic electrostatic energy; five fields (steric, electrostatic, H-bond donor, H-bond acceptor, hydrophobic) are calculated for CoMSIA studies, using the standard parameters.

3.4.3. Statistical Analysis

The relationships between the descriptors and the experimental activity can be studied through various statistical analyses, for instance MLR or PLS regression. Several statistical parameters can then be employed to assess the quality of the model, the most prominent of which are the square of the correlation coefficient (r^2) and the root mean square error (RMSE) (see Note 4) of the predictions. For CoMFA and CoMSIA studies, PLS regression analyses are directly carried out within SYBYL (43), using 4–6 components as independent variables and the experimental pIC_{50} or pK_i values of the molecules as a dependent variable.

3.4.4. Validation of the Model

Models can be validated through the use of cross-validation techniques. Among these, one of the most widespread is the leave-one-out test, which involves taking one molecule out of the training set and predicting its activity on the basis of a model trained with the remaining molecules. The operation is repeated until all the molecules, in turn, have been taken out of the training set one by one. The most used parameter to assess the quality of a leave-one-out cross-validation model is the cross-validated r^2, or q^2—see Golbraikh and Tropsha for further details and caveats (49). Moreover, and most effectively, models can be validated through the evaluation of the activity of a test set, i.e., an additional set of molecules not included in the training set.

*3.4.5. Prediction
of the Activity of Novel
Compounds*

Once a QSAR model has been built and validated, it can be used to predict the activity of newly designed compounds on the basis of the molecular properties associated with them. Following these initial predictions, the compounds can be further modified in order to improve their predicted activity. Eventually, all the designed compounds are ranked on the basis of the QSAR predictions and selected for experimental testing.

*3.5. Prediction
of Biological Activities
Through Docking-
Based Scoring*

In molecular docking, the generated protein–ligand complexes are scored through specific scoring functions, which are based on different principles and endowed with different levels of accuracy (2) (see Note 5). Parallel docking calculations can be run combining various docking algorithms and scoring functions. If experimentally solved protein–ligand complexes are available, they may be conveniently used as controls to assess the accuracy of the docking poses and choose the algorithm the most suitable algorithm to work with a given target.

Docking-based scoring does not require training sets and can be directly used to rank the relative predicted affinity of a set of compounds. The numeric values of the docking scores do not represent free binding energies or affinities and can only be used to estimate the affinity of a compound relativity others. However, if a set of known ligands is available, this can be used as a training set in order to correlate docking scores and the experimental affinities through linear regression analysis. The affinity of novel ligands can then be inferred on the basis of the calculated relationship between docking scores and experimental affinity values. Just as in ligand-based QSAR, the quality of the models can be assessed on the basis of statistical parameters such as r^2 and RMSE of calculated and experimental affinities. Cross-validation and external validation techniques can then be used to validate the models prior to their application to the prediction of the affinity of novel ligands. Moreover, when a training set is available, it can also be used to infer ad hoc "trained" scoring functions through regression analysis, cherry-picking, and combining the components that better correlate with the experimental affinities. However, such ad hoc scoring functions incur the risk of being predictive only within the training set for which they were generated, and deserve a careful and extensive validation through external test sets in order to assess their applicability.

As a caveat, it is worth mentioning that docking scores are rather crude and hence not very effective in the fine ranking of active compounds on the basis of their affinities or potencies. Instead, they are usually suited for distinguishing active from inactive compounds. If a set of known ligands is available, this ability can be monitored through controlled experiments in which the ligands are docked at the target protein together with a large set of decoy compounds. The receiver operating characteristic (ROC) curves obtained in these pilot screenings can be conveniently used to optimize the docking protocol and select the most appropriate scoring functions for a given target.

3.6. Consensus Models   Our method for the construction of consensus models is based on a PLS regression in which the experimental activities are used as the dependent variable, while the activities predicted through different individual models are used as independent variables (see Fig. 2) (5, 48). In particular, the independent variables are converted into one or more components through PLS (see Note 6) and then the regression analysis is performed. As usual, the quality of the model can be monitored on the basis of its r^2 and RMSE values. Moreover, cross-validations and validations with external test sets can be performed.

Consensus modeling can be applied to combine different ligand-based models, different structure-based models, or even ligand-based and structure-based models together. As mentioned in the introduction, since our consensus models originate from PLS regressions, a training set is necessary for their construction, even if the individual components are exclusively structure-based models.

4. Notes

1. The selection of the ligands is a very important step in the construction of a QSAR model. To generate a broadly applicable model, it is fundamental to collect a representative set of diverse ligands.

2. If the protein structure is truncated, i.e., misses a few residues at the N-terminus or the C-terminus, N-acetyl and N-methylamide groups are often used to cap the first-solved N-terminal and last-solved C-terminal residue, respectively. This operation prevents the first- and last-solved residues from being unnaturally electrically charged.

3. In order to avoid overfitting, the model must not be based on an excessive number of descriptors (independent variables). A commonly observed rule prescribes the use of no more than one independent variable per 10 observations (50, 51). For instance, if the training set contains 100 compounds, no more than 10 descriptors should be used. Validating the model through an external test set, i.e., a set of compounds that have not been used to generate the model, is also a good practice to assess the quality of the model, since overfitted models, although very good within training sets, usually perform poorly with external test sets.

4. Many statistical analyses require the variables to follow a normal distribution. Thus, it is recommended that the activity of compounds included in QSAR studies be expressed in logarithmic form, since the use of logarithmic values may help normalizing the distribution of a variable. For instance, the expression of affinities as pK_i (-log K_i) rather than K_i values is preferable. Furthermore, in order to ensure a correct interpretation of QSAR equations and an immediate perception of the weight of each descriptor on the basis of the value of its coefficient, it is also important that the descriptors be standardized. One way of doing this is subtracting to the value of a descriptor calculated for a particular molecule the mean value of the descriptor and dividing the resulting number by the standard deviation of the descriptor (52).

5. The computational time required to score a set of ligands with a scoring function is usually directly correlated to the accuracy of the scoring function. The fastest, less accurate, functions are intended to be used when docking a very high number of molecules; on the contrary, the most accurate and slowest functions are intended to be used when docking a smaller number of molecules.

6. The number of components to be used in a PLS regression should be determined in each case, trying to keep it to a minimum and terminating the introduction of additional components when the last added component barely adds anything to the explanation of the variance of the data.

Acknowledgment

This research was supported by the Intramural Research Program of the NIH, NIDDK.

References

1. Congreve M, Marshall F (2010) The impact of GPCR structures on pharmacology and structure-based drug design. Br J Pharmacol 159:986–996

2. Vilar S, Cozza G, Moro S (2008) Medicinal Chemistry and the Molecular Operating Environment (MOE): Application of QSAR and Molecular Docking to Drug Discovery. Curr Top Med Chem 8:1555–1572

3. Costanzi S, Tikhonova IG, Harden TK, Jacobson KA (2009) Ligand and structure-based methodologies for the prediction of the activity of G protein-coupled receptor ligands. J Comput Aided Mol Des 23:747–754

4. Pierce KL, Premont RT, Lefkowitz RJ (2002) Seven-transmembrane receptors. Nat Rev Mol Cell Biol 3:639–650

5. Vilar S, Karpiak J, Costanzi S (2010) Ligand and structure-based models for the prediction of ligand-receptor affinities and virtual screenings: Development and application to the β_2-adrenergic receptor. J Comput Chem 31:707–720

6. Potemkin V, Grishina M (2008) Principles for 3D/4D QSAR classification of drugs. Drug Discov Today 13:952–959

7. Estrada E (2008) How the parts organize in the whole? A top-down view of molecular

descriptors and properties for QSAR and drug design. Mini Rev Med Chem 8:213–221

8. Cramer RD, Patterson DE, Bunce JD (1988) Comparative Molecular-Field Analysis (CoMFA).1. Effect of Shape on Binding of Steroids to Carrier Proteins. J Am Chem Soc 110:5959–5967

9. Klebe G, Abraham U, Mietzner T (1994) Molecular Similarity Indexes in A Comparative-Analysis (CoMSIA) of Drug Molecules to Correlate and Predict Their Biological-Activity. J Med Chem 37:4130–4146

10. Wold S, Albano C, Dunn WJ, Esbensen K, Hellberg S, Johansson E et al (1984) Modeling data tables by principal components and PLS-class patterns and quantitative predictive relations. Analusis 12:477–485

11. Geladi P, Kowalski BR (1986) Partial least-squares regression: a tutorial. Anal Chim Acta 185:1–17

12. Moro S, Bacilieri M, Ferrari C, Spalluto G (2005) Autocorrelation of molecular electrostatic potential surface properties combined with partial least squares analysis as alternative attractive tool to generate ligand-based 3D-QSARs. Curr Drug Discov Technol 2:13–21

13. Moro S, Bacilieri M, Cacciari B, Spalluto G (2005) Autocorrelation of molecular electrostatic potential surface properties combined with partial least squares analysis as new strategy for the prediction of the activity of human A_3 adenosine receptor antagonists. J Med Chem 48:5698–5704

14. Pastor M, Cruciani G, McLay I, Pickett S, Clementi S (2000) GRid-INdependent descriptors (GRIND): a novel class of alignment-independent three-dimensional molecular descriptors. J Med Chem 43:3233–3243

15. Benedetti P, Mannhold R, Cruciani G, Ottaviani G (2004) GRIND/ALMOND investigations on CysLT1 receptor antagonists of the quinolinyl(bridged)aryl type. Bioorg Med Chem 12:3607–3617

16. Costanzi S, Siegel J, Tikhonova IG, Jacobson KA (2009) Rhodopsin and the others: a historical perspective on structural studies of G protein-coupled receptor. Curr Pharm Des 15:3994–4002

17. Hanson MA, Stevens RC (2009) Discovery of new GPCR biology: one receptor structure at a time. Structure 17:8–14

18. Costanzi S (2008) On the applicability of GPCR homology models to computer-aided drug discovery: a comparison between in silico and crystal structures of the beta2-adrenergic receptor. J Med Chem 51:2907–2914

19. Michino M., Abola E., GPCR Dock 2008 participants, Brooks C.L. 3rd, Dixon J.S., Moult J., Stevens R.C. (2009) Community-wide assessment of GPCR structure modelling and ligand docking: GPCR Dock 2008. *Nat. Rev. Drug Discov.* 8, 455-463

20. Costanzi S (2010) Modeling G protein-coupled receptors: a concrete possibility. Chimica Oggi-Chemistry Today 28:26–31

21. Chipot C, Rozanska X, Dixit SB (2005) Can free energy calculations be fast and accurate at the same time? Binding of low-affinity, non-peptide inhibitors to the SH2 domain of the src protein. J Comput Aided Mol Des 19:765–770

22. Foloppe N, Hubbard R (2006) Towards predictive ligand design with free-energy based computational methods? Curr Med Chem 13:3583–3608

23. Warren GL, Andrews C, Capelli AM, Clarke B, LaLonde J, Lambert MH et al (2006) A critical assessment of docking programs and scoring functions. J Med Chem 49:5912–5931

24. de Graaf C, Rognan D (2008) Selective structure-based virtual screening for full and partial agonists of the β2-adrenergic receptor. J Med Chem 51:4978–4985

25. Reynolds KA, Katritch V, Abagyan R (2009) Identifying conformational changes of the $β_2$-adrenoceptor that enable accurate prediction of ligand/receptor interactions and screening for GPCR modulators. J Comput Aided Mol Des 23:273–288

26. Katritch V, Rueda M, Lam PC, Yeager M, Abagyan R (2010) GPCR 3D homology models for ligand screening: lessons learned from blind predictions of adenosine A_{2A} receptor complex. Proteins 78:197–211

27. Bhattacharya S, Vaidehi N (2010) Computational mapping of the conformational transitions in agonist selective pathways of a G-protein coupled receptor. J Am Chem Soc 132:5205–5214

28. Vilar S, Ferino G, Sharangdhar SP, Berk B, Cavasotto CN, Costanzi S (2011) Docking-based virtual screening for ligands of G protein-coupled receptors: not only crystal structures but also in silico models. J. Mol. Graph. Model 29, 614–623. *Submitted for publication.*

29. Topiol S, Sabio M (2008) Use of the X-ray structure of the $β_2$-adrenergic receptor for drug discovery. Bioorg Med Chem Lett 18:1598–1602

30. Kolb P, Rosenbaum DM, Irwin JJ, Fung JJ, Kobilka BK, Shoichet BK (2009) Structure-based discovery of $β_2$-adrenergic receptor ligands. Proc Natl Acad Sci USA 106: 6843–6848

31. Carlsson J, Yoo L, Gao ZG, Irwin JJ, Shoichet BK, Jacobson KA (2010) Structure-based discovery of A_{2A} adenosine receptor ligands. J Med Chem 53:3748–3755

32. Katritch V, Jaakola VP, Lane JR, Lin J, Ijzerman AP, Yeager M et al (2010) Structure-based discovery of novel chemotypes for adenosine A_{2A} receptor antagonists. J Med Chem 53:1799–1809

33. Vaidehi N, Schlyer S, Trabanino RJ, Floriano WB, Abrol R, Sharma S et al (2006) Predictions of CCR1 chemokine receptor structure and BX 471 antagonist binding followed by experimental validation. J Biol Chem 281:27613–27620

34. Engel S, Skoumbourdis AP, Childress J, Neumann S, Deschamps JR, Thomas CJ et al (2008) A virtual screen for diverse ligands: Discovery of selective G protein-coupled receptor antagonists. J Am Chem Soc 130:5115–5123

35. Tikhonova IG, Sum CS, Neumann S, Engel S, Raaka BM, Costanzi S et al (2008) Discovery of novel agonists and antagonists of the free fatty acid receptor 1 (FFAR1) using virtual screening. J Med Chem 51:625–633

36. Cavasotto CN, Orry AJ, Murgolo NJ, Czarniecki MF, Kocsi SA et al (2008) Discovery of novel chemotypes to a G-protein-coupled receptor through ligand-steered homology modeling and structure-based virtual screening. J Med Chem 51:581–588

37. Bhattacharya S, Subramanian G, Hall S, Lin J, Laoui A, Vaidehi N (2010) Allosteric antagonist binding sites in class B GPCRs: corticotropin receptor 1. J Comput Aided Mol Des 24:659–674

38. Dragon, Talete, SRL, www.talete.mi.it; E-Dragon, Virtual Computational Chemistry Laboratory, www.vcclab.org.

39. The CODESSA PRO project, www.codessa-pro.com.

40. MOE, Chemical Computing Group, Inc., www.chemcomp.com.

41. Schrödinger, LLC, www.schrodinger.com.

42. ICM, MolSoft, LLC, www.molsoft.com.

43. SYBYL, Tripos, Inc., www.tripos.com.

44. GOLD, Cambridge Crystallographic Data Centre, www.ccdc.cam.ac.uk/products/life_sciences/gold.

45. The R project for statistical computing, www.r-project.org.

46. STATISTICA, StatSoft, Inc., www.statsoft.com.

47. Okuno Y, Tamon A, Yabuuchi H, Niijima S, Minowa Y, Tonomura K et al (2008) GLIDA: GPCR–ligand database for chemical genomics drug discovery–database and tools update. Nucleic Acids Res 36:D907–D912

48. Costanzi S, Tikhonova IG, Ohno M, Roh EJ, Joshi BV, Colson AO et al (2007) P2Y₁ antagonists: Combining receptor-based modeling and QSAR for a quantitative prediction of the biological activity based on consensus scoring. J Med Chem 50:3229–3241

49. Golbraikh A, Tropsha A (2002) Beware of q2! J Mol Graph Model 20:269–276

50. Normolle D, Ruffin MT, Brenner D (2005) Design of early validation trials of biomarkers. Cancer Inform 1:25–31

51. Concato J, Peduzzi P, Holford TR, Feinstein AR (1995) Importance of events per independent variable in proportional hazards analysis I. Background, goals, and general strategy. J Clin Epidemiol 48:1495–1501

52. Hill T, Lewicki P (2006) In STATISTICS methods and applications. StatSoft, Tulsa, OK

Chapter 17

Identification of Motions in Membrane Proteins by Elastic Network Models and Their Experimental Validation

Basak Isin, Kalyan C. Tirupula, Zoltán N. Oltvai,
Judith Klein-Seetharaman, and Ivet Bahar

Abstract

Identifying the functional motions of membrane proteins is difficult because they range from large-scale collective dynamics to local small atomic fluctuations at different timescales that are difficult to measure experimentally due to the hydrophobic nature of these proteins. Elastic Network Models, and in particular their most widely used implementation, the Anisotropic Network Model (ANM), have proven to be useful computational methods in many recent applications to predict membrane protein dynamics. These models are based on the premise that biomolecules possess intrinsic mechanical characteristics uniquely defined by their particular architectures. In the ANM, interactions between residues in close proximity are represented by harmonic potentials with a uniform spring constant. The slow mode shapes generated by the ANM provide valuable information on the global dynamics of biomolecules that are relevant to their function. In its recent extension in the form of ANM-guided molecular dynamics (MD), this coarse-grained approach is augmented with atomic detail. The results from ANM and its extensions can be used to guide experiments and thus speedup the process of quantifying motions in membrane proteins. Testing the predictions can be accomplished through (a) direct observation of motions through studies of structure and biophysical probes, (b) perturbation of the motions by, e.g., cross-linking or site-directed mutagenesis, and (c) by studying the effects of such perturbations on protein function, typically through ligand binding and activity assays. To illustrate the applicability of the combined computational ANM—experimental testing framework to membrane proteins, we describe—alongside the general protocols—here the application of ANM to rhodopsin, a prototypical member of the pharmacologically relevant G-protein coupled receptor family.

Key words: Anisotropic network model (ANM), Normal mode analysis (NMA), Structural dynamics, Molecular dynamics (MD) simulations, Structure prediction, Conformational changes, Ensembles of structures, G-protein coupled receptors, Multiscale models and methods

Nagarajan Vaidehi and Judith Klein-Seetharaman (eds.), *Membrane Protein Structure and Dynamics: Methods and Protocols*,
Methods in Molecular Biology, vol. 914, DOI 10.1007/978-1-62703-023-6_17, © Springer Science+Business Media, LLC 2012

1. Introduction

1.1. The Functional Roles of Motions in Membrane Proteins

While the structure of a protein provides some insights into its function, it only yields static information. Detailed information on the dynamics of a protein is necessary for a complete understanding of its function. However, it is not a trivial problem to resolve the functionally relevant motions of biomolecular systems. The timescales involved in different events range from femtoseconds to seconds or even longer, and usually more than a single experimental technique or computational model and method is required to span such a wide range. The spatiotemporal resolutions of the computational models or experimental techniques need to be optimally selected to explore the particular time and length scales of the systems and process of interest.

Computational approaches can visualize molecular events at timescales and resolutions that may not be readily examined by experimental techniques, thus complementing the information acquired from experiments and providing deeper insights into biomolecular mechanisms of function. Not surprisingly, computational biology tools and biophysical theories have been advantageously employed in the last two decades for investigating the dynamics of biomolecular systems. Among them, molecular dynamics (MD) simulations, and normal mode analysis (NMA) found broad utility in biomolecular applications (1–4), and especially in investigating the dynamics of membrane proteins in recent years (5–8). Membrane proteins are particularly important as they serve as targets for the majority of drugs, and their molecular motions remain largely unknown due to scarcity of structural data. It is now widely recognized that the motions of membrane proteins underlie key functional events, such as receptor–ligand binding and signaling, ion channeling and gating, folding and translocation, and the allosteric control of them. An assessment of the type and size of these motions is critically important for designing modulators, agonists, or antagonists of membrane proteins' functions.

1.2. Overview of Structural Data for Membrane Proteins

Obtaining structural data for membrane proteins experimentally is challenging for a number of reasons. First, structural techniques such as X-ray crystallography and NMR spectroscopy require large quantities of proteins purified to homogeneity. For membrane proteins, both the production and purification steps are very difficult. Membrane proteins usually have very low expression levels and are thermally unstable when removed from their native membrane environment and tend to aggregate. Secondly, the need for mimicking membranes requires addition of lipids or detergents that interfere with structural techniques, for example by creating large background signals in NMR and increasing the effective molecular weight of the membrane protein–detergent mixed micellar complex.

Thus, determining membrane protein structures has been a challenge, a fact which is reflected by the limited number of available membrane protein structures resolved and deposited to date in the Protein Data Bank (PDB) (9). As of the end of November 2010 there have been only 1,331 membrane protein structures deposited in the PDB, which corresponds to only 2% of all structures in the PDB (Fig. 1a). Moreover, only 262 out of these 1,331 are unique membrane protein structures, as defined by a sequence cutoff of 95%. Thus, only a small number of unique membrane proteins are available as examples for different functional and structural categories that membrane proteins engage in (Fig. 1b).

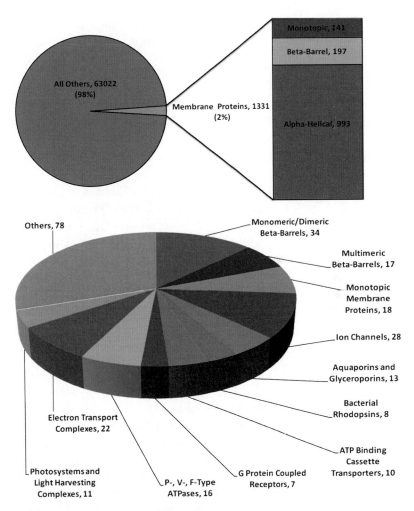

Fig. 1. Fraction and composition of membrane proteins in the PDB. (**a**) The *pie chart* shows that of all structurally reserved proteins, 2% correspond to membrane proteins. The chart on the *right* provides the breakdown of membrane proteins into different structural categories. (**b**) Structural and functional distribution of unique membrane protein structures. Data was retrieved on 26 November 2010 from the curated membrane protein database available at http://blanco.biomol.uci.edu/Membrane_Proteins_xtal.html.

1.3. Need for Complementary Predictive Computational Approaches to Study the Dynamics of Membrane Proteins

The general challenges faced in obtaining structures of membrane proteins experimentally described above, also impact the ability to determine membrane protein motions. Even for those proteins for which a structure exists, it is most often only in one functional state. Thus, there is a great need and opportunity to utilize the protein structures that are available to predict functional motions by computational techniques. These predicted dynamics can then be tested experimentally by targeting specific regions in the protein or types of motions through biophysical approaches. A large battery of techniques is available in which attachment points such as cysteine replacements are introduced at specific sites that allow application of specific techniques, where the choice of technique depends on the timescale and scale of motions to be tested. This approach allows obtaining site-specific information and is ideally suited to target specific predictions made by computational models. In this chapter, we will review this combined experimental–computational approach.

1.4. State of the Art in Computational Modeling of Protein Dynamics: Application to Membrane Proteins

1.4.1. Molecular Dynamics as the Gold Standard for Simulating Motions at Atomic Resolution, but Limited to Short Times and Small Systems

The most widely used atomic-level approach in modeling protein dynamics is molecular dynamics (MD) simulations. The interactions of atoms are described therein by complex potential energy functions and parameters, which define the instantaneous forces exerted on each atom. Knowledge of these forces permits us to calculate the dynamic behavior of the system (composed of the protein with its substrates/ligands, and the environment—mainly water, lipid molecules and ions) using the classical Newton's equations of motion (1). MD simulations thus provide atomic-level detail with high temporal resolution, which is usually viewed as the *gold standard*.

On the other hand, it is also widely known that MD simulations have limitations. First, the trajectories are accurate to the extent that force fields themselves are. Second, the size of the time steps of these simulations, usually 1–2 fs, implies that millions of steps are required to reach the nanosecond timescale; and nanoseconds can provide a good description of local events at best.

The rapid growth in computer power and new developments in parallelizable MD codes designed for high-performance simulations now enable us to simulate large biomolecular systems at hundreds of microseconds or even milliseconds scales (5, 10, 11). These long simulations confirm that the "global" motions of proteins (e.g., interconversions between substates) are several orders of magnitude slower than local motions. For example, the global backbone rearrangements of BPTI, a small protein of 58 residues, occur on a timescale of 10 μs, whereas side-chain motions are faster than 10 ns (11). Clearly, MD simulations on the millisecond timescale can investigate such global motions, including folding and unfolding events for small proteins only. Application to larger proteins, and to membrane proteins in particular, incurs serious convergence problems (6, 12).

A method for exploring global motions accessible under equilibrium conditions is NMA, a technique introduced several decades ago by computational chemists for describing small molecules' vibrational dynamics (13). NMA is a harmonic approximation and assumes that the conformational energy surface can be characterized by a single energy minimum. Classical, full-atomic NMA requires reaching a minimum on the potential energy surface. Hence, a stringent energy minimization of the potential energy function is required before performing NMA, or before evaluating the Hessian matrix (the $3n \times 3n$ matrix of the second derivatives of the potential energy function with respect to the $3n$ degrees of freedom for a structure of n atoms, for atomic NMA). The low-frequency modes obtained by NMA often correlate well with experimentally observed changes between different functional states (e.g., open and closed form of a ligand-free and -bound enzyme). However, energy minimization and numerical evaluation of the Hessian matrix using a full-fledged force field makes full-atomic NMA a computationally expensive method for large biomolecules (1, 14).

The investigation of cooperative motions in large systems requires the adoption of coarse-grained models, and possibly analytical methods, which brings us to the focus of our review, the use of elastic network models (ENMs). NMA has seen a revival in recent years, with the realization that it can be advantageously used in conjunction with simplified models such as ENMs for studying the global motions of proteins (15, 16). A breakthrough in the field was the demonstration by Tirion that the global modes obtained by the detailed NMA could be almost identically reproduced by using a single parameter harmonic potential for the force field (17). This study led to the introduction of the Gaussian Network Model (GNM) (18, 19) as a simplified network representation of the protein structure, amenable to the investigation of its dynamics.

In the GNM, the collective dynamics of the protein (of N residues) is fully defined by an $N \times N$ Kirchhoff matrix that describes the spatial connectivity (or topology) of the structure at the level of α-carbons. The physical basis and the analytical method for examining the statistical mechanics of such a mass-spring network bear close similarities to the elasticity theory of random polymer networks (20).

The same level of coarse-graining, i.e., residue-level description with the position of network nodes identified by that of the α-carbons, has been adopted in the Anisotropic Network Model (ANM) (21–25). The main advantage of performing NMA using the ANM, shortly referred to here as ANM analysis, is the possibility of computing an analytical, unique solution for the collective modes of motion accessible to the examined system. The basic ingredient in the model is the topology of inter-residue contacts in the native structure (as the GNM), which turns out to be a major determinant of equilibrium dynamics. Any deformation from the native state coordinates is resisted by linear springs that associate closely neighboring residues, bonded and nonbonded (Fig. 2). Despite its simplicity, the ANM results proved to be in remarkable agreement with experiments. The global modes obtained with the ANM analysis have been shown in numerous applications to be highly robust (26) and consistent with experimental data in terms of the shapes and mechanisms of motions. This includes, for example, the directions of domain movements and their correlations, the location of hinge-bending sites, the most likely collective rearrangements, or allosteric switches triggered by ligand/substrate binding. The ANM server (27) is being broadly used by the molecular structural and computational biology communities for rapid assessment and visualization of the intrinsic dynamics of known protein structures.

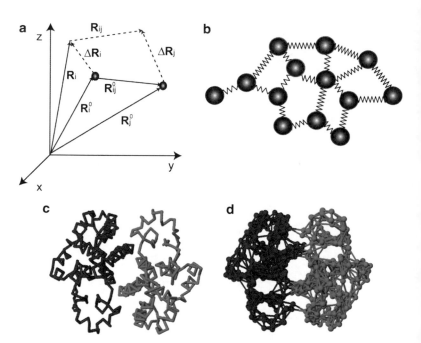

Fig. 2. Representation of a given structure by the ANM. Panel (**a**) displays the equilibrium position of node i, R_i^0 its instantaneous fluctuation ΔR_i, and instantaneous position R_i. The fluctuation from the mean position is given by $\Delta R_i^0 = R_i - R_i^0$. For a pair of interacting residues i and j, we denote the equilibrium inter-node distance vector extending from residue i to residue j as R_{ij}^0 and the instantaneous change in distance vector is $\Delta R_{ij} = R_{ij} - R_{ij}^0 = \Delta R_j - \Delta R_i$. Panel (**b**) displays the node and spring representation where each node represents an amino acid, and springs connect interacting nodes (bonded or nonbonded pairs of amino acids with an interaction cutoff distance of R_c). Panels (**c**) and (**d**) display the backbone (stick), and elastic network (ANM) representations (based on a cutoff distance of 10 Å) for an example protein, glutathione transferase (PDB code 2A2R) (109). Note that the ANM also includes springs at the interface between the two monomers.

However, ENMs, including the GNM and ANM, also have their own limitations. The information provided lacks atomic precision and may in some instances yield unphysical bond lengths or bond angles at the atomic scale (28). Moreover, no absolute timescale and size of motions can be deduced, only the shape of motions. Also, neither the specificity of amino acids nor the effects of a protein's environment (e.g., lipid bilayer in membrane proteins, or water molecules in general) is taken into consideration. In parallel with all NMAs, the approach is in principle limited to movements within a single energy well, and becomes most accurate in the immediate vicinity of the energy well. This means that transitions that involve passage over energy barriers or jumps between minima cannot be predicted. On the other hand, the coarse-graining of the structure and thereby smoothing out of the energy landscape in the ANM presumably help merge substates separated by low energy barriers, thus allowing for sampling of substates separated by low energy barriers.

The basic assumptions and theory underlying the ANM, and its use in predicting structure-based motions as well as its limitations, will be described in detail in Subheading 3.1. From the brief introduction above, it is clear that both extremes, MD simulations and ANM analysis, have limitations, but they also provide complementary information. In particular, they allow for exploring different timescales and events at different resolutions, the former being powerful in examining events in the picoseconds to hundreds of nanoseconds regime, and the latter successfully describing the global machinery of supramolecular systems. Clearly, methods that encompass both regimes are highly desirable, and we will present in Subheading 3.1.3. the development and use of such a method, ANM-steered MD recently applied to rhodopsin (29). First, we provide a brief description of approaches undertaken for extending the range of MD simulations.

1.4.3. Recent Progress in Atomic Simulations: Steered MD, Targeted MD and ANM-Guided MD

Various methods have been developed to accelerate MD simulations and increase their sampling efficiency. Steered MD (SMD) is one of them. The idea is to observe large conformational changes by applying external forces to facilitate such movements (30). A similar approach, Targeted MD (TMD), applies time-dependent, purely geometrical constraints to induce conformational changes for attaining a known target structure at physiological temperatures. Additionally, guiding MD simulations by collective coordinates has proven to be useful for efficient sampling and expansion of the accessible conformational space. Berendsen and coworkers developed a method in which essential dynamics analysis (EDA) or principle component analysis (PCA) of an ensemble of snapshots obtained from MD trajectories are used to generate collective modes (31). These collective modes are used as constraints in MD simulations to sample the conformational space efficiently (31–33).

Collective coordinates have been combined with ensemble sampling by Abheser and Nilges. Their method uses an additional biasing potential defined along the collective modes used as restraints in a set of independent MD trajectories (34). Another technique, Amplified Collective Motion couples the low frequency motions obtained from the ANM to a higher temperature by using a weak coupling method (35).

More recently, we introduced the *ANM-guided MD* methodology (29), a hybrid methodology that will be explained in detail in Subheading 3.1.3. The method, originally introduced for exploring the dynamics of rhodopsin (29) is, in principle, applicable to all membrane proteins. ANM-guided (or *ANM-restrained*) MD is useful for modeling small molecule binding to proteins and flexible docking of proteins in particular. It has also been recently adopted for studying the dynamics of biological systems and protein-protein interactions, in conjunction with MM-PBSA (Molecular Mechanic/Poisson-Boltzmann Surface Area) calculations (36). It also proved useful in establishing the significance of the conformational change induced by thrombosbondin-1 binding to the calreticulin structure, essential to signaling focal adhesion disassembly (36). In the present review, we will provide a detailed description of the methodology, and illustrate its utility in generating conformers that may be advantageously exploited for flexible docking of inhibitors.

1.4.4. Structure-Based Coarse-Grained Computations for Membrane Proteins

Structure-based coarse-grained models have been used for several decades in computational biology with the understanding that these may help explore different time and length scales of biological interest, albeit at low resolution. Pioneering studies in this field are the work of Michael Levitt (37, 38) or the virtual bond model based on α-carbons widely used by Flory and coworkers in delineating the statistical mechanics of biopolymers. Coarse-grained models for membrane proteins, on the other hand, are relatively more recent, presumably due to lack of structural data. As recently reviewed by Sansom and collaborators (39), coarse-grained simulations of membrane proteins are particularly useful for exploring large-scale changes in conformation including the diffusion and reorientation of peptides or insertion of membrane proteins across the lipid bilayer, and self-assembly of membrane protein–detergent micelles. A typical self-assembly of a protein-bilayer requires simulating at least 0.1 µs, and such simulations become prohibitively expensive, if not statistically inaccurate (due to convergence problems), with full atomic models and force fields. In particular, the MARTINI force field introduced by Marrink and coworkers for lipids (40) has found wide applications in recent years and have been successfully extended by the Sansom and Schulten labs to protein–lipid systems (41–43).

In parallel to coarse-grained MD simulations, ANM-based methods have also been exploited for examining the functional motions of membrane proteins, as we have recently reviewed (7, 8). Notable applications include the pore opening of potassium channels (44), the dynamics of the mechanosensitive channel (MscL) (45), the allosteric global torsional motions of nicotinic acetyl choline receptor (46), cooperative domain movements of ABC transporter BtuC (47), and identification of core amino acids and intrinsic dynamics of rhodopsin (44, 48), as will be described in detail below.

2. Materials

Hardware and software to run ANM and ANM-restrained MD. ANM-restrained MD simulations of rhodopsin were performed at four machines each with eight 2.6 GHz Intel Xeon processors and 32 GB memory. The simulations described above complete in approximately 10 h. The molecular dynamics simulations and energy minimization sections of the protocol were run using Nanoscale Molecular Dynamics Software (10). ANM calculations are performed in Fortran 90. The protocol is fully automated using a tcl code that enables the communication of the different sections. This code is available by contacting the authors.

The ANM code is available in both Fortran 90 and C programming languages. ANM calculations of rhodopsin take approximately 1 min on one of the above processors. The Web site implementation of ANM is also available and described in Note 1.

3. Methods

3.1. Methods I: Theory and Modeling

3.1.1. Overview

The diagram in Fig. 3 provides an overview of the overall approach, where modeling of single structures are outlined in the white box and experimental studies in the green boxes. When multiple structures are available, modeling is also used to help in the analysis (green box, right), while other experiments may be driven by the modeling (green box, bottom). The inputs and outputs of these approaches are summarized below and described in detail in the subsequent sections.

Two types of inputs may be used in ANM calculations: a known structure accessible in the PDB, or a structural model generated by the user, e.g., using structure determination techniques or computational modeling such as homology modeling (also see below). For example, the ANM server requires as input either the PDB identifier of the protein of interest (which is directly downloaded from the PDB), or a four-letter filename for the structural model in PDB format (found by browsing and retrieval from the local desktop) (27). In either case, to generate/predict the equilibrium

Modeling   **Experimental validation**

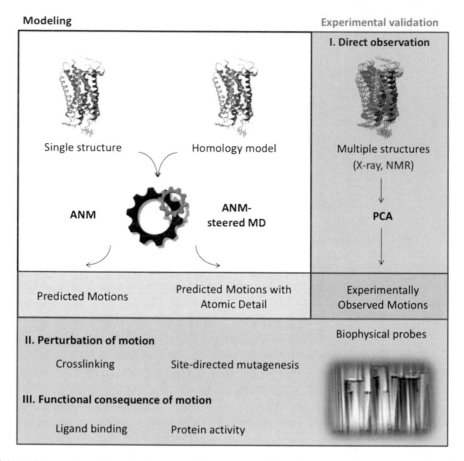

Fig. 3. Schematic overview of the combined experimental–computational approach. Using either single structures or homology (or other) models as an input, ANM or ANM-steered MD simulations are used to predict motions without or with atomic detail, respectively (*white box, left top*). If available, the input into computations can also be multiple structures that can be used to validate predictions as well as help better analyze them ("direct observation," *green box right*). The predictions can be used to design experiments to validate them, through perturbation of motion or study of the functional consequences of motions (*green box, bottom*).

motions of a protein using the ANM, *one* structure file, or the three-dimensional coordinates of the residues forms the input, even at the coarse-grained scale (see below). As a typical example, we will illustrate below the application of ANM to rhodopsin, which permitted us to explain 93% of experimentally observed effects for 119 rhodopsin mutants (49). In the GNM, the input may even be simply the list of "interacting" residue pairs without information on their detailed coordinates. The main ingredient of the model is the $N \times N$ Kirchhoff matrix that describes the presence or absence of contacts between the N residues of the proteins.

On the other hand, if multiple structures are available for a given protein (its mutant, different bound forms, or family members), and if the structures are sufficiently variable/heterogeneous, one can take advantage of the available data to perform a PCA of the ensemble of structures and extract the dominant types of structural variations. This type of information, of purely experimental origin, helps test/validate the theoretically predicted ANM modes. It can also be used to help understand which of the modes predicted by the ANM are exploited by the protein for which types of functional events (50).

If there are no structural data available for the protein of interest, homology (or other) models could be built, which might be used as templates for further refinement and analysis in conjunction with the ANM studies. Currently, there are structural data available for a quarter of all known single-domain families in the protein universe (of all organisms sequenced to date), and >70% of all known sequences can be partially modeled using existing structural data as templates/models (51). Evidently, membrane proteins are the most challenging proteins to predict due to scarcity of structural data ((52); see also Fig. 1). Homology models constructed using ANM modes have been successfully employed to demonstrate the preferential binding of allosteric ligands to active and inactive conformations of metabotropic glutamate receptors (53).

3.1.2. Modeling I: ANM Theory, Assumptions, and Implementation

We present below a detailed description of the steps required for performing the ANM analysis of a given structure. A schematic description is provided in Fig. 4.

ANM Protocol

Input. ANM analysis requires knowledge of α-carbon coordinates. These define the positions of the "nodes" in the elastic network. The coordinates are usually retrieved directly from the PDB file corresponding to the examined structure (e.g., the ANM server directly accepts as input the PDB identifier); or alternatively one can feed the coordinates of a model (e.g., homology model or snapshot from MD simulations) in PDB format. This brings into consideration the first assumption implicit in the ANM analysis: The coordinates in the PDB file are accepted to be the equilibrium coordinates without the need to perform further energy minimization. This is in contrast to classical NMA which requires the energy minimization of the PDB structure or any conformer from simulations prior to performing NMA. The most widely used ANM adaptations use α-carbon coordinates as the node positions, corresponding to a single-site-per-residue description. Other variants use β-carbon coordinates or the lower resolution models with one-site-per-m residues, where m may be as sparse as $m=40$ (54). Such lower resolution models have been successfully used for representing supramolecular systems such as viral capsids and the ribosome. Such larger systems have been reviewed previously (26, 55).

Assessment of the topology of contacts. The topology of contacts here refers to the list of amino acid pairs, or nodes, that are

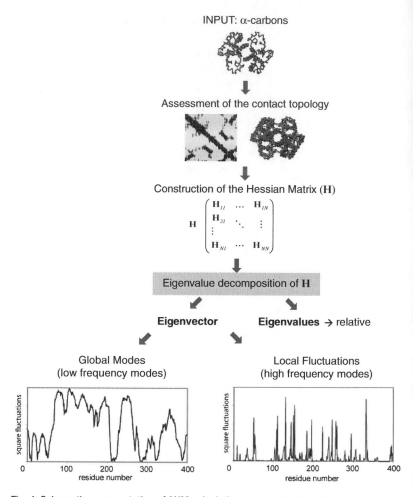

Fig. 4. Schematic representation of ANM calculation steps and released outputs.

"connected" by an elastic spring in the network. By definition, all pairs bonded or nonbonded within a certain cutoff distance R_c are assumed to be connected. Previous analyses demonstrated that $R_c = 12$–15 Å provides a good description of the native contact topology. Thus, the first task is to determine all node pairs within this distance range. This is readily accomplished by taking the difference

$$R_{ij}^0 = R_j^0 - R_i^0 \tag{1}$$

between all pairs of nodes i and j (or all pairs of α-carbons) (Fig. 2a). The pairs of amino acids whose α-carbons are separated by a distance R_{ij}^0 smaller than R_c are thus assumed to interact via a harmonic potential that stabilizes their equilibrium distance R_{ij}^0. At the end of this step we know all pairs of nodes that are connected.

Construction of the Hessian matrix. The Hessian matrix, **H** (Fig. 4), is the only ingredient that the model needs. **H** provides a complete description of the topology of contacts in the original structure. How do we construct **H**? In classical NMA, **H** is by definition a $3N \times 3N$ matrix for a system of N interaction sites, composed of the second derivatives of the total potential energy with respect to the $3N$ degrees of freedom (usually the x-, y-, z- components of the position vectors R_i for all atoms).

Eigenvalue decomposition of **H**. Eigenvalue decomposition of **H** yields $3N-6$ nonzero normal modes, the frequencies and shapes of which are defined by the nonzero eigenvalues and corresponding eigenvectors of **H**. The eigenvalue decomposition may be written as

$$\mathbf{H} = \mathbf{U}\Lambda\mathbf{U}^{\mathrm{T}}. \qquad (2)$$

Here Λ is the diagonal matrix of the eigenvalues. Λ is composed of six zero eigenvalues (due to translational and rotational invariance of **H**) and $3N-6$ nonzero eigenvalues $\{\lambda_1, \lambda_2, ..., \lambda_{3N-6}\}$ organized in ascending order (i.e., $\lambda_1 \le \lambda_2 \le .... \le \lambda_{3N-6}$). **U** is a $3N \times 3N$ matrix composed of $3N$-dimensional eigenvectors written as columns u_1, $u_2, ..., u_{3N-6}$ (preceded by the six eigenvectors corresponding to the zero eigenvalues).

What is the physical meaning of eigenvectors and eigenvalues? Each eigenvector represents a direction of motion in the $3N$-dimensional conformational space. It may also be viewed as a supervector, the super-elements of which are each three-dimensional vectors corresponding to the displacements of individual nodes, listed in sequential order along the protein sequence. The kth eigenvector, for example, is composed of N blocks (each being a three-dimensional vector) representing the relative displacement of the N α-carbons along the kth mode. The kth eigenvalue, on the other hand, provides a measure of the frequency, and size, of the motion as it scales with the square frequency of the kth mode and the size of the fluctuations are inversely proportional to the frequency. Note that the eigenvectors provide information on the relative movements of residues, not their absolute size, since each eigenvector is normalized (sum of elements squared is equal to zero). The size of the motion scales with the reciprocal square root of eigenvalues, as will be further clarified below.

Eigenvalue decomposition is the most time-consuming step of ANM. A suitable subroutine, e.g., BLZPACK (http://crd.lbl. gov/~osni/marques.html#BLZPACK) which is an implementation of the block Lanczos algorithm to solve large, sparse matrices, is used for this purpose. For molecules with 300 residues represented by a 900×900 Hessian matrix, the solution is obtained within 2 s. For proteins with thousands of residues it is obtained within a few minutes.

Mode shapes. The kth mode shape, also called k^{th} mode profile, refers to the relative square displacements of residues, $[(\Delta R_i)^2]k = [(\Delta X_i)^2]k + [(\Delta Y_i)^2]k + [(\Delta Z_i)^2]k$ along mode k as a function of residue index $1 \le i \le N$. This profile is readily evaluated from the trace (sum of the diagonal elements) of each diagonal super element of the matrix $[u_k \, u_k^T]$. The mode shape is normalized, i.e., the terms plotted as a function of residue index i sum up to unity. Therefore, the kth mode shape represents the distribution of residue motions driven by mode k.

Global modes, hinge sites, and recognition sites. Global motions usually involve large portions of the molecule, i.e., they are highly cooperative, and have been shown in numerous applications to be relevant to biological function (16). The slowest mode shape indeed provides valuable information on the global dynamics of the molecule. Minima refer to sites that serve as hinge sites, or anchors: these sites play a key role in mediating the collective dynamics of the entire molecule. Not surprisingly, residues located at such sites are usually conserved. Maxima in the global mode shapes, on the other hand, usually refer to substrate recognition sites. Their intrinsic mobility facilitates optimal recognition and binding of substrates. Both, hinge regions as well as substrate recognition sites are in general critical for biological function. For example, the highly conserved shallow pockets that serve as receptor-binding sites in influenza virus hemagglutinin A, or the antigen binding hypervariable loops of immunoglobulins, form maxima in the slowest mode shape, whereas linkers or interfacial regions between domains subject to anticorrelated motions form minima (15, 28, 55, 56) It is important to note that the global modes are insensitive to the details of the models and energy parameters used in normal mode analyses; they are essentially defined by the distribution of inter-residue contacts, which is rigorously accounted for in the ANM.

High frequency modes. In contrast to global modes, the high frequency modes are highly localized, and as such they are sensitive to the detailed interactions at the atomic level. They usually contain white noise contributions that need to be filtered out in order to extract physically meaningful information. Not surprisingly, these modes have been referred to as "uninteresting modes" (31) when extracted from MD trajectories. They usually drive *isolated* fluctuations, as opposed to the *correlated* ones that underlie the intramolecular communication.

The ANM results differ from those extracted from MD simulations or from full atomic NMA, in that they are devoid of random noise effects; and they are uniquely determined for the given topology of native contacts. The high frequency modes identified by the GNM, in particular, proved to be "interesting": they indicate the most strongly constrained sites in the presence of the intricate coupling between *all* residues. The peaks emerging in these mode shapes are usually implicated in folding nuclei, or key tertiary contacts stabilizing the overall fold. As a consequence, they are evolutionarily conserved among different members of a given family (57, 58).

Application of ANM to Rhodopsin

To illustrate the application of ANM to membrane proteins, we describe here its application to rhodopsin, the vertebrate dim-light photoreceptor. Rhodopsin is the prototypic member of the largest known superfamily of cell surface membrane receptors, the G-protein coupled receptors (GPCRs) (59). These receptors perform diverse functions including responses to light, odorant molecules, neurotransmitters, hormones and a variety of other signals. All GPCRs contain a bundle of seven transmembrane (TM) helices (H1-H7) (60). In rhodopsin, this TM bundle contains the chromophore, 11-*cis*-retinal, covalently bound to the ε-amino group of Lys296 in H7. The capture of a photon by rhodopsin isomerizes 11-*cis*-retinal into all-*trans*-retinal. The structural perturbation in the TM helical bundle results in formation of a series of distinct photointermediates that ultimately lead to tertiary structure changes in the cytoplasmic (CP) domain that are the hallmark of the active state of rhodopsin, Meta II (61, 62). Meta II binds the heterotrimeric G-protein, transducin, which is activated via exchange of GDP to GTP (63).

The aim of the protocol described in this chapter is to integrate experimental and computational data to better understand the structural changes that lead to the Meta II state. The global mode profile generated by the ANM for rhodopsin is presented in Fig. 5a. Panel b compares the theoretically predicted (red, ANM) and experimentally observed (X-ray crystallographic; black) *B*-factors $B_i = (8\pi^2 / 3) < (\Delta R_i)^2 >$. Note that the relative displacements of the different structural elements become more distinctive upon extraction of the global mode (panel a). Figure 5c maps the latter information onto a color-coded ribbon diagram. Minima in the mode shape curve are colored blue, and maxima, orange. Seven minima are identified and their centers are labeled in Fig. 5. The ribbon diagram reveals that they all occupy a position in the middle of the TM helices, about halfway between the extracellular (EC) and CP sides. As such, the loci of minima serve as a hinge plane about which the two halves of the molecule undergo anticorrelated motions.

Largest amplitude motions are predicted for the loop regions between the TM helices and the CP loops tend to undergo larger size motions than the EC loops (Fig. 5a). The EC loop 2 between H4 and H5 is much less mobile than the shorter CP loops 2 and 3 between H3 and H4, and H5 and H6, respectively. This suggests there is a possibly larger conformational motion upon retinal isomerization on the CP side than the EC side.

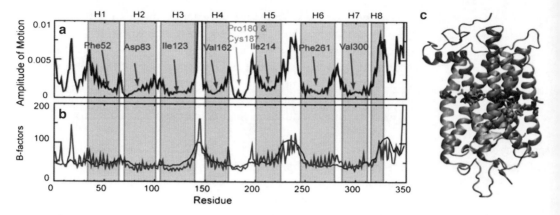

Fig. 5. Rhodopsin motions predicted by the ANM. (**a**) The distribution of square displacements of residues along ANM mode 1. The non-TM regions exhibit higher mobilities in general, especially CL2 (between H3 and H4) and CL3 (between H5 and H6). Residues acting as hinge centers (minima) are labeled. Most lie at the TM helices centers, except for two additional minima: Pro180 and Cys187 near the EC entrance to the chromophore binding pocket. (**b**) Experimental (*black*) and pre-dicted B-factors from the ANM (*dashed blue*). (**c**) Ribbon diagram of rhodopsin color-coded according to the relative motions in panel (**a**) in the order of increasing mobility *blue* (lowest mobility), *cyan, green, yellow, orange* (highest mobility). Side chains are shown for the seven hinge residues labeled in panel (**a**) and 11-*cis*-retinal is shown in *light blue* space-filling representation. The hinge site divides the protein into two anti-correlated regions, one on the CP side and the other containing the chromophore binding pocket on the EC side. The image was generated using VMD (110). More details are described in (49).

The ANM analysis of rhodopsin led to the identification of three groups of residues of interest overall summing to 61 residues implicated in functional dynamics (49): (1) those participating in the global hinge region (colored red in Fig. 6), (2) those directly affected by retinal isomerization (blue), and (3) those emerging as peaks in ANM fast modes, i.e., distinguished in the high frequency modes (green). The complete list of residues in the three categories is given in Table 1 of our previous work (49).

3.1.3. Modeling II: ANM-Guided MD

ANM-Guided MD Protocol

The intrinsic dynamics of proteins has been shown in many applications to be dominated by the protein architecture itself (15, 64, 65) and the same property appears to hold for many membrane proteins as well (8). On the other hand, the interactions with the environment such as the lipid bilayer, water molecules, ions and other ligands may also play a role in defining the detailed mechanisms of structural changes in membrane proteins. We developed a method referred to as ANM-restrained-MD, or ANM-guided MD, for explicitly taking into account the specificities of residues and their interactions with the environment. We applied the methodology to rhodopsin (29). The idea is to use iteratively the deformations derived from ANM analysis as restraints in MD trajectories. This permits us to guide the MD trajectories so as to be able to sample the cooperative motions that are otherwise beyond reach.

In ANM-guided-MD, an ensemble of ANM modes is used in an iterative scheme, as depicted in Fig. 7. Essentially, the algorithm

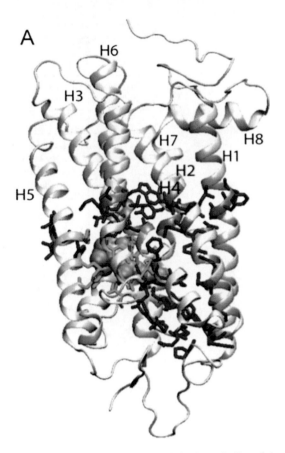

Fig. 6. Residues identified by ANM to play a key role in the activation of rhodopsin. Three sets of residues are highlighted: global hinge sites in red, amino acids affected by retinal isomerization in *blue*, and peaks in high frequency modes in *green* (Adapted from Isin et al., 2006) (49).

consists of two loops: The first (inner) loop generates a succession of conformations using ANM modes as harmonic restraints in MD runs, succeeded in each case by a short energy minimization algorithm to allow the molecule to settle in a local energy minimum (The details of mode selection criteria are given in Notes 2). To this aim, we select from a pool of low frequency ANM modes. Then, for each mode we define two target conformations and we run short MD simulations (20 ps) in the presence of harmonic restraints that favor these target structures. Since the restraints may lead to unrealistic strains in the structure, we perform a short energy minimization succeeding each run and choose from the two alternative structures the one that is energetically favored. After a sufficient number of iterations by screening all modes selected by collectivity and eigenvalue dispersion criteria (detailed in the following sections) a new cycle (outer loop) is initiated with the updated ANM modes corresponding to the structure reached by the end of the first cycle, and this procedure is terminated after

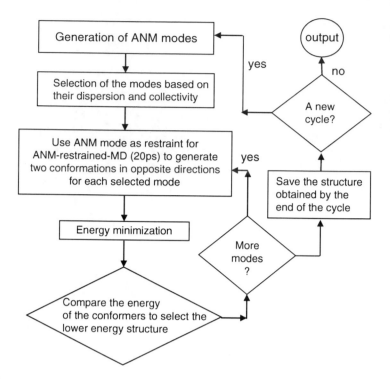

Fig. 7. Schematic description of the ANM-guided MD methodology.

a certain number of cycles when the targeted conformation with experimentally identified residue interactions or the final acceptable root-mean-square deviation (RMSD) from the target is reached. The underlying assumption in this protocol is that ANM-derived restraints drive the excursion of the molecule toward a direction that would otherwise be naturally selected at much longer times.

We developed a fully automated Tcl code for the implementation of ANM-restrained-MD protocol in the NAMD software. The following sections provide the details on the individual steps of the protocol.

Selection of distinctive and cooperative modes. The lowest frequency modes of ANM are chosen to drive the MD simulation at each cycle as they represent the functional motions. Two criteria were considered: mode frequency dispersion (or eigenvalue distribution) and degree of collectivity to further determine the most relevant lowest frequency modes. The details of these criteria are given in the Notes section see Note 2.

Preparing the target conformations. Since each mode corresponds to a fluctuation between two oppositely directed motions, both directions being equally probable, two sets of deformations are

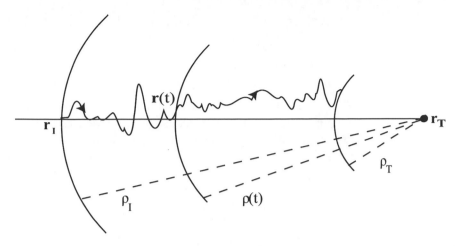

Fig. 8. Applying harmonic restraints along the ANM modes in MD. Two dimensional schematic representation of a pathway that the molecule follows while applying harmonic restraints along the ANM modes in MD. The distance $r(t)$ from the target structure is gradually decreased during the course of simulation. *Dashed lines* show the distances of the molecule from the target structure at the initial conformation (r_I), at time t ($r(t)$), and at the final (r_T) conformation.

considered for each mode, referred to as "plus" or "minus" displacements along the particular mode axis. The corresponding "target" conformation at a given cycle is required to have an RMSD in C^{α}-atom positions lower than 1.5 Å compared to the starting conformer, if the reconfiguration is performed along mode 1. The allowed upper RMSD becomes $1.5(\lambda_i/\lambda_1)^{1/2}$ Å for mode i, consistent with the relative size $(\lambda_i/\lambda_1)^{1/2}$ of mode i with respect to mode 1.

Application of harmonic restraints along the ANM modes in MD. Each C^{α}-atom is harmonically restrained to approach the target conformations for 20 ps. Figure 8 schematically describes the application of harmonic restraints in MD simulations. The instantaneous distance $\rho(t)$ of each configuration from the target configuration can be written as:

$$\rho(t) = |\mathbf{r}(t) - \mathbf{r}_T| = \left(\sum (\mathbf{r}_i(t) - \mathbf{r}_{Ti})^2\right)^{1/2}, \qquad (3)$$

where $\mathbf{r}(t)$ designates the $3N$-dimensional instantaneous conformation vector of the biomolecule, $\mathbf{r}_T$ is its target conformation; $\mathbf{r}_i(t)$ is the 3×1 instantaneous position vector of the atom i at time t; and $\mathbf{r}_{Ti}$ is its target conformation.

The implementation of the restraints is achieved by the following steps:

1. Set the distance ρ_I between the initial and target conformations as $\rho_I = |\mathbf{r}_I - \mathbf{r}_T|$

2. Solve the equation of motion containing the additional *force* for the restrains, by assigning initial coordinates using the initial conformation and an appropriate set of initial velocities.

3. At each succeeding timestep, decrease the distance by $\Delta\rho = (\rho_{\mathrm{I}} - \rho_{\mathrm{T}})\dfrac{\Delta t}{t_{\mathrm{s}}}$ where t_{s} is the simulation time and ρ_{T} is the distance of the conformation from the target structure at the end of the simulation, which should be as small as possible.

Energy minimization. After reaching the two target structures for a given mode, both are subjected to 1,000 steps of energy minimization using the steepest descent algorithm to relieve possible unrealistic distortions and to select the lower energy conformer among the two.

Application of ANM-Guided MD to Rhodopsin

ANM-guided MD was applied to rhodopsin to obtain atomic detail on the simulation (29). The results helped confirm and further refine the precise location of the previously identified hinge site (highlighted in red Fig. 6a). We observed that the hinge residues at H1, H2, and H7 are relatively closer to the CP region where two water molecules are found to further stabilize the hinge site, details that are not possible to obtain with ANM alone. Inclusion of explicit water in our model thus contributed to refine the precise location of the global hinge region, as well as identify a hydrogen bond network consolidated by water molecules. Several new interactions were also observed to contribute to the mechanism of signal propagation from the retinal-binding pocket to the G-protein binding sites in the CP domain (29).

Figure 9a displays the rhodopsin ribbon diagram color-coded by the RMSDs observed by the end of the ANM-guided-MD simulations, from red (least mobile) to blue (most mobile). Except for Pro23, all residues that exhibit minimal displacements (colored red in Fig. 9a) are clustered in two regions, the chromophore binding pocket and the CP end of H1, H2, and H7, which we shortly refer to as chromophore-binding and CP sites. These two sites are enclosed by ellipses and displayed from different perspectives in the respective panels b and c. Below we present more details on these two sites.

CP ends of H1, H2 and H7. There exists an extensive interhelical hydrogen-bond network between H1, H2 and H7. This network includes highly conserved residues in the GPCR family including the interhelical N–D pair (Asn55 on H1, Asp83 on H2) and the NPXXY motif between Asn302 and Tyr306 on H7. We have found that two residues belonging to the NPXXY motif on H7, Asn302, and Tyr306, are connected to H1 and H2 through water molecules located in the cavity between helices H1, H2 and H7. Notably, in the resulting conformation, Asn55 (H1), Asp83 (H2), and Asn302 (H7) are hydrogen bonded to a central water molecule (Fig. 9b); a second water molecule interacts closely with Thr62 (H1), Asn73 (H2), and Tyr306 (H7) (Fig. 9b). Overall ~20 water molecules interacting with H1, H2, H3, H4, and H7 residues

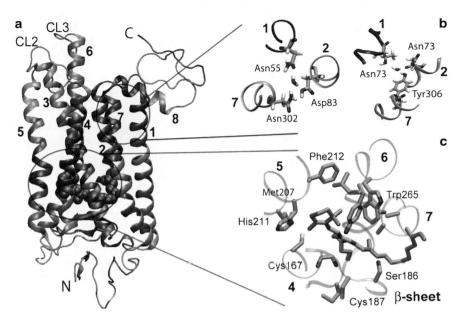

Fig. 9. Rhodopsin dynamics and key interactions near all-*trans*-retinal obtained by ANM-guided MD. (**a**) Ribbon diagram of the structure color-coded by the mobilities observed in simulations (*red*, most constrained; *green/blue*, most mobile). Non-TM regions exhibit higher mobilities in general, especially CL2 (between H3 and H4) and CL3 (between H5 and H6). The all-*trans* retinal is colored *brown*. Residues that exhibit the lowest RMSDs are clustered in two regions: around the chromophore, and in the CP portion of helices 1, 2, and 7. The CP ends of H3, H4, H5, and H6, including the loops CP2 and CP3, exhibit high RMSDs. (**b**) CP end of H1, H2 and H7, viewed from the CP region. Two water molecules are found be connected to highly conserved residues throughout the simulations. The first forms hydrogen bonds with D55 (H1), N302 (H7) and D83 (H2) (*left*), and the second forms hydrogen bonds with T62 (H1), N73 (H2) and Y306 (H7) (*right*). (**c**) Hinge residues in the vicinity of the chromophore viewed from the CP regions. These are residues distinguished by their high stability (low mobility). They include Cys167 on H4, Ser186 and Cys187 on the β-sheet, Met207, His211, and Phe212 on H5, and Trp265 on H6. All-*trans*-retinal is shown in orange (Adapted from Isin et al., 2008).

were observed in our simulations to span the helical bundle from the EC to the CP region and some exchanged neighbors during the simulations.

Chromophore Binding Pocket, and Rearrangements to Accommodate All-*Trans*-Retinal

The chromophore binding pocket is highly packed. Therefore, small conformational changes in the retinal are sufficient to significantly affect interactions within the binding site. We have observed that the isomerization of the retinal to all-*trans* state and the accompanying flip of the β-ionone ring cause steric clashes with the surrounding residues (49). These clashes are relieved upon rearrangements in the positions and orientations of the surrounding helices, which in turn induce a redistribution of contacts in the chromophore binding pocket. We note in particular that the number of atom–atom contacts between Trp265 and all-*trans* retinal (based on interatomic distance cutoff of 4.5 Å) is significantly lower than those made in the *cis* form (Fig. 9c). Furthermore, Phe261, Tyr268 and Ala269 on H6 make contacts with the β-ionone-ring to stabilize 11-*cis*-retinal in the dark state (not shown), while these contacts are lost during ANM-guided MD runs. Instead, new amino acids, Cys167 on H4 and two residues Phe203, Met207 and His211 on H5, and Thr118 on H3 line the chromophore binding pocket.

3.2. Methods II: Experimental Validation

3.2.1. Experimental Validation I. PCA of Multiple Structures

PCA Protocol

The PCA of structural ensembles permits us to visualize the major structural differences by identifying the dominant directions of conformational variations. PCA may be performed for an ensemble of PDB structures resolved for the same protein in the presence of different substrates, for NMR models obtained for the same protein (66, 67), or MD snapshots from essential dynamics analysis (EDA). The first principal component axis displays the greatest variance in the dataset, and followed by the second principal component axis, and so on. Briefly, PCA is achieved by designating for each conformer *s* in the ensemble of *M* conformers, the conformation vector

$$\mathbf{q}^{(s)} = ((R_1^{(s)})^T, ..., (R_N^{(s)})^T)^T = \left(x_1^{(s)}, y_1^{(s)}, ..., x_N^{(s)}, y_N^{(s)}, z_N^{(s)}\right)^T \quad (4)$$

or the $3N$-dimensional fluctuation vector

$$\Delta\mathbf{q}^{(s)} = \mathbf{q}^{(s)} - \mathbf{q}^0 \quad (5)$$

that describes the departure $\Delta R_i^{(s)} = R_i^{(s)} - R_i^0$ in the position vectors of the N sites from their equilibrium position R_i^0. The equilibrium positions are identified by the average over all conformers (i.e., optimally superimposed PDB structures or snapshots from MD trajectories). The covariance matrix is expressed in terms of these fluctuations as

$$C_{PCA} = M^{-1}\Sigma_k\Delta\mathbf{q}^{(k)}\Delta\mathbf{q}^{(k)T}. \quad (6)$$

Comparison of ANM results with those disclosed by the PCA of experimental data thus reduces to the comparison of the covariances from the two approaches, or the comparison of the two sets of eigenvalues/eigenvectors that describe the dominant motions (ANM) or structural variations (PCA).

Application of PCA to Rhodopsin

Recent application of this type of comparison to GPCR family proteins yielded the results presented in Fig. 10 (7, 8). In this study, we used all available rhodopsin structures in the ground and photoactivated states and two opsin structures (7, 8). In panel a, these 16 known structures that project onto the conformational subspace spanned by the two dominant PCA axes (corresponding to PCA modes 1 and 2) are displayed. These two modes account for 62% and 12% of the structural variability in the dataset (see Eq. 3.14). Comparison of panels c and d demonstrates that the dominant structural changes observed in experiments (PCA mode 1) for rhodopsin agree well with the conformational changes predicted by the ANM, in accord with the results obtained (50) for other proteins.

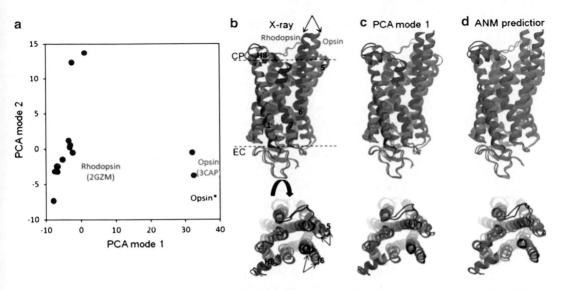

Fig. 10. PCA and ANM calculations for rhodopsin. (a) Distribution of 16 X-ray structures in the conformation subspace spanned by the PCA modes 1 and 2. PCA1 differentiates the inactive and (putative) activated structures; PCA2 further differentiates between the structures in the cluster of inactive rhodopsins (b) Superimposition of experimentally determined rhodopsin and opsin structures, indicated by the labels on panel (a). (c) Rhodopsin structure generated by deforming the opsin structure along PCA1. (d) Rhodopsin conformation predicted by deforming the opsin structure along the 20 lowest frequency ANM modes. Calculations have been performed for C^α atoms; the remaining backbone atoms were reconstructed with BioPolymer module of Sybyl 8.3 (Tripos). ANM calculations were performed using a relatively short cutoff distance ($R_c = 8$ Å) so as to release interhelical constraints.

3.2.2. Experimental Validation II. Biophysical Probes

When direct structure determination is not feasible, introduction of biophysical probes can be a suitable alternative to probe the local environment at a particular site, especially when combined with ANM predictions. The predictions allow the design of specific experiments, where individual sites can be chosen with more insight. In the absence of any predictions, a large numbers of mutants need to be screened to identify the regions of conformational change. With the predictions, biophysical probes can be introduced site-specifically to probe for changes in the environment at sites where such changes have been predicted by ANM.

Biophysical Probes Protocols

A simple and very specific way to introduce biophysical probes is through cysteine mutagenesis followed by reaction with derivatization reagents that carry the functional group to be investigated with the goal of probing its environment. The free sulfhydryl group of the cysteine is amenable for chemical derivatization with different reagents which can then be characterized by different spectroscopic methods. For example, cysteine accessibility can be studied by 4,4′-dithiodipyridine (4-PDS) labeling. 4-PDS reacts with free sulfhydryl groups in cysteines to produce stoichiometric amounts of 4-thiopyridone that absorbs at 323 nm (68). The reaction rates and extents of a cysteine in a protein with 4-PDS can be used as a biophysical probe for the local environment of that cysteine. A step-by-step protocol for 4-PDS labeling of purified rhodopsin can be found in (69).

Another popular approach is application of EPR spectroscopy to a spin-labeled protein. This is a highly informative approach to study conformational changes in a protein. The native cysteines or the introduced cysteines in the proteins are typically derivatized with a nitroxide label. The most commonly used nitroxide label is methanethiosulfonate and the corresponding side chain derivatized is designated as R1 (70). EPR can be used to measure (1) dynamics of the side chain by the spectral line shape, (2) distance between the paramagnetic labels via dipolar interactions and (3) solvent accessibility of side chains via accessibility parameters. For studying relative domain movements, two R1 labels are suitably introduced in the proteins and the distance between the labels is measured under different conditions. A step-by-step protocol for EPR studies of rhodopsin can be found in (71).

Cysteine accessibility and EPR spectroscopy are only examples of the methods that can be used, others include fluorescence spectroscopy, NMR spectroscopy, and others.

Application of Biophysical Probes to Rhodopsin Identified by ANM and ANM-Guided MD Simulations

To understand the conformational changes accompanying the activation of rhodopsin and other GPCRs, extensive cysteine-scanning mutagenesis experiments were conducted in combination with site-directed spin labeling followed by EPR analysis of mobility, accessibility, spin-spin interactions, sulfhydryl reactivity, and disulfide cross-linking rates (62, 72–79). Upon isomerization, the mobility of spin-labeled side chains at the buried surfaces of H1, H2, H3, H6, and H7 were found to increase. Furthermore, experiments suggest that the CP ends of helices, especially H3, H5, and H6, need to be highly flexible to bind and activate the G-protein The increased accessibility of Val250 and Thr251 upon activation (72) and the changes in the spin-spin distances with respect to Val139Cys (77) were attributed to the movement of H6 away from the helical bundle such that its interaction with H3 and H7 was weakened. These observations were later confirmed by the recent crystal structures in the active forms of rhodopsin and other GPCRs such as β2- and β1-adrenergic receptors (80–82). We have previously shown that the first two global modes of ANM drive the relative rearrangements of helices H3, H4, H5, and H6 (49); the ANM-guided MD simulations closely reproduce these modes at the CP ends of these helices. On the other hand, the CP ends of H1, H2, and H7 are presently found to closely maintain their dark state structure (Fig. 9a, b), consistent with the relatively smaller conformational changes observed experimentally in this region (73, 74, 79, 83–86).

Interaction and stabilization of conserved motifs through water molecule network in the transmembrane region. It has been noted that water molecules within the TM region could play critical roles in regulating the activity of GPCRs and the spectral sensitivity in visual pigments (87). One of the two water molecules that mediates the hydrogen bond network between the NPXXY motif on H7 and N–D pair on H1 and H2 is already in contact with these residues in the X-ray structure (87). Both water molecules remain at the same position despite the implementation of ten cycles of ANM-restraints in MD simulations. Later, the role of water molecules facilitating the hydrogen bond network within the TM region of another GPCR, β2-adrenergic receptor, was also shown by the MD simulations (88–90) and the recent crystal structures of this protein (80, 81).

Activation of rhodopsin by redistribution of residue contacts with its ligand. Cross-linking experiments by using photoactivatable analogs of 11-*cis*-retinal (91, 92) showed that in the dark state, 11-*cis* retinal cross-links to Trp265, while the all-*trans*-retinal cross-links to Ala169 instead of Trp265. Further investigation by high-resolution solid-state NMR measurements also showed that Trp126 and Trp265 interact more weakly with retinal in the active state (93). Additionally, the NMR data also determined that both the side chain of Glu122 and the backbone carbonyl of His211 are disrupted by the orientation of the β-ionone ring of all-*trans*-retinal in Meta II (93). The authors in (93) further proposed that the contact of the ionone ring with H5 near His211 moves H5 to an active state orientation.

These findings are consistent with the redistribution of contacts between the chromophore and rhodopsin we have observed during our simulations. We have found that the β-ionone ring of 11-*cis*-retinal in the dark state is almost parallel to the aromatic ring of Trp265 on H6. Furthermore, Phe261, Tyr268, and Ala269 on H6 make contacts with the β-ionone-ring to stabilize 11-*cis*-retinal in the dark state (not shown). These contacts are either lost or drastically reduced during ANM-guided MD simulations, and new contacts with other amino acids are made, including Cys167 on H4 and Phe203, Met207 and His211 on H5, and Thr118 on H3 (Fig. 9c). We note that these new interactions also provide insights about the activation mechanism of other GPCRs. Site-directed mutagenesis studies of the β_2-adrenergic receptor have previously showed that Ser203 and Ser207 on H5 are critically important for binding catechol hydroxyl moieties from the aromatic ring of the agonists and hence activation of the protein (94).

3.2.3. Experimental Validation III: Perturbation of Motions

Perturbation of Motions Protocols

A convenient method to identify if motions at specific domains or regions of proteins play a functional role is to restrict these motions and measure the effect of this restriction on ligand binding or on activation. One way to achieve this is by suitably placing a pair of cysteines on the protein and to monitor the formation of a disulfide bond between them. Once a disulfide bond has formed, the relative motion between the two cysteines and the regions connecting them is restricted. The step-by-step protocol for measuring disulfide bond formation in rhodopsin can be found elsewhere (84). Site-directed mutagenesis is a technique to selectively introduce or delete specific amino acid(s) in protein by mutating bases in the corresponding nucleotide sequence. Many protocols have been prepared, and commercial kits are available.

Application of Perturbation of Motions to Rhodopsin

Disulfide bonds have been strategically introduced in rhodopsin to study their effects on G protein activation, Meta II decay and phosphorylation by rhodopsin kinase (95). Different effects on function were observed, depending on where the disulfide bonds were formed, although all regions were shown by biophysical probes to undergo conformational changes upon light activation. In particular, a disulfide tethering helix VIII to the first CP loop does not affect any of these functional characteristics, in contrast to a disulfide bond cross-linking the ends of helices III and VI, while a disulfide bond tethering the C-terminus to the CP loops actually enhanced G protein activation, while fully abolishing phosphorylation by rhodopsin kinase. These results show that some motions are dispensable for function.

3.2.4. Experimental Validation IV: Functional Consequences of Motions

Functional Consequences of Motions Protocols

Introducing mutations in proteins or restricting their motions through disulfide bond formation (described in Subheading 3.2.3., Perturbation of motions) can be used to investigate the significance of a particular motion for the function of a protein. The functional assay best suited to study different proteins depends on what the function of the protein is. In the case of GPCRs, binding and activation of the G protein is a hallmark of its function. For receptors other than GPCRs, other signaling proteins affected by receptor activation may be suitable. For receptors with enzymatic activity such as kinase activity, auto-phosphorylation may be the best assay. For other membrane proteins like ion channels yet other functional assays are available. Because the functional assay differs for each protein under study, we do not provide specific protocols here.

In addition to G protein activation, another common functional assay in receptors relate to binding of ligand. In rhodopsin, the retinal binding assayed by absorbance spectroscopy is the method of choice. Radioligand assays are more generally used to study other GPCRs. Retinal binding in rhodopsin is often used to not only assess ligand binding itself, but also as an indirect readout to study folding and misfolding, which is relevant to the retinal degeneration disease, Retinitis pigmentosa. Finally, loss of retinal binding after rhodopsin activation is also used as a functional assay, referred to as "Meta II decay." The half-life of the increase in tryptophan fluorescence upon retinal loss is used as a measure of stability of the activated state.

Application of Functional Consequences of Motions to Rhodopsin Found by ANM and ANM-Guided MD Simulations

G protein activation. In order for the G protein to optimally bind and be activated by rhodopsin, the CP loops and ends of helices have to become more accessible to the aqueous environment and thus the G protein. The ANM predictions can thus be validated by surface accessibility measurements. Indeed, the CP ends of H3, H4, H5, and H6, and the connecting loops CL2 and CL3 at the CP region, are highly mobile as evident by their high RMS deviation values (Figs. 5 and 9), in agreement with spin labeling and cross-linking experiments (77, 78, 91, 92, 96). The surface accessibility of the G-protein contact site near the ERY motif is found to increase during the simulations.

Mutations associated with autosomal dominant retinitis pigmentosa. Two members of residues identified to take part in the global hinge site, Cys187 and Cys110, form a disulfide bridge, critical for folding and stability of rhodopsin (97). Mutation of both residues was also found to be associated with autosomal dominant retinitis pigmentosa (ADRP), a hereditary progressive blinding disease (98, 99). This is most probably caused by the destabilization of the opsin structure near the chromophore binding site in Meta II and dark state of rhodopsin. During ANM-guided MD simulations, the positions of Cys187 and Cys110 do not change and these residues have the lowest RMS deviation from the original structure.

In addition to the two regions indicated in Fig. 9a, another deep minimum is observed which contains only one residue, Pro23, near the soluble N-terminus at the extracellular region. It has been shown that Pro23His mutation results in severe misfolding of the entire protein, made irreversible by the formation of a wrong disulfide bond (100). This mutation has been associated with the most frequently occurring form of ADRP (101). The pathogenicity of human mutant Pro23His causing retinal degeneration was also confirmed by transgenic mice strain experiments (102, 103).

Meta II decay rates. Meta II stability of rhodopsin mutants is characterized by quantifying Meta II decay rates (104). This method has been useful to estimate the role of a given amino acid in structure and function of the protein. We used our compiled comprehensive list for Meta II decay rates of rhodopsin mutants (49) to determine the effect of the hinge residues on Meta II stability. 97 of reported 228 amino acid replacements corresponded to unique positions in the rhodopsin sequence. In all cases, Meta II decay was determined using the fluorescence assay developed by Farrens and Khorana (104). ANM correctly predicts 93% of the experimentally observed effects in rhodopsin mutants for which the decay rates have been reported. The observed validation with experimental data, whenever available, strongly supports the use of GNM/ANM for guiding experiments. In particular we note among them a few aromatic residues, Tyr301, Trp161, and Phe212, which appear to play a central role in stabilizing the global hinge region (Fig. 11).

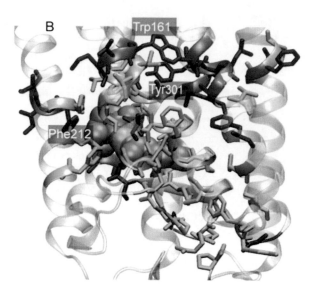

Fig. 11. Comparison of ANM results with experimental data. Key residues predicted by the ANM (see Fig. 6, colored residues) and confirmed by Meta II and folding experiments to play a critical role are colored *green*, and the rest, *purple*. Both structures refer to the dark state conformation with 11-*cis* retinal shown in space-filled model colored *cyan*.

4. Notes

1. Web site implementations of GNM and ANM
   We have developed the following software and database servers for performing and retrieving the types of calculations described in the present chapter: ANM server (27) for ANM calculations; GNM server (105) and GNM database (106) for performing online GNM calculations and releasing precomputed data, respectively; PCA_NEST server for performing PCA of NMR structures (67), and ProDy software (107) for performing PCA of ensembles of structures and comparison with ANM predictions.

2. Choosing ANM modes for guiding MD simulations
   Two criteria are used to select the ANM modes for the ANM-guided MD simulations. These are the mode frequency dispersion and the degree of collectivity. The mode frequency dispersion is examined to identify a subset that has distinctive frequencies. The degree of collectivity (108), on the other hand, is calculated using

$$\kappa = \frac{1}{N}\exp(-\sum_{i}^{N}\alpha\Delta R_i^2 \log \alpha\Delta R_i^2).$$

where α is a normalization factor to obtain $\sum_{i}^{N}\alpha\Delta R_i^2 = 1$ (25) to ascertain that the selected modes are sufficiently cooperative where most of residues are engaged in the protein motion. The degree of collectivity, κ, takes values between 0 and 1. The value zero means that there is no collective motion in the corresponding mode while in a mode with collectivity 1 all residues contribute the conformational change. This criterion is useful for eliminating the cases where the low frequency modes induce a motion in a loosely coupled chain segment only (e.g., the N- or C-terminus). In the case of rhodopsin, the lowest frequency modes were also observed to be the most cooperative ones, and the frequency distribution indicated that the subset of the first three, or first seven, modes were separable since the frequency values of the remaining modes are distinctively higher.

Acknowledgments

This work was in part supported by the National Science Foundation grant CCF-1144281, a CAREER grant CC044917 and by the National Institutes of Health Grant 5R01LM007994-06 (IB) and NIAID U01 AI070499 (ZNO).

References

1. Becker OM, MacKerell ADJ, Roux B, Wanatabe M (2001) Computaional biochemistry and biophysics. Marcel Dekker, New York

2. Cui Q, Bahar I (2006) Normal mode analysis. Theory and applications to biological and chemical systems. CRC Press, Taylor & Francis Group, Boca Raton, FL

3. Leach AR (2001) Molecular modelling: principles and applications. Prentice Hall, Upper Saddle River, NJ

4. Schlick T (2002) Molecular modeling and simulation: an interdisciplinary guide. Springer, New York

5. Dror RO, Jensen MO, Borhani DW, Shaw DE (2010) Exploring atomic resolution physiology on a femtosecond to millisecond timescale using molecular dynamics simulations. J Gen Physiol 135:555–562

6. Grossfield A, Zuckerman DM (2009) Quantifying uncertainty and sampling quality in biomolecular simulations. Annu Rep Comput Chem 5:23–48

7. Bahar I (2010) On the functional significance of soft modes predicted by coarse-grained models for membrane proteins. J Gen Physiol 135:563–573

8. Bahar I, Lezon TR, Bakan A, Shrivastava IH (2010) Normal mode analysis of biomolecular structures: functional mechanisms of membrane proteins. Chem Rev 110:1463–1497

9. Berman HM, Westbrook J, Feng Z, Gilliland G, Bhat TN, Weissig II, Shindyalov IN, Bourne PE (2000) The protein data bank. Nucleic Acids Res 28:235–242

10. Phillips JC, Braun R, Wang W, Gumbart J, Tajkhorshid E, Villa E, Chipot C, Skeel RD, Kale L, Schulten K (2005) Scalable molecular dynamics with NAMD. J Comput Chem 26: 1781–1802

11. Shaw DE, Maragakis P, Lindorff-Larsen K, Piana S, Dror RO, Eastwood MP, Bank JA, Jumper JM, Salmon JK, Shan Y, Wriggers W (2010) Atomic-level characterization of the structural dynamics of proteins. Science 330:341–346

12. Grossfield A, Feller SE, Pitman MC (2007) Convergence of molecular dynamics simulations of membrane proteins. Proteins 67:31–40

13. Brooks B, Karplus M (1985) Normal modes for specific motions of macromolecules: application to the hinge-bending mode of lysozyme. Proc Natl Acad Sci U S A 82:4995–4999

14. Berendsen HJ, Hayward S (2000) Collective protein dynamics in relation to function. Curr Opin Struct Biol 10:165–169

15. Bahar I, Rader AJ (2005) Coarse-grained normal mode analysis in structural biology. Curr Opin Struct Biol 15:586–592

16. Bahar I, Lezon TR, Yang LW, Eyal E (2010) Global dynamics of proteins: bridging between structure and function. Annu Rev Biophys 39: 23–42

17. Tirion MM (1996) Large amplitude elastic motions in proteins from a single-parameter, atomic analysis. Phys Rev Lett 77:1905–1908

18. Bahar I, Atilgan AR, Erman B (1997) Direct evaluation of thermal fluctuations in proteins using a single-parameter harmonic potential. Fold Des 2:173–181

19. Haliloglu T, Bahar I, Erman B (1997) Gaussian dynamics of folded proteins. Phys Rev Lett 79:3090–3093

20. Flory PJ (1976) Statistical thermodynamics of random networks. Proc R Soc London A 351:351–380

21. Atilgan AR, Durell SR, Jernigan RL, Demirel MC, Keskin O, Bahar I (2001) Anisotropy of fluctuation dynamics of proteins with an elastic network model. Biophys J 80:505–515

22. Doruker P, Atilgan AR, Bahar I (2000) Dynamics of proteins predicted by molecular dynamics simulations and analytical approaches: application to alpha-amylase inhibitor. Proteins 40:512–524

23. Isin B, Doruker P, Bahar I (2002) Functional motions of influenza virus hemaglutinin: a structure-based analytical approach. Biophys J 82:569–581

24. Keskin O, Durrell SR, Bahar I, Jernigan RL, Covell DG (2002) Relating molecular flexibility to function. A case study of tubulin. Biophys J 83:663–680

25. Tama F, Sanejouand YH (2001) Conformational change of proteins arising from normal mode calculations. Protein Eng 14:1–6

26. Tama F, Brooks CL (2006) Symmetry, form, and shape: guiding principles for robustness in macromolecular machines. Annu Rev Biophys Biomol Struct 35:115–133

27. Eyal E, Yang LW, Bahar I (2006) Anisotropic network model: systematic evaluation and a new web interface. Bioinformatics 22:2619–2627

28. Ma J (2005) Usefulness and limitations of normal mode analysis in modeling dynamics of biomolecular complexes. Structure 13: 373–380

29. Isin B, Schulten K, Tajkhorshid E, Bahar I (2008) Mechanism of signal propagation upon retinal isomerization: insights from molecular dynamics simulations of rhodopsin restrained by normal modes. Biophys J 95: 789–803

30. Isralewitz B, Baudry J, Gullingsrud J, Kosztin D, Schulten K (2001) Steered molecular dynamics investigations of protein function. J Mol Graph Model 19:13–25

31. Amadei A, Linssen AB, Berendsen HJ (1993) Essential dynamics of proteins. Proteins 17: 412–425

32. Amadei A, Linssen AB, de Groot BL, van Aalten DM, Berendsen HJ (1996) An efficient method for sampling the essential subspace of proteins. J Biomol Struct Dyn 13:615–625

33. de Groot BL, Amadei A, Scheek RM, van Nuland NA, Berendsen HJ (1996) An extended sampling of the configurational space of HPr from E. coli. Proteins 26:314–322

34. Abseher R, Nilges M (2000) Efficient sampling in collective coordinate space. Proteins 39:82–88

35. Zhang Z, Shi Y, Liu H (2003) Molecular dynamics simulations of peptides and proteins with amplified collective motions. Biophys J 84:3583–3593

36. Yan Q, Murphy-Ullrich JE, Song YH (2010) Structural insight into the role of thrombospondin-1 binding to calreticulin in calreticulin-induced focal adhesion disassembly. Biochemistry 49:3685–3694

37. Levitt M, Warshel A (1975) Computer simulation of protein folding. Nature 253: 694–698

38. Levitt M (1976) A simplified representation of protein conformations for rapid simulation of protein folding. J Mol Biol 104:59–107

39. Sansom MS, Scott KA, Bond PJ (2008) Coarse-grained simulation: a high-throughput computational approach to membrane proteins. Biochem Soc Trans 36:27–32

40. Marrink SJ, Risselada HJ, Yefimov S, Tieleman DP, de Vries AH (2007) The MARTINI force field: coarse grained model for biomolecular simulations. J Phys Chem B 111:7812–7824

41. Bond PJ, Holyoake J, Ivetac A, Khalid S, Sansom MS (2007) Coarse-grained molecular dynamics simulations of membrane proteins and peptides. J Struct Biol 157:593–605

42. Psachoulia E, Fowler PW, Bond PJ, Sansom MS (2008) Helix-helix interactions in membrane proteins: coarse-grained simulations of glycophorin a helix dimerization. Biochemistry 47:10503–10512

43. Shih AY, Freddolino PL, Arkhipov A, Schulten K (2007) Assembly of lipoprotein particles revealed by coarse-grained molecular dynamics simulations. J Struct Biol 157:579–592

44. Shrivastava IH, Bahar I (2006) Common mechanism of pore opening shared by five different potassium channels. Biophys J 90: 3929–3940

45. Valadie H, Lacapcre JJ, Sanejouand YH, Etchebest C (2003) Dynamical properties of the MscL of Escherichia coli: a normal mode analysis. J Mol Biol 332:657–674

46. Taly A, Delarue M, Grutter T, Nilges M, Le NN, Corringer PJ, Changeux JP (2005) Normal mode analysis suggests a quaternary twist model for the nicotinic receptor gating mechanism. Biophys J 88:3954–3965

47. Weng J, Ma J, Fan K, Wang W (2008) The conformational coupling and translocation mechanism of vitamin B12 ATP-binding cassette transporter BtuCD. Biophys J 94:612–621

48. Rader AJ, Anderson G, Isin B, Khorana HG, Bahar I, Klein-Seetharaman J (2004) Identification of core amino acids stabilizing rhodopsin. Proc Natl Acad Sci U S A 101: 7246–7251

49. Isin B, Rader AJ, Dhiman HK, Klein-Seetharaman J, Bahar I (2006) Predisposition of the dark state of rhodopsin to functional changes in structure. Proteins 65:970–983

50. Bakan A, Bahar I (2009) The intrinsic dynamics of enzymes plays a dominant role in determining the structural changes induced upon inhibitor binding. Proc Natl Acad Sci U S A 106: 14349–14354

51. Levitt M (2009) Nature of the protein universe. Proc Natl Acad Sci U S A 106:11079–11084

52. Ganapathiraju M, Jursa CJ, Karimi HA, Klein-Scctharaman J (2007) TMpro web server and web service: transmembrane helix prediction through amino acid property analysis. Bioinformatics 23:2795–2796

53. Yanamala N, Tirupula KC, Klein-Seetharaman J (2008) Preferential binding of allosteric modulators to active and inactive conformational states of metabotropic glutamate receptors. BMC Bioinformatics 9(Suppl 1):S16

54. Doruker P, Jernigan RL, Bahar I (2002) Dynamics of large proteins through hierarchical levels of coarse-grained structures. J Comput Chem 23:119–127

55. Chennubhotla C, Rader AJ, Yang LW, Bahar I (2005) Elastic network models for understanding biomolecular machinery: from enzymes to supramolecular assemblies. Phys Biol 2:S173–S180

56. Ming D, Kong YF, Lambert MA, Huang Z, Ma JP (2002) How to describe protein motion without amino acid sequence and atomic coordinates. Proc Natl Acad Sci U S A 99:8620–8625

57. Bahar I, Atilgan AR, Demirel MC, Erman B (1998) Vibrational dynamics of folded proteins: significance of slow and fast motions in relation to function and stability. Phys Rev Lett 80:2733–2736

58. Demirel MC, Atilgan AR, Jernigan RL, Erman B, Bahar I (1998) Identification of kinetically hot residues in proteins. Protein Sci 7: 2522–2532

59. Sakmar TP, Menon ST, Marin EP, Awad ES (2002) Rhodopsin: insights from recent structural studies. Annu Rev Biophys Biomol Struct 31:443–484

60. Gether U (2000) Uncovering molecular mechanisms involved in activation of G protein-coupled receptors. Endocr Rev 21:90–113

61. Klein-Seetharaman J (2002) Dynamics in rhodopsin. Chembiochem 3:981–986

62. Meng EC, Bourne HR (2001) Receptor activation: what does the rhodopsin structure tell us? Trends Pharmacol Sci 22:587–593

63. Lambright DG, Sondek J, Bohm A, Skiba NP, Hamm HE, Sigler PB (1996) The 2.0 angstrom crystal structure of a heterotrimeric G protein46. Nature 379:311–319

64. Changeux JP, Edelstein SJ (2005) Allosteric mechanisms of signal transduction. Science 308:1424–1428

65. Eisenmesser EZ, Millet O, Labeikovsky W, Korzhnev DM, Wolf-Watz M, Bosco DA, Skalicky JJ, Kay LE, Kern D (2005) Intrinsic dynamics of an enzyme underlies catalysis. Nature 438:117–121

66. Lange OF, Lakomek NA, Fares C, Schroder GF, Walter KF, Becker S, Meiler J, Grubmuller H, Griesinger C, de Groot BL (2008) Recognition dynamics up to microseconds revealed from an RDC-dcrivcd ubiquitin ensemble in solution. Science 320:1471–1475

67. Yang LW, Eyal E, Bahar I, Kitao A (2009) Principal component analysis of native ensembles of biomolecular structures (PCA_NEST): insights into functional dynamics. Bioinformatics 25:606–614

68. Grassetti DR, Murray JF Jr (1967) Determination of sulfhydryl groups with 2,2′- or 4,4′-dithiodipyridine. Arch Biochem Biophys 119:41–49

69. Dutta A, Tirupula KC, Alexiev U, Klein-Seetharaman J (2010) Characterization of membrane protein non-native states. 1. Extent of unfolding and aggregation of rhodopsin in the presence of chemical denaturants. Biochemistry 49:6317–6328

70. Hubbell WL, Gross A, Langen R, Lietzow MA (1998) Recent advances in site-directed spin labeling of proteins. Curr Opin Struct Biol. 8:649–656

71. Resek JF, Farahbakhsh ZT, Hubbell WL, Khorana HG (1993) Formation of the meta II photointermediate is accompanied by conformational changes in the cytoplasmic surface of rhodopsin. Biochemistry 32: 12025–12032

72. Altenbach C, Yang K, Farrens DL, Farahbakhsh ZT, Khorana HG, Hubbell WL (1996) Structural features and light-dependent changes in the cytoplasmic interhelical E-F loop region of rhodopsin: a site-directed spin-labeling study. Biochemistry 35:12470–12478

73. Altenbach C, Klein-Seetharaman J, Hwa J, Khorana HG, Hubbell WL (1999) Structural features and light-dependent changes in the sequence 59–75 connecting helices I and II in rhodopsin: a site-directed spin-labeling study. Biochemistry 38:7945–7949

74. Altenbach C, Cai K, Khorana HG, Hubbell WL (1999) Structural features and light-dependent changes in the sequence 306–322 extending from helix VII to the palmitoylation sites in rhodopsin: a site-directed spin-labeling study. Biochemistry 38:7931–7937

75. Cai K, Langen R, Hubbell WL, Khorana HG (1997) Structure and function in rhodopsin: topology of the C-terminal polypeptide chain in relation to the cytoplasmic loops. Proc Natl Acad Sci U S A 94:14267–14272

76. Farahbakhsh ZT, Ridge KD, Khorana HG, Hubbell WL (1995) Mapping light-dependent structural changes in the cytoplasmic loop connecting helices C and D in rhodopsin: a site-directed spin labeling study. Biochemistry 34:8812–8819

77. Farrens DL, Altenbach C, Yang K, Hubbell WL, Khorana HG (1996) Requirement of rigid-body motion of transmembrane helices for light activation of rhodopsin. Science 274:768–770

78. Hubbell WL, Altenbach C, Hubbell CM, Khorana HG (2003) Rhodopsin structure, dynamics, and activation: a perspective from crystallography, site-directed spin labeling, sulfhydryl reactivity, and disulfide cross-linking. Adv Protein Chem 63:243–290

79. Klein-Seetharaman J, Hwa J, Cai K, Altenbach C, Hubbell WL, Khorana HG (2001) Probing the dark state tertiary structure in the cytoplasmic domain of rhodopsin: proximities between amino acids deduced from spontaneous disulfide bond formation between Cys316 and engineered cysteines in cytoplasmic loop 1. Biochemistry 40:12472–12478

80. Rasmussen SG, Choi HJ, Rosenbaum DM, Kobilka TS, Thian FS, Edwards PC, Burghammer M, Ratnala VR, Sanishvili R, Fischetti RF, Schertler GF, Weis WI, Kobilka BK (2007) Crystal structure of the human beta2 adrenergic G-protein-coupled receptor. Nature 450:383–387

81. Rosenbaum DM, Zhang C, Lyons JA, Holl R, Aragao D, Arlow DH, Rasmussen SG, Choi HJ, Devree BT, Sunahara RK, Chae PS, Gellman SH, Dror RO, Shaw DE, Weis WI, Caffrey M, Gmeiner P, Kobilka BK (2011) Structure and function of an irreversible agonist-beta(2) adrenoceptor complex. Nature 469:236–240

82. Warne T, Moukhametzianov R, Baker JG, Nehme R, Edwards PC, Leslie AG, Schertler GF, Tate CG (2011) The structural basis for agonist and partial agonist action on a beta(1)-adrenergic receptor. Nature 469:241–244

83. Altenbach C, Cai K, Klein-Seetharaman J, Khorana HG, Hubbell WL (2001) Structure and function in rhodopsin: mapping light-dependent changes in distance between residue 65 in helix TM1 and residues in the sequence 306–319 at the cytoplasmic end of helix TM7 and in helix H8. Biochemistry 40:15483–15492

84. Cai K, Klein-Seetharaman J, Farrens D, Zhang C, Altenbach C, Hubbell WL, Khorana HG (1999) Single-cysteine substitution mutants at amino acid positions 306–321 in rhodopsin, the sequence between the cytoplasmic end of helix VII and the palmitoylation sites: sulfhydryl reactivity and transducin activation reveal a tertiary structure. Biochemistry 38:7925–7930

85. Cai K, Klein-Seetharaman J, Altenbach C, Hubbell WL, Khorana HG (2001) Probing the dark state tertiary structure in the cytoplasmic domain of rhodopsin: proximities between amino acids deduced from spontaneous disulfide bond formation between cysteine pairs engineered in cytoplasmic loops 1, 3, and 4. Biochemistry 40:12479–12485

86. Klein-Seetharaman J, Hwa J, Cai K, Altenbach C, Hubbell WL, Khorana HG (1999) Single-cysteine substitution mutants at amino acid positions 55–75, the sequence connecting the cytoplasmic ends of helices I and II in rhodopsin: reactivity of the sulfhydryl groups and their derivatives identifies a tertiary structure that changes upon light-activation. Biochemistry 38:7938–7944

87. Okada T, Fujiyoshi Y, Silow M, Navarro J, Landau EM, Shichida Y (2002) Functional role of internal water molecules in rhodopsin revealed by X- ray crystallography. Proc Natl Acad Sci U S A 99:5982–5987

88. Dror RO, Arlow DH, Borhani DW, Jensen MO, Piana S, Shaw DE (2009) Identification of two distinct inactive conformations of the

beta2-adrenergic receptor reconciles structural and biochemical observations. Proc Natl Acad Sci U S A 106:4689–4694

89. Han DS, Wang SX, Weinstein H (2008) Active state-like conformational elements in the beta2-AR and a photoactivated intermediate of rhodopsin identified by dynamic properties of GPCRs. Biochemistry 47:7317–7321

90. Romo TD, Grossfield A, Pitman MC (2010) Concerted interconversion between ionic lock substrates of the beta(2) adrenergic receptor revealed by microsecond timescale molecular dynamics. Biophys J 98:76–84

91. Borhan B, Souto ML, Imai H, Shichida Y, Nakanishi K (2000) Movement of retinal along the visual transduction path. Science 288:2209–2212

92. Nakayama TA, Khorana HG (1990) Orientation of retinal in bovine rhodopsin determined by cross-linking using a photoactivatable analog of 11-cis-retinal. J Biol Chem 265:15762–15769

93. Patel AB, Crocker E, Eilers M, Hirshfeld A, Sheves M, Smith SO (2004) Coupling of retinal isomerization to the activation of rhodopsin. Proc Natl Acad Sci U S A 101:10048–10053

94. Strader CD, Candelore MR, Hill WS, Sigal IS, Dixon RA (1989) Identification of two serine residues involved in agonist activation of the beta-adrenergic receptor. J Biol Chem 264:13572–13578

95. Cai K, Klein-Seetharaman J, Hwa J, Hubbell WL, Khorana HG (1999) Structure and function in rhodopsin: effects of disulfide cross-links in the cytoplasmic face of rhodopsin on transducin activation and phosphorylation by rhodopsin kinase. Biochemistry 38:12893–12898

96. Bourne HR (1997) How receptors talk to trimeric G proteins. Curr Opin Cell Biol 9:134–142

97. Hwa J, Reeves PJ, Klein-Seetharaman J, Davidson F, Khorana HG (1999) Structure and function in rhodopsin: further elucidation of the role of the intradiscal cysteines, Cys-110, -185, and -187, in rhodopsin folding and function. Proc Natl Acad Sci U S A 96:1932–1935

98. Richards JE, Scott KM, Sieving PA (1995) Disruption of conserved rhodopsin disulfide bond by Cys187Tyr mutation causes early and severe autosomal dominant retinitis pigmentosa. Ophthalmology 102:669–677

99. Vaithinathan R, Berson EL, Dryja TP (1994) Further screening of the rhodopsin gene in patients with autosomal dominant retinitis pigmentosa. Genomics 21:461–463

100. Hwa J, Klein-Seetharaman J, Khorana HG (2001) Structure and function in rhodopsin: mass spectrometric identification of the abnormal intradiscal disulfide bond in misfolded retinitis pigmentosa mutants. Proc Natl Acad Sci U S A 98:4872–4876

101. Berson EL (1993) Retinitis pigmentosa. The Friedenwald lecture. Invest Ophthalmol Vis Sci 34:1659–1676

102. Olsson JE, Gordon JW, Pawlyk BS, Roof D, Hayes A, Molday RS, Mukai S, Cowley GS, Berson EL, Dryja TP (1992) Transgenic mice with a rhodopsin mutation (Pro23His): a mouse model of autosomal dominant retinitis pigmentosa. Neuron 9:815–830

103. Wang M, Lam TT, Tso MO, Naash MI (1997) Expression of a mutant opsin gene increases the susceptibility of the retina to light damage. Vis Neurosci 14:55–62

104. Farrens DL, Khorana HG (1995) Structure and function in rhodopsin. Measurement of the rate of metarhodopsin II decay by fluorescence spectroscopy. J Biol Chem 270:5073–5076

105. Yang LW, Liu X, Jursa CJ, Holliman M, Rader AJ, Karimi HA, Bahar I (2005) iGNM: a database of protein functional motions based on Gaussian Network Model. Bioinformatics 21:2978–2987

106. Yang LW, Rader AJ, Liu X, Jursa CJ, Chen SC, Karimi HA, Bahar I (2006) oGNM: online computation of structural dynamics using the Gaussian Network Model. Nucleic Acids Res 34:W24–W31

107. Bakan A, Meireles LM, Bahar I (2011) ProDy: protein dynamics inferred from theory and experiments. Bioinformatics 27(11):1575–1577

108. Bruschweiler R (1995) Collective protein dynamics and nuclear-spin relaxation. J Chem Phys 102:3396–3403

109. Tellez-Sanz R, Cesareo E, Nuccetelli M, Aguilera AM, Baron C, Parker LJ, Adams JJ, Morton CJ, Lo BM, Parker MW, Garcia-Fuentes L (2006) Calorimetric and structural studies of the nitric oxide carrier S-nitrosoglutathione bound to human glutathione transferase P1-1. Protein Sci 15:1093–1105

110. Humphrey W, Dalke A, Schulten K (1996) VMD: visual molecular dynamics. J Mol Biol 14:33–38

Chapter 18

Modeling the Structural Communication in Supramolecular Complexes Involving GPCRs

Francesca Fanelli

Abstract

This article describes a computational strategy aimed at studying the structural communication in G-Protein Coupled Receptors (GPCRs) and G proteins. The strategy relies on comparative Molecular Dynamics (MD) simulations and analyses of wild-type (i.e., reference state) vs. mutated (i.e., perturbed state), or free (i.e., reference state) vs. bound (i.e., perturbed state) forms of a GPCR or a G protein. Bound forms of a GPCR include complexes with small ligands and/or receptor dimers/oligomers, whereas bound forms of heterotrimeric GDP-bound G proteins concern the complex with a GPCR. The computational strategy includes structure prediction of a receptor monomer (in the absence of high-resolution structure), a receptor dimer/oligomer, and a receptor–G protein complex, which constitute the inputs of MD simulations. Finally, the analyses of the MD trajectories are instrumental in inferring the structural/dynamics differences between reference and perturbed states of a GPCR or a G protein. In this respect, focus will be put on the analysis of protein structure networks and communication paths.

Key words: GPCRs, Constitutively active mutants, Comparative modeling, Dimerization, Protein–protein docking, Molecular dynamics, Protein structure network

1. Introduction

Intramolecular and intermolecular communications inside a protein molecule and among the components of protein networks are at the basis of signal transduction and functioning mechanisms of membrane proteins, in particular, G Protein-Coupled Receptors (GPCRs) (Reviewed in ref. 1). GPCRs regulate most aspects of cell activity by transmitting extracellular signals inside the cell (reviewed in refs. 2, 3). GPCRs share an up-and-down bundle of seven transmembrane helices connected by three intracellular (IL1, IL2, and IL3) and three extracellular (EL1, EL2, and EL3) loops, an extracellular N-terminus and an intracellular C-terminus. Upon activation by extracellular signals, the receptors activate the α-subunit in

Nagarajan Vaidehi and Judith Klein-Seetharaman (eds.), *Membrane Protein Structure and Dynamics: Methods and Protocols*, Methods in Molecular Biology, vol. 914, DOI 10.1007/978-1-62703-023-6_18, © Springer Science+Business Media, LLC 2012

heterotrimeric guanine nucleotide binding proteins (G proteins) by catalyzing the exchange of bound GDP for GTP, i.e., they act as Guanine Nucleotide Exchange Factors (GEFs). Thus, GPCRs are allosteric proteins that transform extracellular signals into promotion of nucleotide exchange in intracellular G proteins. These receptors have regions of low and high flexibility that communicate with each other, even if distal. GPCRs exist as complex statistical conformation ensembles (4, 5). Their functional properties are related to the distribution of states within the native ensemble, which is differently affected by ligands, oligomeric state (i.e., monomer, homo- and heterodimers, and oligomers), interacting proteins, and amino acid mutations. Regulated protein–protein interactions are key features of many aspects of GPCR function and there is increasing evidence that these receptors act as part of multicomponent units comprising a variety of signaling and scaffolding molecules (3, 6).

Unraveling the GPCR functioning mechanisms at the atomic level is difficult due to the limited high resolution structural information, which, is presently limited to: a) rhodopsin in its dark (inactive), constitutively active, and photoactivated states (7–11); b) β_2- and β_1-adrenergic receptors (ARs) bound to agonists, partial agonists, inverse agonists, antagonists, a nanobody, and heterotrimeric Gs (12–20); c) A_{2A} adenosine receptor ($A_{2A}R$) bound to an antagonist, a number of agonists, and an allosteric inverse agonist antibody (21–25); d) squid rhodopsin bound to 11-cis retinal (26); e) CXCR4 chemokine receptor bound to small molecule and cyclic peptide antagonists (27); f) D3 dopamine receptor in complex with a D2/D3 selective antagonist (28); g) the H1 histamine receptor (H_1R) in complex with an antagonist (29); h) the m2 and m3 muscarinic receptors bound to antagonists (30, 31); i) the δ-, κ- and μ- opioid receptos bound to antagonists (32–34); j) the nociceptin/orphanin FQ receptor in complex with a peptide mimetic antagonist (35); and k) a lipid GPCR in complex with a sphingolipid antagonist (36). In spite of the extraordinary advance in structure determination of GPCRs since year 2007, there is still little knowledge of the likely architecture of GPCR dimers/oligomers, the effects of ligand binding, site-directed mutagenesis and dimerization on the intrinsic dynamics of the receptor, as well as the receptor impact on the intrinsic dynamics of the G protein, which makes GPCRs act as GEFs.

This article describes a strategy addressing the allosteric structural communication involving GPCRs by integrating different computational approaches in a comparative framework. These approaches include, building of the structural model of the receptor, prediction of the likely architecture of GPCR homo- and heterodimers/oligomers, prediction of the ligand-receptor and the receptor–G protein complexes that constitute the functional unit, and simulating receptor and G protein dynamics in their native, mutated, free, and differently bound states. Ultimately, structural/dynamic differences between reference and perturbed states of a GPCR or G protein are inferred by comparing time-averaged features, essential motions, structure networks, and communication paths.

2. Materials

1. The MODELLER software is employed for comparative modeling of the initial structural model of a given receptor or G protein, not yet solved at the atomic level of detail (http://www.salilab.org/modeller/ (37, 38)).

2. Protein quality check and automatic adjustment of side chain rotamers are done through the "Protein Health" and "Protein Design" modules of the Quanta molecular graphics package (www.accelrys.com), which is not freely available. However, amino acid side chain rotamer libraries and software for side chain conformation prediction can be freely downloaded from the Dunbrack's Web site (http://dunbrack.fccc.edu/bbdep/index.php). Moreover, many of the freely available protein modeling programs (e.g., Chimera (http://www.cgl.ucsf.edu/chimera/), Pymol (http://www.pymol.org/), and Bodil (http://users.abo.fi/bodil/)) allow modifications of amino acid side chains according to rotamer libraries. On the other hand, freely available software for checking the stereochemical quality of a protein structure include PROCHECK (http://www.ebi.ac.uk/thornton-srv/software/PROCHECK/) and WHATIF (http://swift.cmbi.ru.nl/whatif/).

3. Rigid body protein–protein docking for predicting receptor quaternary structures or receptor–G protein complexes is carried by means of the ZDOCK program (http://zlab.bu.edu/zdock/ (39, 40)).

4. Reorientation of a receptor structure according to the membrane topology can be done by means of the FiPD software, which serves essentially to analyze the outputs of the ZDOCK program (visit "Software" at http://www.csbl.unimore.it (41)).

5. MD simulations are carried out by means of the CHARMM software (http://www.charmm.org/html/info/intro.html (42)).

6. All MD analyses are done by means of the Wordom software, which is freely downloadable from: http://wordom.sourceforge.net/index.html (43).

7. Cluster analysis of the docking solutions can be done by means of the FiPD software.

8. All drawings shown in Figs. 1–4 were done using the software PyMOL 1.1r1 (http://www.pymol.org/).

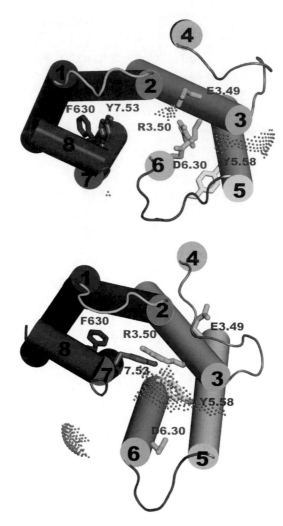

Fig. 1. Cytosolic ends of wild-type luteinizing hormone receptor (LHR) (inactive state, *top*) and of D578[(6.44)]H constitutively active LHR mutant (*bottom*). The numbering in parentheses follows the arbitrary scheme by Ballesteros and Weinstein (58). Details on the building of the structural models, which are average minimized structures, are published elsewhere (46). The structures are seen from the cytosolic side in a direction perpendicular to the membrane surface. Helices 1, 2, 3, 4, 5, 6, 7, and 8 are, respectively, colored *blue*, *orange*, *green*, *pink*, *yellow*, *cyan*, *violet*, and *red*. IL 1, 2, and 3 are, respectively, *lime*, *gray*, and *magenta*. Details of the interaction of selected conserved amino acids are shown in *stick representation*, which mark the structural differences between inactive (reference) and active (perturbed) states. *Gray dots* are the solvent accessible surface computed over R464[(3.50)], V467[(3.53)], T468[(5.54)], K566[(6.32)], I567[(6.33)], and K570[(6.36)]. The relative SASA in wild-type and mutated LHR forms is 89 and 255 Å², respectively. This figure illustrates how selected interaction patterns and size–shape descriptors such as SASA can be used to mark structural differences between functionally different states.

Fig. 3. (continued) of the circle is proportional to the number of links made by the considered node, with the lowest value corresponding to one link. Link color refers to the frequency of the link. In this respect, *cyan*, *green*, *yellow*, *orange*, and *red* correspond, respectively, to the following frequency (F) ranges: $50 \leq F \leq 60$ %, $60 \leq F \leq 70$ %, $70 \leq F \leq 80$ %, $80 \leq F \leq 90$ % and $F \geq 90$ %. This figure illustrates how a perturbation such as a point mutation may result in a different composition of the structure network that characterizes the wild-type form.

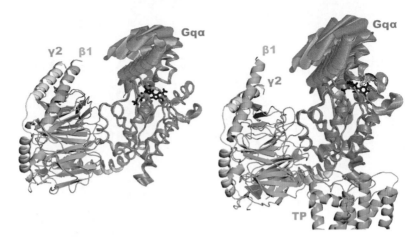

Fig. 2. Cα-displacements of the Gq α-subunit in its free (*left*) state and in complex with the thromboxane A2 receptor (TP) (*right*) along the second eigenvector from a PCA run on a concatenated trajectory made of 12,000 frames. A number of conformations of the α-subunit were generated between the minimum and maximum projection on the selected eigenvector. These pictures refer to a work published elsewhere (45). The G protein α-, β-, and γ-subunits are, respectively, *gray*, *cyan* and *yellow*, whereas TP is *green*. The GDP molecule is *red*. *Red dots* indicate the SAS of the nucleotide. This figure illustrates how a perturbation such as receptor binding induces changes in the essential motions of the G protein α-subunit. In this respect, comparative PCA may serve to highlight divergences in the intrinsic dynamics of a reference system (i.e., receptor-free α-subunit) and a perturbed system (i.e., receptor-bound α-subunit).

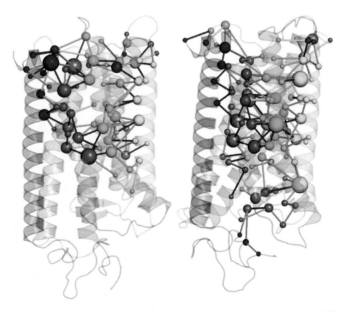

Fig. 3. 3D PSG concerning wild-type LHR (i.e., inactive state, *left*) and in D578$^{(6.44)}$H constitutively active LHR mutant (*right*). These pictures result from a work published elsewhere (46). The *spheres* centered on the Cα-atoms concern node pairs in the PSG, which are linked in more than 50 % of frames in a 10 ns trajectory. Nodes are colored according to their location (i.e., according to the different receptor regions). In this respect, helices 1, 2, 3, 4, 5, 6, 7, and 8 are, respectively, *blue*, *orange*, *green*, *pink*, *yellow*, *cyan*, *violet*, and *red*. IL1 and EL1 are *lime*, IL2 and EL2 are *gray*, and IL3 and EL3 are *magenta*. The diameter

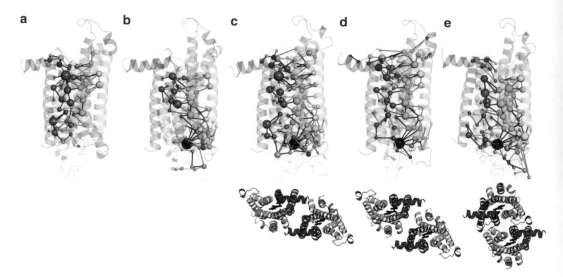

Fig. 4. Average minimized structures concerning the $A_{2A}R$ in its free (**a**) and antagonist-bound (i.e., bound to the antagonist ZM241385 (ZMA)) (**b**) monomeric forms, as well as in three different ZMA-bound dimeric forms (**c–e**). The architectures of the three different dimers are shown at the *bottom* of *panels* **c–e**. For the color coding of the different receptor portions, see the legend to Fig. 3. Pictures refer to results published elsewhere (47). The *top images* show the effects of ligand binding (*panel* **b**) and of dimerization (*panels* **c–e**) on the PSG characterizing the apo-form (*panel* **a**) (see the legend to Fig. 3 for a description of 3D PSG). This figure illustrates how PSN analysis can be used to infer differences in the intramolecular and intermolecular communication following ligand binding or receptor dimerization.

3. Methods

3.1. Modeling the GPCR Structure

1. The initial structure modeling of a given GPCR is achieved by comparative modeling by the MODELLER software, which is based on the satisfaction of spatial restraints (37, 38). The inputs to the program are restraints on the spatial structure of the amino acid sequence(s) and ligands to be modeled. The output is a 3D structure that satisfies these restraints as well as possible. Restraints can in principle be derived from a number of different sources. These include related protein structures (comparative modeling), NMR experiments (NMR refinement), rules of secondary structure packing (combinatorial modeling), secondary structure predictions, cross-linking experiments, fluorescence spectroscopy, image reconstruction in electron microscopy, site-directed mutagenesis, intuition, and residue–residue and atom–atom potentials of mean force. The restraints can operate on distances, angles, dihedral angles, pairs of dihedral angles, and some other spatial features defined by atoms or pseudo atoms. By defaults, MODELLER automatically derives the restraints only from the known related structures and their alignment with the target sequence. A 3D model is obtained by optimization of a molecular probability

density function (PDF). The molecular PDF for comparative modeling is optimized with the variable target function procedure in Cartesian space that employs methods of conjugate gradients and MD with simulated annealing. The goodness of a comparative model depends on the goodness of the sequence alignment and on the structural quality of the template(s). External restraints can be added by the user. Nonhomologous regions, such as loops, can be modeled following an *ab initio* approach (see Note 1) (38).

2. As for comparative modeling of GPCRs, selection of template structures from one or more of the GPCR crystal structures available so far must privilege sequence similarity between target and template proteins. Fully automated modeling of an entire GPCR sequence is rarely feasible even for members of the same subfamily, and human intervention is necessary to modify the selected template, to add extra restraints (e.g., to elongate α-helices or add β-strands), to add patches (i.e., disulphide bridges), to refine loops *ab initio* (i.e., by the "loop-model" option). A number of different sequence alignments are worth probing. For each alignment, it is worth generating multiple models (i.e., more than 100 for each alignment) by randomizing all the Cartesian coordinates of standard residues in the initial model (i.e., "randomize.xyz" option). A high degree of model refinement within MODELLER is recommended. From each run, the top ten models are selected, characterized by the lowest values of the MODELLER objective function (which means lowest degree of restraint violation).

3. The set of selected models are subjected to quality checks, which verify the correctness of the main chain conformation, leading to selection of one or more model(s) (see Note 2 and Subheading 2).

4. The side chains conformations of selected model(s) are finally optimized by using rotamer libraries (see Note 3). We use the side chain modeling tool within the Quanta commercial software (www.accelrys.com), since it allows the use of three different backbone-independent and backbone-dependent rotamer libraries either on the whole protein or on a wide variety of amino acid selections. However, a number of rotamer libraries and software for side chain modeling are freely available and worth using (see Subheading 2).

5. Further refinement of such model(s) by taking average minimized structures from MD simulations is frequently required especially in those cases, in which sequence identity between template and target is low (e.g., ~20 % in the transmembrane domains and higher in the loops and in the N- and C-termini). Time-averaged structures (i.e., the structures obtained by averaging the Cartesian coordinates of the structural models over

the frames that constitute the MD trajectory) are worth computing over the initial, central, last as well as all frames of the MD trajectory. In cases of low Cα-RMSD values concerning the trajectory frames, the different average structures will be quite similar. It is worth noting that for any average structure, energy minimization is mandatory to eliminate physically meaningless bond lengths, bond angles and torsion angles.

3.2. Prediction of GPCR Dimer/ Oligomer Architecture

1. Prediction of likely architectures of GPCRs homo- and heterodimers follows a computational approach developed for quaternary structure predictions of transmembrane (TM) α-helical proteins (41). The approach consists of rigid-body docking using a version of the ZDOCK program devoid of desolvation as a component of the docking score (39). It does not employ symmetry constraints either for improving sampling or in the filtering step. Furthermore, there are no size limitations for the systems under study, which are not limited to the TM regions but include the loop regions as well. The only requirement with this approach is the structural model of the monomer and the knowledge of a set of Cα-atoms, which lie at the two lipid/water interfaces, defining two parallel planes (see Note 4). In the target monomer, these two planes must be parallel to the xy plane and, hence, perpendicular to the z-axis, to make the membrane topology filter work properly (see Note 5). If these planes are parallel to the *xy* plane and, hence, perpendicular to the z-axis, the orientation of the monomer is considered good and no reorientation is needed. In contrast, if such planes are not parallel to the *xy* plane, the monomer needs a reorientation. The latter can be done through the FiPD software (visit "Software" at http://www.csbl.unimore.it).

   The structural models subjected to docking simulations can be either crystal structures or minimized averages from MD simulations on selected models from comparative modeling.

2. In the case of homodimerization, two identical copies of the structural receptor model are docked together, i.e., one monomer is used as a fixed protein (target) and the other as a mobile protein (probe). For predicting heterodimers, the structural model of receptor A is taken as a target, whereas the structural model of receptor B is taken as a probe and/or vice versa. A rotational sampling interval of 6° is recommended, (i.e., dense sampling, "-D" option), and the best 4,000 solutions are retained and ranked according to the ZDOCK score (i.e., "-N 4000" setup).

3. The best docked solutions by shape complementarity as selected by the docking program are then filtered using the FiPD software (41), the "membrane topology" filter, which discards all

the solutions that violate the membrane topology requirements (see Note 5). The membrane topology filter, indeed, discards all the solutions characterized by a deviation angle from the original z-axis, i.e., tilt angle, and a displacement of the geometrical center along the z-axis, i.e., z-offset, above defined threshold values. In the case of GPCRs, tilt angle and z-offset thresholds of 0.4 rad ("-a 0.4" setup) and 6.0 Å ("-o 6.0" setup) are worth using (see Note 6). Following such filtering, discarded solutions generally constitute more than 94 % of the solutions selected by the docking program.

4. The filtered solutions from each run are merged with the target protein, leading to an equivalent number of dimers that are clustered using a $C\alpha$-RMSD threshold of 3.0 Å for each pair of superimposed dimers. All the amino acid residues in the dimer are included in $C\alpha$-RMSD calculations. Cluster analysis is based on a Quality Threshold-like clustering algorithm (44) implemented both in the FiPD and Wordom software (41, 43) (see Note 7). Since the filtering cutoffs of the membrane topology parameters are intentionally quite permissive (see Note 6), inspection of the cluster centers (i.e., the solutions with the highest number of neighbors in each cluster) often serves as a final filter to discard remaining false positives, leading to a reduction of the reliable solutions to about 1 % of the total 4,000 solutions. The best scored docking solution(s) from the most populated and reliable cluster(s) is(are) finally chosen.

5. If necessary, adjustment of the torsion angles of interface amino acid side chains involved in steric clashes is done by using rotamer libraries (see Note 3).

6. Predictions of higher order oligomers can be done either by selecting solutions from reliable clusters (i.e., which can coexist in a multimeric complex), or by running a new docking by using the predicted dimer as a target and the monomer as a probe. Filtering and cluster analysis then follow the same procedure as that employed for predicting dimers.

3.3. Prediction of Receptor–G Protein Complex Structures

1. Prediction of the structure of the protein complex between a given GPCR and the cognate G protein is achieved following docking simulations using the ZDOCK version that calculates desolvation penalty, pairwise shape complementarity, and electrostatics to score the docked complexes. The receptor is used as a target, whereas heterotrimeric G protein is the probe. To improve sampling efficiency, only the cytosolic portions of the receptor are taken into account in docking simulations (i.e., by "blocking" atoms that cannot be part of the interface). Dense sampling and final retrieval of the best 4,000 solutions are done.

2. To filter the most reliable solutions among the 4,000 best scored docked GPCR–G protein complexes, a distance cutoff

of 20 Å is used between the Cα-atom of the fully conserved arginine of the E/DRY motif of the receptor and the Cα-atom of the last amino acid on the G protein α-subunit. Since such distance-based filtering is intentionally quite permissive to avoid unwanted elimination of reliable solutions, the filtered solutions are, then, subjected to cluster analysis, by using the QT-like algorithm with a Cα-RMSD cutoff of 4.0 Å. Finally, visual inspection of the cluster centers serves to discard the remaining solutions that violate the expected membrane topology of the G protein (see Note 8). The best scored solution(s) from the most populated and reliable (i.e., that accomplish the expected membrane topology of the G protein) clusters are finally selected and subjected to side chain rotamer optimization (see Note 3 and Subheading 2).

3.4. MD Simulations

1. MD simulations can be used to study the allosteric communication (i.e., intramolecular and intermolecular) involving GPCRs. Comparative analyses of MD trajectories of wild-type (i.e., reference state) vs. mutated (i.e., perturbed state) or free (i.e., reference state) vs. bound (i.e., perturbed state) forms of a GPCR or a G protein (1, 45–47) would provide such information. Bound forms of a GPCR include complexes with small ligands and receptor dimers/oligomers. The allosteric communication can be, hence, inferred by comparing the essential motions as well as selected time-dependent features of reference and perturbed states. In this context, time-dependent features may include structure networks and shortest communication paths.

2. Strategies to reduce the system's degrees of freedom and detect meaningful motions are worth using. They include the employment of implicit water/membrane models like the GBSW (48) implemented in the CHARMM molecular simulation software (42). Such a solvation model is used with the all-hydrogen parameter set PARAM22/CMAP. The surface tension coefficient (representing the nonpolar solvation energy) is set to 0.03 kcal/(mol Å^2) and the membrane thickness centered at $Z=0$ is set to 30.0 Å with a membrane smoothing length of 5.0 Å ($w_m=2.5$ Å). Additional ways to reduce the degrees of freedom may include the use of intra-helical restraints between the oxygen atom of residue i and the backbone nitrogen atom of residue $i+4$, except for prolines. Noncanonical α-helix conformations are excluded from the intra-backbone restraints as well. The scaling factor of such restraints is 10 and the force constant at 300 K is 10 kcal/mol Å. The structural model must have an appropriate membrane topology, i.e., the Cα-atoms predicted to be at the membrane/water interface must lie in the xy plane. Prior to MD simulations, the potential energy of the simulation system should be minimized using steepest descent followed by Adopted Basis Newton-Raphson (ABNR)

minimization, until the root mean square gradient is less than 0.001 kcal/mol Å.

3. As for the setup of equilibrium MD simulations, the bond lengths involving the hydrogen atoms are restrained by the SHAKE algorithm, allowing for an integration time step of 0.002 ps. The system is slowly heated to 300 K with 7.5 K rises every 2.5 ps per 100 ps, by randomly assigning velocities from a Gaussian distribution. After heating, the system is allowed to equilibrate for a system-dependent time period. Equilibration is followed by the MD production phase that consists of constant temperature (300 K) simulations in the nanosecond/submicrosecond time scale (see Note 9).

3.5. Comparative MD Analyses

1. The structural differences between reference and perturbed states of a GPCR or a G protein can be inferred by comparing properties either computed on the average structures, or plotted as time series (i.e., property vs. time plots). The selected structural properties must be the same for all the compared systems (Note 9).

   Useful time-averaged properties include inter-residue main chain and/or side chain distances accounting for relative motions of selected segments of the protein, and/or formation or breakage of networks of interactions between conserved residues (Fig. 1), and/or changes in the solvent accessible surface area (SASA) of selected receptor segments in response to the initial perturbation (i.e., small molecule/protein binding or point mutation). In particular, as also demonstrated by the crystal structures of dark and photoactivated (Meta II) rhodopsin, as well as constitutively active opsin (7–9), in GPCRs of the rhodopsin family, the SASA index computed over an amino acid set comprising sometimes the fully conserved arginine of the E/DRY motif and selected surrounding amino acids (see Note 10) can be used as an indicator of presence or absence of activation (Fig. 1). Other time averages may include the Cα-RMSD (computable by the "RMSD" module of Wordom (43)) accounting for differences in the average backbone flexibility between inactive and active states. Furthermore, the comparative analyses of the Cα-atom Root Mean Square Fluctuations (Cα-RMSFs, computable by the "RMSF" module of Wordom (43)) concerning reference and perturbed states can be used as well, which provide differences in deviation between the position of the Cα-atoms in each trajectory frame and the relative average position over the whole trajectory. Comparisons of the Cα-RMSF plots would provide information on the flexibility of selected regions of the protein in response to a perturbation occurring at a distal site. This is the case of comparative Cα-RMSF analysis of free and receptor-bound forms of the G protein α-subunit (45).

2. Other comparative analyses done on equilibrium MD trajectories include the Principal Component Analysis (PCA) on the Cα-atoms (i.e., the "PCA" module in Wordom (43)). This is a strategy to isolate and identify low frequency, high amplitude movements in the dynamics, thus separating meaningful concerted motions (i.e., essential motions) from noise and high frequency oscillations (49) (see Note 11). Comparing the essential motions of reference and perturbed states of a GPCR or a G protein provides an understanding of the intrinsic dynamics associated with the functioning mechanism of the considered protein (Fig. 2).

3. The structural differences between reference and perturbed states of a GPCR or a G protein resulting from MD simulations can be also expressed as, (a) changes in the parameters that define the structure network of the reference protein and, (b) changes in the dynamic distribution of the shortest communication paths connecting two amino acids. This is achieved through the Protein Structure Network (PSN) analysis. The PSN analysis implemented in the Wordom software is (43) a product of graph theory applied to protein structures (50). A graph is defined by a set of points (nodes) and connections (edges) between them. In a protein structure graph (PSG), each amino acid is represented as a node and these nodes are connected by edges based on the strength of noncovalent interactions between nodes (I_{ij}) (51, 52) (see Note 12). I_{ij} are calculated for all nodes, excluding $i \pm n$, where n is a given neighbor cutoff of 3 ("--PROXIMITY 3" setup in the "PSN" module). An interaction strength cutoff I_{min} is then chosen and any residue pair ij for which $I_{ij} \geq I_{min}$ is considered to be interacting and hence is connected in the PSG. Selection of the proper I_{min} for the system under study requires calculation of a number of PSGs over an I_{min} range (e.g., from 0 to 5 by 0.1 steps: "--INTMIN 0.0:5.0:0.1" setup in the "PSN module"). In a given PSG, the residues making zero edges are termed as orphans and those that make four or more edges are referred to as hubs at that particular I_{min}. Node interconnectivity is finally used to highlight cluster-forming nodes, where a cluster is a set of connected amino acids in a graph (see Note 13). Cluster size, i.e., the number of nodes constituting a cluster, varies as a function of the I_{min}, and the size of the largest cluster is used to calculate the I_{critic} value. The latter is defined as the I_{min} at which the size of the largest cluster is half the size of the largest cluster at $I_{min} = 0.0$ %. At $I_{min} = I_{critic}$ weak node interactions are discarded, emphasizing the effects of stronger interactions on PSN properties. Thus, the first I_{min} value higher than the I_{critic} is recommended, which results from an I_{min} scanning from 0 to 5 by 0.1 steps. Comparative PSN analyses allow to search for differences in the network parameters (e.g., identity

and number of nodes, hubs, links-connecting nodes and links-connecting hubs) between reference and perturbed GPCR (Figs. 3 and 4) or G protein. Such differences are linked to intramolecular or intermolecular communication mechanisms.

4. Following calculation of the PSN-based connectivities (by the "PSN" module) and of correlated Cα-atom motions (i.e., by using the Linear Mutual Information (LMI) method ("--LMI" option in the "CORR" module of Wordom)), for each frame, the procedure to search for the shortest path(s) between each residue pair (i.e., by the "PSNPATH" module) consists of: (a) searching for the shortest path(s) between each selected amino acid pair based upon the PSN connectivities, and (b) selecting the shortest path(s) that contain(s) at least one residue correlated (i.e., with a LMI cross-correlation ≥ 0.3: "--CUTOFF 0.3" setup in the PSN module) with either one of the two extremities (i.e., the first and last amino acids in the path). The path search implemented in Wordom relies on the Dijkstra's algorithm (52). Once the shortest paths have been found, calculation of the path frequencies, i.e., number of frames containing the selected path divided by the total number of frames in the trajectory, is done. All those paths characterized by frequency values above given thresholds, which depend on the simulated system, are then subjected to cluster analysis by means of the QT algorithm, according to a similarity score (53, 54) (see Note 14). Meaningful paths can be then selected as the centers of the most populated clusters. In addition, coarse path representations can be used to compare the structural communication in reference and perturbed systems (54). These coarse paths consist in the representation of the most recurrent nodes and links either in the entire set of paths (above a frequency threshold) or among the members of the most populated clusters (54). Thus in a comparative framework, path analysis on MD trajectories represent a further way to highlight differences in the structural communication between reference and perturbed GPCR or G protein (46, 47).

4. Notes

1. *Ab initio* modeled loops are less reliable than the loop structures modeled by satisfying the spatial restraints derived from the sequence alignment of target and template proteins. (i.e., by comparative modeling).

2. In some cases, it is worth considering selection of more than one model, showing comparable quality check parameters but with some significant structural differences (e.g., in loops conformations).

3. The Dunbrack and Karplus (55), Ponder and Richards (56), and Sutcliffe (57) rotamer libraries are used, by automatically assigning the conformation that gets consensus among the three with no steric clashes. In general, more than one receptor conformer results from this operation. Such conformers are all worth considering for further calculations.

4. Membrane topology predictors from a single predictor such as PRODIV-TMHMM_0.91 or consensus prediction from more than one source are worth using to individuate the interface Cα-atoms lying in the *xy* plane.

5. For the membrane topology filter to work properly, the two docked structural models must have the appropriate orientation with respect to the putative membrane. This is due to the fact that ZDOCK expresses its docking solutions in terms of a *x,y,z*-translation and a RzRxRz-rotation of the probe. If both target and probe are properly oriented, i.e., with the interface Cα-atoms lying in the *xy* plane, the translation along the *z*-axis can be considered as an offset out of the membrane, and the Rx component of the rotation as a deviation from the original orientation in the membrane. Wrong membrane topology of the docked proteins generate errors in the filtering stage.

6. Selection of quite permissive filtering cutoffs is necessary to avoid removal of reliable solutions as a consequence of possible small input deviations from the proper membrane topology.

7. The QT-like algorithm first calculates the Cα-RMSD for each superimposed pair of dimers/oligomers and then it computes the number of neighbors for each dimer/oligomer by using a threshold Cα-RMSD. The dimer/oligomer with the highest number of neighbors is considered as the center of the first cluster. All the neighbors of this configuration are removed from the ensemble of configurations to be counted only once. The center of the second cluster is then determined in the same way as for the first cluster, and this procedure is repeated until each structure is assigned to a cluster.

8. It is worth noting that the criteria for evaluating the correctness of the G protein membrane topology are quite rough. Indeed, acceptable membrane topologies were considered those characterized by the main axis of the N-terminal helix of the α-subunit almost parallel and close enough to the membrane surface to allow the post-translational hydrophobic modifications of the α- and γ-subunits to insert into the membrane. In this respect, visual inspection serves to discard obvious deviations.

9. Final results must be representative of a number of MD simulations differing in the input structural models and/or in the input setup.

10. Selection of the amino acids that contribute to SASA must be customized ad hoc on the considered system and privilege maximal difference in SASA between functionally different states (i.e., inactive and active).

11. A covariance matrix is constructed by using the Cartesian coordinates of the Cα-atoms as variable set and the trajectory frames as data set. PCA is carried out both on single or concatenated Cα-trajectories of the molecular systems under comparison. According to the Essential Dynamics analysis protocol, the diagonalization of the covariance matrix produces a set of eigenvectors and eigenvalue pairs, which indicate, respectively, directions and amplitudes of motions. Eigenvectors characterized by high eigenvalues describe motions with great atomic displacements. The motions along the most significant eigenvectors can be obtained by projecting each frame of the original trajectories over a number of eigenvectors, which describe the essential subspace of the system.

12. The strength of interaction between residues i and j (Iij) is evaluated as a percentage given by equation above: $I_{ij} = \dfrac{n_{ij}}{\sqrt{N_i N_j}} \times 100$ where Iij is the percentage interaction between residues i and j; nij is the number of atom–atom pairs between the side chains of residues i and j within a distance cutoff (4.5 Å); Ni and Nj are normalization factors for residue types i and j, which take into account the differences in size of the side chains of the residue types and their propensity to make the maximum number of contacts with other amino acid residues in protein structures.

13. Node clusterization procedure is such that nodes are iteratively assigned to a cluster if they can establish a link with at least one node in such a cluster. A node not linkable to existing clusters initiates a novel cluster and so on until the node list is exhausted.

14. Path clusterization relies on a similarity score (S) between paths a and b, computed according to the following equation:

$$S_{a,b} = \left(\frac{2C_N}{N_a + N_b} \right) \cdot 0.15 + \left(\frac{2 Max(C_P)}{N_a + N_b} \right) \cdot 0.4 + \left(\frac{2C_L}{L_a + L_b} \right) \cdot 0.45$$

Where (a) C_N is the number of common nodes in both paths; (b) N_a and N_b are the number of nodes in paths a and b, respectively; (c) $Max(C_P)$ is the greatest number of nodes at the same position in the path as obtained by sliding the nodes of path a over the nodes of path b by one position at a time and then inverting the two paths (i.e., sliding path b over path a); (d) C_L is the number of common links in both paths (i.e., those links connecting pairs of identical nodes); and (e) L_a and L_b are the number of links in path a and b, respectively. The similarity score ranges from 0, for two totally different paths, and 1, for two identical paths.

The similarity score is also used to compute the cluster centers. A cluster center is, indeed, the path with the highest average S among all the paths in the cluster, i.e., the path with the highest number of neighbors in the cluster.

Acknowledgment

This study was supported by a Telethon-Italy grant n. S00068TELU and S00068TELC.

Michele Seeber, Angelo Felline, Francesco Raimondi, and Daniele Casciari deserve acknowledgment for their valuable contribution to method development.

References

1. Fanelli F, De Benedetti PG (2011) Update 1 of: computational modeling approaches to structure-function analysis of G Protein-coupled receptors. Chem Rev 111:438–535

2. Lefkowitz RJ (2000) The superfamily of heptahelical receptors. Nat Cell Biol 2:133–136

3. Pierce KL, Premont RT, Lefkowitz RJ (2002) Seven-transmembrane receptors. Nat Rev Mol Cell Biol 3:639–650

4. Onaran HO, Scheer A, Cotecchia S, Costa T (2000) In: Kenakin T, Angus J (eds) Handbook of experimental pharmacology, vol 148, Springer, Heidelberg, pp 217–280

5. Kenakin T (2002) Efficacy at G-protein-coupled receptors. Nat Rev Drug Discov 1:103–110

6. Brady AE, Limbird LE (2002) G protein-coupled receptor interacting proteins: emerging roles in localization and signal transduction. Cell Signal 14:297–309

7. Palczewski K (2006) G protein-coupled receptor rhodopsin. Annu Rev Biochem 75: 743–767

8. Choe HW, Kim YJ, Park JH, Morizumi T, Pai EF, Krauss N, Hofmann KP, Scheerer P, Ernst OP (2011) Crystal structure of metarhodopsin II. Nature 471:651–655

9. Park JH, Scheerer P, Hofmann KP, Choe HW, Ernst OP (2008) Crystal structure of the ligand-free G-protein-coupled receptor opsin. Nature 454:183–187

10. Scheerer P, Park JH, Hildebrand PW, Kim YJ, Krauss N, Choe HW, Hofmann KP, Ernst OP (2008) Crystal structure of opsin in its G-protein-interacting conformation. Nature 455:497–502

11. Standfuss J, Edwards PC, D'Antona A, Fransen M, Xie G, Oprian DD, Schertler GF (2011) The structural basis of agonist-induced activation in constitutively active rhodopsin. Nature 471:656–660

12. Rasmussen SG, Choi HJ, Rosenbaum DM, Kobilka TS, Thian FS, Edwards PC, Burghammer M, Ratnala VR, Sanishvili R, Fischetti RF, Schertler GF, Weis WI, Kobilka BK (2007) Crystal structure of the human beta(2) adrenergic G-protein-coupled receptor. Nature 450:383–387

13. Cherezov V, Rosenbaum DM, Hanson MA, Rasmussen SG, Thian FS, Kobilka TS, Choi HJ, Kuhn P, Weis WI, Kobilka BK, Stevens RC (2007) High-resolution crystal structure of an engineered human beta2-adrenergic G protein-coupled receptor. Science 318:1258–1265

14. Warne T, Serrano-Vega MJ, Baker JG, Moukhametzianov R, Edwards PC, Henderson R, Leslie AG, Tate CG, Schertler GF (2008) Structure of a beta1-adrenergic G-protein-coupled receptor. Nature 454:486–491

15. Hanson MA, Cherezov V, Griffith MT, Roth CB, Jaakola VP, Chien EY, Velasquez J, Kuhn P, Stevens RC (2008) A specific cholesterol binding site is established by the 2.8 A structure

of the human beta2-adrenergic receptor. Structure 16:897–905

16. Wacker D, Fenalti G, Brown MA, Katritch V, Abagyan R, Cherezov V, Stevens RC (2010) Conserved binding mode of human beta(2) adrenergic receptor inverse agonists and antagonist revealed by X-ray crystallography. J Am Chem Soc 132:11443–11445

17. Rasmussen SG, Choi HJ, Fung JJ, Pardon E, Casarosa P, Chae PS, Devree BT, Rosenbaum DM, Thian FS, Kobilka TS, Schnapp A, Konetzki I, Sunahara RK, Gellman SH, Pautsch A, Steyaert J, Weis WI, Kobilka BK (2011) Structure of a nanobody-stabilized active state of the beta(2) adrenoceptor. Nature 469:175–180

18. Rosenbaum DM, Zhang C, Lyons JA, Holl R, Aragao D, Arlow DH, Rasmussen SG, Choi HJ, Devree BT, Sunahara RK, Chae PS, Gellman SH, Dror RO, Shaw DE, Weis WI, Caffrey M, Gmeiner P, Kobilka BK (2011) Structure and function of an irreversible agonist-beta(2) adrenoceptor complex. Nature 469:236–240

19. Warne T, Moukhametzianov R, Baker JG, Nehme R, Edwards PC, Leslie AG, Schertler GF, Tate CG (2011) The structural basis for agonist and partial agonist action on a beta(1)-adrenergic receptor. Nature 469:241–244

20. Rasmussen SG, Devree BT, Zou Y, Kruse AC, Chung KY, Kobilka TS, Thian FS, Chae PS, Pardon E, Calinski D, Mathiesen JM, Shah ST, Lyons JA, Caffrey M, Gellman SH, Steyaert J, Skiniotis G, Weis WI, Sunahara RK, Kobilka BK (2011) Crystal structure of the beta(2) adrenergic receptor-Gs protein complex. Nature 469:175–181

21. Jaakola VP, Griffith MT, Hanson MA, Cherezov V, Chien EY, Lane JR, Ijzerman AP, Stevens RC (2008) The 2.6 angstrom crystal structure of a human A2A adenosine receptor bound to an antagonist. Science 322:1211–1217

22. Xu F, Wu H, Katritch V, Han GW, Jacobson KA, Gao ZG, Cherezov V, Stevens RC (2011) Structure of an agonist-bound human A2A adenosine receptor. Science 332:322–327

23. Lebon G, Warne T, Edwards PC, Bennett K, Langmead CJ, Leslie AG, Tate CG (2011) Agonist-bound adenosine A_{2A} receptor structures reveal common features of GPCR activation. Nature 474:521–525

24. Dore AS, Robertson N, Errey JC, Ng I, Hollenstein K, Tehan B, Hurrell E, Bennett K, Congreve M, Magnani F, Tate CG, Weir M, Marshall FH (2011) Structure of the adenosine A(2A) receptor in complex with ZM241385 and the xanthines XAC and caffeine. Structure 19:1283–1293

25. Hino T, Arakawa T, Iwanari H, Yurugi-Kobayashi T, Ikeda-Suno C, Nakada-Nakura Y, Kusano-Arai O, Weyand S, Shimamura T, Nomura N, Cameron AD, Kobayashi T, Hamakubo T, Iwata S, Murata T (2012) G-protein-coupled receptor inactivation by an allosteric inverse-agonist antibody. Nature 482:237–240

26. Murakami M, Kouyama T (2008) Crystal structure of squid rhodopsin. Nature 453:363–367

27. Wu B, Chien EY, Mol CD, Fenalti G, Liu W, Katritch V, Abagyan R, Brooun A, Wells P, Bi FC, Hamel DJ, Kuhn P, Handel TM, Cherezov V, Stevens RC (2010) Structures of the CXCR4 chemokine GPCR with small-molecule and cyclic peptide antagonists. Science 330:1066–1071

28. Chien EY, Liu W, Zhao Q, Katritch V, Han GW, Hanson MA, Shi L, Newman AH, Javitch JA, Cherezov V, Stevens RC (2010) Structure of the human dopamine D3 receptor in complex with a D2/D3 selective antagonist. Science 330:1091–1095

29. Shimamura T, Shiroishi M, Weyand S, Tsujimoto H, Winter G, Katritch V, Abagyan R, Cherezov V, Liu W, Han GW, Kobayashi T, Stevens RC, Iwata S (2011) Structure of the human histamine H1 receptor complex with doxepin. Nature 475:65–70

30. Haga K, Kruse AC, Asada H, Yurugi-Kobayashi T, Shiroishi M, Zhang C, Weis WI, Okada T, Kobilka BK, Haga T, Kobayashi T (2012) Structure of the human M2 muscarinic acetylcholine receptor bound to an antagonist. Nature 482:547–551

31. Kruse AC, Hu J, Pan AC, Arlow DH, Rosenbaum DM, Rosemond E, Green HF, Liu T, Chae PS, Dror RO, Shaw DE, Weis WI, Wess J, Kobilka BK (2012) Structure and dynamics of the M3 muscarinic acetylcholine receptor. Nature 482:552–556

32. Granier S, Manglik A, Kruse AC, Kobilka TS, Thian FS, Weis WI, Kobilka BK (2012) Structure of the delta-opioid receptor bound to naltrindole. Nature 485:400–404

33. Wu H, Wacker D, Mileni M, Katritch V, Han GW, Vardy E, Liu W, Thompson AA, Huang XP, Carroll FI, Mascarella SW, Westkaemper RB, Mosier PD, Roth BL, Cherezov V, Stevens RC (2012) Structure of the human kappa-opioid receptor in complex with JDTic. Nature 485:327–332

34. Manglik A, Kruse AC, Kobilka TS, Thian FS, Mathiesen JM, Sunahara RK, Pardo L, Weis WI, Kobilka BK, Granier S (2012) Crystal structure of the micro-opioid receptor bound to a morphinan antagonist. Nature 485:321–326

35. Thompson AA, Liu W, Chun E, Katritch V, Wu H, Vardy E, Huang XP, Trapella C, Guerrini R, Calo G, Roth BL, Cherezov V, Stevens RC (2012) Structure of the nociceptin/orphanin FQ receptor in complex with a peptide mimetic. Nature 485:395–399

36. Hanson MA, Roth CB, Jo E, Griffith MT, Scott FL, Reinhart G, Desale H, Clemons B, Cahalan SM, Schuerer SC, Sanna MG, Han GW, Kuhn P, Rosen H, Stevens RC (2012) Crystal structure of a lipid G protein-coupled receptor. Science 335:851–855

37. Sali A, Blundell TL (1993) Comparative protein modelling by satisfaction of spatial restraints. J Mol Biol 234:779–815

38. Fiser A, Do RK, Sali A (2000) Modeling of loops in protein structures. Protein Sci 9:1753–1773

39. Chen R, Li L, Weng Z (2003) ZDOCK: an initial-stage protein-docking algorithm. Proteins 52:80–87

40. Chen R, Weng Z (2003) A novel shape complementarity scoring function for protein-protein docking. Proteins 51:397–408

41. Casciari D, Seeber M, Fanelli F (2006) Quaternary structure predictions of transmembrane proteins starting from the monomer: a docking-based approach. BMC Bioinformatics 7:340

42. Brooks BR, Bruccoleri RE, Olafson BD, States DJ, Swaminathan S, Karplus M (1983) Charmm: a program for macromolecular energy, minimization and dynamics calculations. J Comput Chem 4:187–217

43. Seeber M, Felline A, Raimondi F, Muff S, Friedman R, Rao F, Caflisch A, Fanelli F (2011) Wordom: a user-friendly program for the analysis of molecular structures, trajectories, and free energy surfaces. J Comput Chem 32:1183–1194

44. Heyer LJ, Kruglyak S, Yooseph S (1999) Exploring expression data: identification and analysis of coexpressed genes. Genome Res 9:1106–1115

45. Raimondi F, Seeber M, Benedetti PG, Fanelli F (2008) Mechanisms of inter- and intramolecular communication in GPCRs and G proteins. J Am Chem Soc 130:4310–4325

46. Angelova K, Felline A, Lee M, Patel M, Puett D, Fanelli F (2011) Conserved amino acids participate in the structure networks deputed to intramolecular communication in the lutropin receptor. Cell Mol Life Sci 68:1227–1239

47. Fanelli F, Felline A (2011) Dimerization and ligand binding affect the structure network of A(2A) adenosine receptor. Biochim Biophys Acta 1808:1256–1266

48. Im W, Feig M, Brooks CL 3rd (2003) An implicit membrane generalized born theory for the study of structure, stability, and interactions of membrane proteins. Biophys J 85:2900–2918

49. Amadei A, Linssen AB, Berendsen HJ (1993) Essential dynamics of proteins. Proteins 17:412–425

50. Vishveshwara S, Brinda KV, Kannan N (2002) Protein structure: insights from graph theory. J Theor Comput Chem 1:187–211

51. Vishveshwara S, Ghosh A, Hansia P (2009) Intra and inter-molecular communications through protein structure network. Curr Protein Pept Sci 10:146–160

52. Dijkstra EW (1959) A note on two problems in connexion with graphs. Numer Math 1:269–271

53. Heyer LJ, Kruglyak S, Yooseph S (1999) Exploring expression data: identification and analysis of coexpressed genes. Genome Res 9:1106–1115

54. Raimondi F, Felline A, Portella G, Orozco M, Fanelli F (2012) Light on the structural communication in Ras GTPases. J Biolmol Struct Dyn (in press)

55. Dunbrack RL Jr, Karplus M (1993) Backbone-dependent rotamer library for proteins. Application to side-chain prediction. J Mol Biol 230:543–574

56. Ponder JW, Richards FM (1987) Tertiary templates for proteins. Use of packing criteria in the enumeration of allowed sequences for different structural classes. J Mol Biol 193:775–791

57. Sutcliffe MJ, Hayes FR, Blundell TL (1987) Knowledge based modelling of homologous proteins, Part II: Rules for the conformations of substituted sidechains. Protein Eng 1:385–392

58. Ballesteros JA, Weinstein H (1995) Integrated methods for the construction of three-dimensional models and computational probing of structure-function relations in G protein-coupled receptors. Methods Neurosci 25:366–428

Chapter 19

Exploring Substrate Diffusion in Channels Using Biased Molecular Dynamics Simulations

James Gumbart

Abstract

Substrate transport and diffusion through membrane-bound channels are processes that can span a range of time scales, with only the fastest ones being amenable to most atomic-scale equilibrium molecular dynamics (MD) simulations. However, the application of forces within a simulation can greatly accelerate diffusion processes, revealing important structural and energetic features of the channel. Here, we demonstrate the use of two methods for applying biases to a substrate in a simulation, using the ammonia/ammonium transporter AmtB as an example. The first method, steered MD, applies a constant force or velocity constraint to the substrate, permitting the exploration of potential substrate pathways and the barriers encountered, although typically far outside of equilibrium. On the other hand, the second method, adaptive biasing forces, is quasi-equilibrium, permitting the derivation of a potential of mean force, which characterizes the free energy of the substrate during transport.

Key words: Steered molecular dynamics, Adaptive biasing forces, Potential of mean force, AmtB, Ammonia/ammonium transport

1. Introduction

Living cells are defined in part by the possession of a membrane surrounding their perimeter, thereby delineating the border between the inside and the outside. Membranes additionally surround the organelles of eukaryotic cells, providing different environments for different functions within a cell. The membrane barrier must selectively allow permeation of some molecules but not others; to accomplish this feat, a variety of membrane protein channels and transporters have evolved, at least one for nearly every type of molecule that must be imported or exported (1).

The existence of a membrane barrier and the proteins that mediate exchange of ions and other substrates across it leads immediately to the question of how transport across the membrane,

Nagarajan Vaidehi and Judith Klein-Seetharaman (eds.), *Membrane Protein Structure and Dynamics: Methods and Protocols*,
Methods in Molecular Biology, vol. 914, DOI 10.1007/978-1-62703-023-6_19, © Springer Science+Business Media, LLC 2012

whether it be active or passive, is achieved and, in particular, how its done selectively. Although some common mechanisms exist, each protein has evolved specific features to optimally transport its particular substrate. A substrate pathway along with associated conformational changes in the protein typically characterizes the transport process. These changes are often small or even nonexistent for passive channels, but can be quite large for active transporters. The substrate pathway can also be characterized by the substrate's free energy along it, known as the potential of mean force (PMF), which dictates the speed of transport and the selectivity.

While X-ray crystallography and other experimental structural techniques can provide snapshots of membrane proteins, even in multiple states, they cannot display the dynamics connecting those states. Computational methods, namely, molecular dynamics (MD) simulations, provide a means of animating the static structures, allowing for, e.g., the visualization of an entire permeation event of a substrate through a channel or a transporter. However, the relevant time scales for many transport processes are beyond that afforded by all-atom MD, which is typically limited to a few microseconds currently. As such, methods to accelerate the process of interest in a simulation have been developed. One example is steered molecular dynamics (SMD) (2, 3), a method in which forces are applied to part of the system, e.g., the substrate, to drive it along a predefined direction. Commonly implemented using a constant force or a constant velocity, this method is useful for exploring possible permeation pathways through a membrane protein. However, because the forces imposed are usually orders of magnitude greater than would be experienced in a living system, interpretation of the results is often limited to a qualitative description of the possible behavior of the substrate and protein. For a quantitative picture of transport, one must turn to more advanced methods. For example, the results of multiple SMD simulations combined through application of Jarzynski's equality provide a means of recovering an equilibrium PMF from nonequilibrium events (4–8). Alternatively, adaptive biasing forces (ABF) generate quasi-equilibrium trajectories from which the PMF can be deduced (9–12).

Numerous channels and transporters have been studied using MD-based methods. SMD has been used to address substrate binding and/or transport in the water channel aquaporin (13–16), ADP/ATP carrier AAC (17), neurotransmitter transporters LeuT (18), sugar transporter LacY (19, 20), and many others. In addition, SMD has been used to probe the conformational changes underlying function for many membrane proteins, such as the vitamin B_{12} transporter BtuB (21) and the protein-conducting channel SecY (22–24). ABF has been used to find the free-energy profile for, e.g., glycerol in the channel GlpF (25), ADP in AAC (26), and ions in the nicotinic acetylcholine and glycine receptors (27). Both

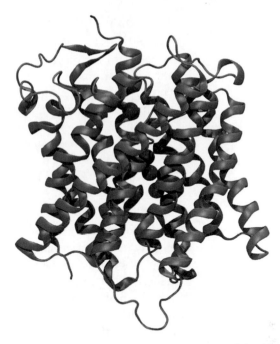

Fig. 1. Crystal structure of AmtB. The *top* of the figure is the extracellular side of the channel and the *bottom* is the cytoplasmic side. The four crystallographically resolved NH_3/NH_4^+ molecules, denoted Am1 to Am4 from *top* to *bottom*, are also seen in *blue*.

methods, SMD and ABF, are clearly well established, but still require care to ensure their proper application.

In this chapter, applications of SMD and ABF to the ammonium transporter AmtB will be explored (see Fig. 1). AmtB has been well studied by computational methods, including those used to calculate free energies (28–33). SMD will be used first to gain an approximate knowledge of the permeation pathway and the barriers along it. Then the PMF for ammonia in the central region of the channel will be calculated using ABF. The appropriate choice of parameters and potential difficulties will also be discussed.

2. Materials

The investigations of AmtB described below will utilize the 1.35-Å structure solved in 2004 (34). Prior to carrying out SMD and ABF simulations, a membrane-water system containing AmtB should be built and equilibrated.

2.1. Programs

1. Obtain the latest version of the molecular visualization and analysis program VMD (35) from http://www.ks.uiuc.edu/Research/vmd/.

2. For running MD simulations, download NAMD (version 2.7 or higher) from http://www.ks.uiuc.edu/Research/namd/.

3. A program for plotting and analyzing the data resulting from the simulations is also required. The free program xmgrace (unix) is recommended. It can be downloaded from http://plasma-gate.weizmann.ac.il/Grace/. Other useful graphing programs are Mathematica, http://www.wolfram.com/, Matlab, http://www.mathworks.com/, and gnuplot, http://www.gnuplot.info/. For Windows, Microsoft Excel can also be used.

2.2. Files

1. Download the structure of AmtB from the Protein Data Bank at http://www.pdb.org/ (PDB code: 1U7G).

2. A fully built and equilibrated system can be downloaded from http://www.ks.uiuc.edu/Training/Tutorials/science/channel/channel-tutorial-files/. Example input files for running the simulation as well as output can also be found here. Alternatively, using the tools within VMD, one can build the system starting from the PDB.

3. Standard CHARMM parameters (36, 37) will be used for the simulations. The full parameter set can be downloaded at http://mackerell.umaryland.edu/CHARMM_ff_params.html.

2.3. Hardware

1. A desktop or a laptop computer running any common operating system (Linux, Mac OS, or Windows) is required. For best visualization, a dedicated graphics card is recommended.

2. For most production NAMD simulations, including ones used in this chapter, running on a cluster or a supercomputer is necessary. Information on obtaining an account and time on NSF's Teragrid system can be found at https://www.xsede.org/using-xsede.

3. Methods

Before investing computer time in relatively lengthy free-energy calculations, SMD simulations will be used to gain a qualitative picture of the permeation pathway for ammonia in the AmtB channel. The crystallized ammonium ion on the extracellular side of the channel, mutated to a neutral ammonia molecule, will be pulled toward the cytoplasmic side at a constant velocity. You can then analyze the resulting output.

3.1. Setting Up and Running the SMD Simulation

1. The equilibrated system will be used as the starting point for the SMD simulations. Set up a configuration file for a standard equilibrium simulation, which will be modified accordingly.

2. To use the SMD feature in NAMD, the following parameters need to be set in the configuration file:

```
SMD on
SMDFile smd01.ref
SMDk 5
SMDVel .00001 ;# .00001 A/timestep=10 A/ns
(1 fs timesteps used)
SMDDir 0 0 -1
SMDOutputFreq 100
```

The parameter *SMDVel* defines a pulling speed of 10 Å/ns, which is fast enough to traverse the channel, defined by *SMDDir* to be in the –z direction, in an ~5-ns simulation (see Note 1). The file, *smd01.ref*, is almost identical to the PDB of the entire system and controls to which atoms the SMD force will be applied. A nonzero value in the occupancy column of this file indicates that forces will be applied, a "0" not. For pulling a single ammonia molecule, only the nitrogen of the ammonia will be set to 1. Finally, *SMDk* specifies the force constant for the spring connecting the SMD atom to the imaginary atom moving at constant velocity (see Note 2).

3. To prevent the protein, or even the entire system, from drifting under the applied force, restraints must be applied to counterbalance it. However, it is important that these restraints do not limit any potential conformational changes of the protein during the simulation. The following parameters apply positional restraints to a limited number of atoms:

```
constraints on
consexp 2
consref rest6.ref
conskfile rest6.ref
conskcol O
selectConstraints on
selectConstrX off
selectConstrY off
selectConstrZ on
```

Of particular note is the restriction to apply restraints in only the z direction, which corresponds to the SMD direction. Similar to the *SMDFile* above, the *conskfile* specifies which atoms to apply restraints to as well as the force constant for those harmonic restraints; in this case, the Cα atoms of residues 7, 97, 149, 225, 333, and 348, which are at the periphery of the extracellular side of AmtB, are chosen with a force constant of 5.0 kcal/mol Å^2 (for comparison, 1.0 kcal/mol Å^2 is also shown in Fig. 2e, f). The file in *consref* specifies the positions to which those atoms should be restrained, and many

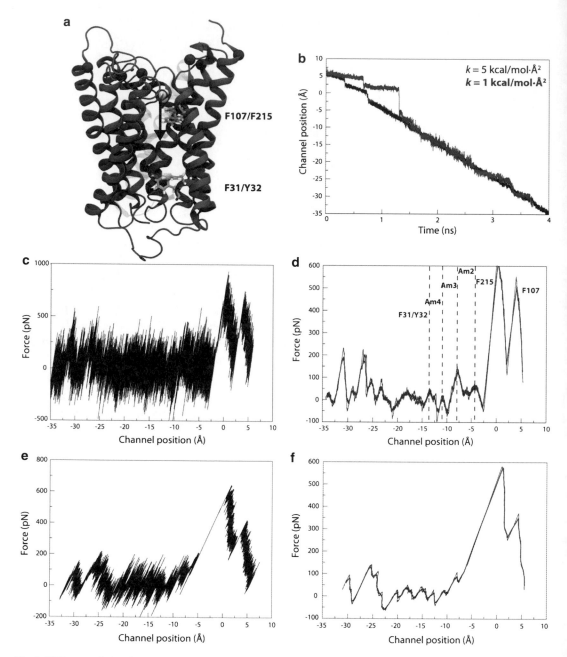

Fig. 2. SMD as applied to NH$_3$ in AmtB. (**a**) Simulation setup. The NH$_3$ molecule begins above the first gate and is pulled in the z direction toward the second gate and beyond. *Red spheres* near the *top* indicate residues restrained to negate any net motion arising from the SMD forces. (**b**) Position of NH$_3$ during the SMD simulations for two different values of the SMD spring constant k. Notice how a larger k (*black*) enforces better the constant velocity of the pulled atom than the smaller k (*red*). (**c**) Force vs. position for SMD with $k=5$ kcal/mol Å^2. (**d**) Running average of force vs. position with averages taken over 0.5-ns (*light, red line*) and 1-ns (*heavy, blue line*) windows. Force peaks are correlated with specific channel features. (**e**, **f**) Same as (**c**, **d**) but with $k=1$ kcal/mol Å^2.

times is identical to the input PDB file. See Note 3 for an alternative way to apply restraints.

4. In some cases, the molecule being pulled may find it easier to move laterally outside of the channel before continuing translocation than to cross a large barrier. To limit such a possibility, restraints in the *xy* plane can be applied. However, these restraints should only affect the molecule if it tries to move outside the channel, and otherwise be zero. One way to implement such a restraint in NAMD is through *tclBCforces*. Additionally, the *colvars* (short for collective variables) module is useful for defining restraints, as well as for calculating free energies (see below).

   In the NAMD configuration file, an external file containing the colvars setup is referenced:

```
colvars on
colvarsConfig restrain.in
```

In restrain.in, you will find the following:

```
colvarsTrajFrequency 1000
colvarsRestartFrequency 1000
colvar {
 name restrain01
 width 0.5
 lowerboundary 0.0
 upperboundary 8.0
 lowerwallconstant 100.0
 upperwallconstant 100.0
 distanceXY {
  main {
   atomnumbers { <list of ammonia atom(s)> }
  }
  ref {
   atomnumbers { <list of protein atoms> }
  }
  axis ( 0.0, 0.0, 1.0 )
 }
}
```

The *distanceXY* collective variable defines a restraint that, in this case, limits the motion of the ammonia molecule to a cylinder within 8 Å of the central axis of the protein (see Note 4).

5. Run the simulation. For a 5-ns simulation, it may take 1–3 days to complete, depending on available computational resources. Sample output is provided.

3.2. Analysis of the SMD Simulation

1. For a constant-velocity simulation, it is particularly useful to look at the force as a function of position in the channel. To extract this information from the log file, tools such as awk or grep (Linux)

can be used. Alternatively, one can write a script in VMD to extract the required information and output it to a new file. For the SMD simulation run here, the content of the script reads as follows:

```
### Open the log file for reading and the out-
put .dat file for writing
set file [open "AmtB-SMD01.log" r]
set output [open "FvsP.dat" w]
### Loop over all lines of the log file
while { [gets $file line] !=-1 } {
### Determine if a line contains SMD output.
If so, write the
### current z position followed by the force
along z scaled by
### by the direction of pulling (0,0,-1)
 if {[string range $line 0 3] == "SMD "} {
   puts $output "[expr [lindex $line 4]] [expr
   -1*[lindex $line 7]]"
 }
}
### Close the log file and the output file
close $file
close $output
```

Run the script in VMD and plot the resulting output in, e.g., Grace or Excel (see Fig. 2). Right now, the data appears very noisy. To clean it up, calculate a running average with a window size of 500. This can be done in Grace by going to the menu *Data→Transformations→Running Averages* or in Excel by typing in an adjacent column of the first row ($B1:$B500), and then using the "fill down" function (location depends on Excel version) to fill the remaining cells.

2. Examine the simulation trajectory in VMD and try to correlate the force peaks in the plot with specific events. For example, the first two large peaks correspond to breaching the first two hydrophobic gates in AmtB, formed by Phe107 and Phe215. The three internal binding sites for NH_3 are also revealed by this plot. Compare them to the location of those found in the crystal structure. Finally, the other gate formed by Tyr32 and Phe31 can also be identified.

Now with a basic understanding of the NH_3 permeation pathway and the barriers along it, a more quantitative characterization can be carried out. We will use the adaptive biasing forces method in NAMD (10, 38) to determine the PMF along the channel axis.

3.3. Setting Up and Running the ABF Simulation

1. First, an appropriate reaction coordinate must be defined. While an obvious choice would be just to use the absolute position of NH_3 along the channel axis, this does not account

for any fluctuations of the channel in *z*. A more apt choice is to relate the position of the molecule with that of the channel's center of mass; this also has the benefit of negating the need for additional restraints. Next, appropriate bounds for this coordinate must be defined. Using VMD, load the equilibrated PDB file for the AmtB system. Make two atom selections, one for NH_3 (i.e., "resname AMM1 and noh") and one for the protein Cα atoms (i.e., "protein and name CA"). You can get the distance separating them with the *tcl* command

```
expr [lindex [measure center $AMM1sel] 2] -
[lindex [measure center$protCAsel] 2]
```

which is, approximately, the beginning of your reaction coordinate. We will use the two hydrophobic gates to more explicitly define the range, i.e., residues 107 and 215 on one side and 31 and 32 on the other. Again, using the *tcl* command above, but with $upperCAsel and $lowerCAsel instead of $AMM1sel, find the relevant region of the channel.

2. To enhance the efficiency of the ABF algorithm, the span of the reaction coordinate will be subdivided into equally spaced windows. The range found above is approximately [-13, 5], while the NH_3 is initially positioned 9 Å above the center of the protein. Therefore, we will use four windows, each 5-Å wide, over the range (−13,7]. Create four directories, one for each window. Using the trajectory from the SMD simulation, find frames in which NH_3 is somewhere in each of the windows, i.e., between 2 and 7 Å above the protein's center of mass and so on. Write out a PDB for the entire system for each representative frame named, e.g., *win1start.pdb*. These will be the starting points for the four individual ABF simulations.

3. The NAMD configuration file used for the SMD simulations can be used again here with a few minor edits. First, the keyword *coordinates* should reference the appropriate starting frame for a given window. Remove any reference to SMD and constraints, but leave that for colvars.

4. As in the SMD section above, colvars will be used for the ABF calculations. The colvars input is similar, with a few important differences:

```
colvarsTrajFrequency 1000
colvarsRestartFrequency 1000
colvar {
 name Translocation
 width 0.1
 lowerboundary --z1--
 upperboundary --z2--
 lowerwallconstant 100.0
 upperwallconstant 100.0
 distanceZ {
```

```
main {
  atomnumbers { <list of ammonia atom(s)> }
}
ref {
  atomnumbers { <list of protein atoms> }
}
axis ( 0.0, 0.0, 1.0 )
}
}

abf {
  colvars Translocation
  fullSamples 1000
}
```

In this case, instead of *distanceXY* used to restrain the off-axis movement, the *distanceZ* colvar is used. The width of 0.1 Å defines the bin size and is sufficiently small to generate a smooth PMF. The biasing forces and PMF calculation are invoked by the *abf* block. Create a colvars input file for each window, defining the lower and upper boundaries appropriately.

5. Run the simulation for each window. To restart an ABF simulation, in the subsequent configuration files, add the line

```
colvarsInput    <inputname>.restart.colvars.
state
```

which references the output of the previous run (see Note 5).

3.4. Analysis of the ABF Simulations

Because some windows may require significant sampling to obtain a reasonably accurate PMF, you may also use the provided output for further analysis.

1. To recombine the output from the separate windows in a single PMF, we will again use ABF. Create a new directory called *merge* and place in there the last *grad* and *count* files from each window. Also copy a NAMD configuration file and a colvars input file.

2. In the configuration file, set it to *run* for 0 steps. Ensure that it still properly references PDB, PSF, and restart files. It is not important which input files you use for this simulation.

3. You will need to modify the colvars input file in two ways. First, change the upper and lower boundaries so that they encompass the entire range, i.e., −13 and 7. Then, add the following line to the *abf* block:

```
inputPrefix <win1name> <win2name> <win3name>
<win4name>
```

where *win_name* refers to the prefix for the output files of that window.

4. Run the new simulation. NAMD will read in the gradients and counts from the provided output of the individual windows

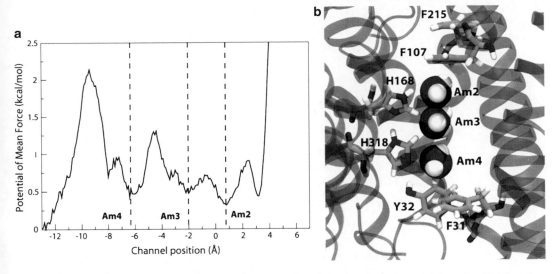

Fig. 3. ABF as applied to NH$_3$ in AmtB. (**a**) PMF as a function of NH$_3$ position relative to the protein's center. (**b**) Close-up on the interior of the channel. The locations of the three minima in the PMF are indicated by the *blue* and *white* NH$_3$ molecules, with the *darker red spheres* showing the crystallized positions. Notice the near-perfect overlap. Key residues at the gates and in the channel are also highlighted.

and generate a new set of outputs, including a new PMF, covering the entire reaction coordinate.

5. Plot the final PMF. Through comparison to the simulation, try to correlate the various extrema to specific locations in the channel. The minima should correspond to the locations of the crystallized ammonia molecules in the PDB 1U7G, as shown in Fig. 3. See Note 6.

4. Notes

1. The pulling velocity should be as slow as possible in order to maximize sampling of the channel environment. Choose a velocity that allows you to complete the simulation in a reasonable amount of time given the computer power available.

2. The force constant affects how well the constant velocity of the pulled atom is maintained. It should be high enough to ensure that the potential in the channel is accurately measured but not so high that the measurement is dominated by noise; this is also known as the "stiff-spring approximation" (7). The appropriate value will depend on the size of the barriers encountered, although a good rule of thumb is that the deviation due to thermal noise, i.e., $\Delta z = \sqrt{k_B T / k}$, is at least less than 0.5 Å. See Fig. 2 for a comparison of the resulting force and distance profiles using different force constants.

3. To restrain the center of mass of a large group of atoms in a single dimension (*x*, *y*, or *z*), SMD can also be used. Set the *SMDFile* to one containing the atoms to restrain, the *SMDDir* to 0 0 1, and *SMDVel* to 0. Because of the way SMD is formulated, non-axial directions cannot be used in this manner. Also, if SMD is used for restraints, an alternative method will be needed to apply the steering forces, such as *tclForces*.

4. In *colvars*, as in NAMD, atom numbering begins from 1, whereas in VMD, it begins from 0. An easy way to get the 1-based atom numbers in VMD is to use the keyword *serial* instead of *index*.

5. It is difficult a priori to define the amount of sampling required to obtain a converged PMF. One way to assess convergence is to plot sequential PMFs and observe any changes as the simulation progresses. After two or more runs for a given window, plot the last two PMFs and compare. You can also plot the counts, which measure the number of times each bin has been sampled. When the PMF is converged, NH_3 should diffuse freely along the reaction coordinate, as the biasing forces cancel out any barriers and wells in the free energy that are encountered. Therefore, the counts should also increase uniformly for all bins after convergence.

6. Notice the relatively large barriers at the hydrophobic gates. It has been suggested that ammonium becomes deprotonated at the first gate, transits the channel as ammonia, and is reprotonated at the second gate (34). The large barriers at the gates support this suggestion, as NH_3 cannot easily enter or leave the channel. As an additional exercise, you can repeat the PMF calculation for NH_4^+ to determine if it can overcome the barriers more easily.

Acknowledgements

This work was supported by the National Institutes of Health (P41-RR005969). Simulations were run at the National Center for Supercomputing Applications (MCA93S028).

References

1. Khalili-Araghi F et al (2009) Molecular dynamics simulations of membrane channels and transporters. Curr Opin Struct Biol 19: 128–137

2. Izrailev S et al (1997) Molecular dynamics study of unbinding of the avidin-biotin complex. Biophys J 72:1568–1581

3. Sotomayor M, Schulten K (2007) Single-molecule experiments in vitro and in silico. Science 316:1144–1148

4. Hummer G, Szabo A (2001) Free energy reconstruction from nonequilibrium single-molecule pulling experiments. Proc Natl Acad Sci U S A 98:3658–3661

5. Jarzynski C (1997) Equilibrium free-energy differences from nonequilibrium measurements: a master equation approach. Phys Rev E 56:5018–5035

6. Jarzynski C (1997) Nonequilibrium equality for free energy differences. Phys Rev Lett 78:2690–2693

7. Park S et al (2003) Free energy calculation from steered molecular dynamics simulations using Jarzynski's equality. J Chem Phys 119: 3559–3566

8. Park S, Schulten K (2004) Calculating potentials of mean force from steered molecular dynamics simulations. J Chem Phys 120: 5946–5961

9. Darve E, Pohorille A (2001) Calculating free energies using average force. J Chem Phys 115:9169–9183

10. Darve E, Rodríguez-Gómez D, Pohorille A (2008) Adaptive biasing force method for scalar and vector free energy calculations. J Chem Phys 128:144120

11. Hénin J, Chipot C (2004) Overcoming free energy barriers using unconstrained molecular dynamics simulations. J Chem Phys 121: 2904–2914

12. Hénin J et al (2010) Exploring multidimensional free energy landscapes using time-dependent biases on collective variables. J Chem Theor Comp 6:35–47

13. Chen H et al (2007) Charge delocalization in proton channels. I. The aquaporin channels and proton blockage. Biophys J 92:46–60

14. Ilan B et al (2004) The mechanism of proton exclusion in aquaporin channels. Proteins 55: 223–228

15. Jensen MØ et al (2002) Energetics of glycerol conduction through aquaglyceroporin. GlpF. Proc Natl Acad Sci U S A 99: 6731–6736

16. Wang Y, Schulten K, Tajkhorshid E (2005) What makes an aquaporin a glycerol channel: a comparative study of AqpZ and GlpF. Structure 13:1107–1118

17. Wang Y, Tajkhorshid E (2008) Electrostatic funneling of substrate in mitochondrial inner membrane carriers. Proc Natl Acad Sci U S A 105:9598–9603

18. Celik L, Schiott B, Tajkhorshid E (2008) Substrate binding and formation of an occluded state in the leucine transporter. Biophys J 94:1600–1612

19. Jensen MØ et al (2007) Sugar transport across lactose permease probed by steered molecular dynamics. Biophys J 93:92–102

20. Yin Y et al (2006) Sugar binding and protein conformational changes in lactose permease. Biophys J 91:3972–3985

21. Gumbart J, Wiener MC, Tajkhorshid E (2007) Mechanics of force propagation in TonB-dependent outer membrane transport. Biophys J 93:496–504

22. Gumbart J, Schulten K (2006) Molecular dynamics studies of the archaeal translocon. Biophys J 90:2356–2367

23. Gumbart J, Schulten K (2007) Structural determinants of lateral gate opening in the protein translocon. Biochemistry 46:11147–11157

24. Gumbart J, Schulten K (2008) The roles of pore ring and plug in the SecY protein-conducting channel. J Gen Physiol 132:709–719

25. Henin J et al (2008) Diffusion of glycerol through Escherichia coli aquaglyceroporin GlpF. Biophys J 94:832–839

26. Dehez F, Pebay-Peyroula E, Chipot C (2008) Binding of ADP in the mitochondrial ADP/ATP carrier is driven by an electrostatic funnel. J Am Chem Soc 130:12725–12733

27. Ivanov I et al (2007) Barriers to ion translocation in cationic and anionic receptors from the cys-loop family. J Am Chem Soc 129:8217–8224

28. Bostick DL, Brooks CL III (2007) Deprotonation by dehydration: the origin of ammonium sensing in the AmtB channel. PLoS Comput Biol 3:e22

29. Lamoureux G, Klein ML, Bernèche S (2007) A stable water chain in the hydrophobic pore of the AmtB ammonium transporter. Biophys J 92:L82–L84

30. Lin Y, Cao Z, Mo Y (2006) Molecular dynamics simulations on the Escherichia coli ammonia channel protein AmtB: mechanism of ammonia/ammonium transport. J Am Chem Soc 128:10876–10884

31. Luzhkov VB et al (2006) Computational study of the binding affinity and selectivity of the bacterial ammonium transporter AmtB. Biochemistry 45:10807–10814

32. Nygaard TP et al (2006) Ammonium recruitment and ammonia transport by E. coli ammonia channel AmtB. Biophys J 91:4401–4412

33. Yang H et al (2007) Detailed mechanism for AmtB conducting NH_4^+/NH_3: molecular dynamics simulations. Biophys J 92:877–885

34. Khademi S et al (2004) Mechanism of ammonia transport by Amt/MEP/Rh: structure of AmtB at 1.35 Å. Science 305:1587–1594

35. Humphrey W, Dalke A, Schulten K (1996) VMD—visual molecular dynamics. J Mol Graph 14:33–38

36. MacKerell AD Jr et al (1998) All-atom empirical potential for molecular modeling and dynamics studies of proteins. J Phys Chem B 102:3586–3616

37. MacKerell AD Jr, Feig M, Brooks CL III (2004) Extending the treatment of backbone energetics in protein force fields: limitations of gas-phase quantum mechanics in reproducing protein conformational distributions in molecular dynamics simulations. J Comp Chem 25:1400–1415

38. Phillips JC et al (2005) Scalable molecular dynamics with NAMD. J Comp Chem 26:1781–1802

Nagarajan Vaidehi and Judith Klein-Seetharaman (eds.), *Membrane Protein Structure and Dynamics: Methods and Protocols*,
Methods in Molecular Biology, vol. 914, DOI 10.1007/978-1-62703-023-6, © Springer Science+Business Media, LLC 2012